SON OF A HUNDRED KINGS

BOOKS BY

THOMAS B. COSTAIN

Son of a Hundred Kings

The Conquerors: The Pageant of England

High Towers

The Moneyman

The Black Rose

Ride with Me

For My Great Folly

Joshua: A Study in Leadership
(in collaboration with Rogers MacVeagh)

SON OF A HUNDRED KINGS

A Novel of the Nineties

by

THOMAS B. COSTAIN

DOUBLEDAY & COMPANY, INC.
Garden City, New York, 1950

PRINTED IN THE UNITED STATES
AT
THE COUNTRY LIFE PRESS, GARDEN CITY, N.Y.

FIRST EDITION

To

THOMAS COSTAIN STEINMETZ

My Grandson

Author's Note

A few readers will recognize, no doubt, the general background against which this story is told. A still smaller number may identify certain traits in the characters who play parts in it. I hasten to state, therefore, that the story itself is completely fictitious, that it does not grow out of anything which occurred in the town in question or, so far as I know, in any other part of the world; except that many years ago a case was reported of a boy being sent alone across the Atlantic without funds and with a sign on his back. There have been feuds between brothers since the days of Cain and Abel, and the involvement of innocent bystanders in crime goes almost as far back.

Although I have drawn on a few experiences of my own, I disclaim autobiographical intent and desire to explain that, with a single exception (and that exception easily recognizable to a few), no relative of mine appears on the stage. Perhaps I should be more explicit and say that Aunt Tilly, the life partner of the one exception, bears no resemblance whatever to the gentle little woman who was chiefly responsible for my advent into this world. Old friends and acquaintances may catch occasional flashes of themselves: the characters in the story nevertheless are composite figures, as in practically all fiction, a rag from one quarter, a bone from another, a hank of hair from somewhere else, conceived and constructed to play certain necessary parts.

THOMAS B. COSTAIN

Lakeville, Conn.
April 27, 1950.

Book One

Chapter I

I

There was so much happening on this cold, clear New Year's Day that it is difficult to decide where the story should begin. There was, first and foremost, the arrival of the boy from England. Boys were arriving from England all the time, but there were quite special circumstances in connection with this one. There was also the tragedy which occurred late in the afternoon. As the annals of the small city of Balfour were plain and simple in the main, any death by violence was bound to create complications and much trouble. In addition there were several lesser happenings, any one of which might do well enough as a starting point. A tussle broke out between two small boys during a Sunday-school party at the ivy-covered home of the minister, of all places. Drygoodsman McGregor, sitting proudly in his office immediately behind the life-sized statue of himself which adorned the front of the store and projected out above his famous sign, *Lockie McGregor, The Remarkable Man,* made a resolution to which he would adhere, being a man of stern fiber, through thick and thin. Finally, there was Police Magistrate Jenkinson, grumbling in front of his fire and dictating the whole course of the story by delaying the special session of his court until late afternoon.

Perhaps, therefore, it would be best to start with the day itself. It was one of considerable importance, being the first day of the year 1890 and so the official start of a fabulous period which would come to be called in course of time the Gay Nineties. It may have been that Nature knew in advance about this absurd but rather pleasant decade which was getting under way and felt a sense of responsibility. At any rate, this first day had been given a setting such as never could have been achieved by sentimental Christmas card or sugary valentine. A heavy snow had turned the world white, and there was a sun as cold and brilliant as a finely cut diamond (and which, at a later hour, would paint the horizon as red as the nose of Santa Claus). It was indeed a perfect day to usher in the years when things would be so proper on the surface and so naughty underneath, when taste would be represented by curlicues and architectural gingerbread, when great moist tears would drip from the notes of popular songs; when, finally, people would be firmly convinced that civilization had achieved the absolute peak of perfection and nothing would ever change again.

It was strange that this conviction of a static world should have been so general, for these years were to see the start of the greatest changes the poor old globe had ever experienced. South of the border, in a workshop filled with curious machines, a silent man was putting the final touches to a magic cylinder which would hum and rasp and squeak with a human voice. In a very short time thereafter two brothers would be running a bicycle shop and dreaming of getting much closer than Icarus to the sun and stars. Strips of celluloid were being made which would flow through a contraption like a glorified magic lantern and would cause figures on a screen to move. In a very few years there would be a race of horseless carriages from Paris to Bordeaux and back at an average speed of fifteen miles an hour!

Still more fateful, still more threatening to the peace and content of this seemingly unchangeable world, the gay ten years were to see underground movements starting, hatreds festering along Slavic rivers and Balkan mountains, black smoke belching from the tall chimneys of Essen and Skoda, and Pomeranian grenadiers goose-stepping with feverish haste. There was no hint of all this on the surface, nor would there be until after the turn of the century. People were to go on singing, laughing, eating three huge meals a day, and sleeping like tops, without any suspicion at all that dark buds were beginning to sprout on the vines of wrath.

There was less suspicion perhaps in Balfour than in any other part. This busy western Ontario city had a life so completely and passionately its own that a meeting of the Bicycling Club or a fire on Holbrook Street was of much more concern than all the rumors circulating in all the chancelleries of Europe.

2

The conductor of the express train from Montreal walked through a succession of close-smelling Pullman cars until he reached the first of the day coaches. It was a few minutes past seven only, but most of the passengers had aroused themselves already from their uneasy slumbers on the straight-up-and-down seats and were gazing with apathetic eyes at the white landscape. The whining voices of children and the scolding tones of harassed mothers filled the coach. He did not pause until he reached a seat occupied by a small boy and a small satchel of red carpet.

"Well, Ludar," he said in a cheerful voice. "And how are you this morning?"

"I'm well, sir," answered the boy in a high, English voice.

He was a thin little fellow, between six and seven years old, and wearing on his white and pinched face an expression of the utmost unhappiness. His clothes consisted of a belted coat and knickerbockers of gray corduroy, badly rumpled and travel-stained, and heavy shoes with brass on the toes to prevent scuffing. Pulling himself together out of deference to authority, he sat up very straight in the seat.

"But you were sick during the night, I hear."

The boy nodded. It was apparent that he feared this great man in peaked cap and shiny braid would decide on some form of punishment for him. "The food came up again," he explained. In an attempt at extenuation he added, "I get sick very easy, sir."

He might have gone on to say that it had been like this ever since he got off the boat at that big city called Halifax and started taking trains. There had always been people around him, asking questions and staring as though he had come right out of a zoo; mostly ladies with faces which were stern though they tried to be friendly, telling him what to do and waiting severely until he obeyed. Generally it had been about food. Everyone assumed that he was on the point of starvation and that it was their duty to take him off the train at the first opportunity and ply him with dishes he did not like. It did not matter how much he protested that he was not hungry. He must be hungry, a boy traveling alone like this without anyone to look after him. So, off the train he would be marched, to sit on a high stool in a station restaurant. The food was always the same: ham, which he disliked; eggs, which made him sick; and greasy fried potatoes, which turned his stomach at the first glance—all of it washed down by a bitter-tasting drink called coffee, which scalded his throat. Eating had never been a pleasure anyway because most foods disagreed with him. However, he would swallow as much as he could, which was never enough to satisfy the persistent benevolence which was paying fifteen or twenty cents to stuff him full. He would be led back to the coach in a disapproving silence which made it clear that he had been ungrateful.

As soon as the train started, of course, he would be sick. Unfortunately his weak stomach was not considerate enough to manifest its symptoms gradually so that things could be taken care of in time. Instead there would be a violent retching and all the unwelcome food would be transferred to the world around him, most particularly to his own clothes. A man in uniform, with a disgusted look on his face and a pail of water and a mop in his hands, would come along and say: "Again, eh! You dirty little devil, you!" While the work of cleaning up proceeded, Ludar would lie back on the hard seat, pale and sick and so ashamed of himself that he would turn away when anyone passed.

The conductor squatted on the opposite seat, holding in the air the small gadget which served as the badge of his trade and clicking it busily as he talked, as though punching a rapid succession of tickets. "You're going to meet your father in Balfour, eh?" he began. "Well, you're a lucky boy, I guess. It's a fine city. It's jam-full of factories and growing like sixty. There's plenty of chances there for boys who want to get out early and make a little money. And you'll be having a good start, Ludar. The money the passengers on the boat collected for you has hardly been touched because so many people wanted to be kind to you and buy your meals themselves. I have a matter of eleven dollars and sixty-five cents right here in my pocket for you which I'll hand over to the conductor of the train at T'ronto. He'll take you on to Balfour. It will be on the Grand Trunk—but I'm afraid that can't be helped. It's the only way to get to that town. *He* will hand it over to your father when you arrive. I guess eleven dollars will come in handy, eh?"

"Yes, sir," said the boy. He had heard the word dollar used a great deal since arriving in this country and knew it had to do with money. He wondered why these people never spoke of shillings and pence.

"You'll get there around noon," went on the conductor, nodding his head and continuing to use the tone he would employ with another adult. "Your father will be right there on the platform, waiting for you. One of the high mucky-mucks in Toronto will send him a telegram today, saying you're due to arrive. I guess you'll be glad to see him again."

"I've never seen my father, sir."

The conductor looked startled. "Never seen him? Well, that *does* beat the Dutch! Coming out all this distance alone to find a father you've never as much as laid an eye on! You *are* a game little rooster!"

Ludar had been thinking over what the man in the peaked cap had told him. "Won't I be on trains any more after today, sir?"

"That's exactly what I mean, Ludar. You'll be home at noon or thereabouts today."

"Will I be able to sleep in a bed again, sir? And have baths like I used to?"

"Yep. You'll sleep in a fine warm bed and have as many baths as you want. What do you think of that?"

What the boy thought was apparent from the look of delight in his eyes. They were blue eyes and, because he had lost weight during his long journey, they seemed to overshadow the rest of his face.

"And I think, my lad," went on the official, "that you really need that last item very much by this time."

On the seat beside the conductor were a few presents which well-intentioned passengers had given the lonely traveler. There were a tin pail and a wooden shovel which an elderly lady had contributed, saying that they would be useful "on one of his little holidays." Ludar did not know what holidays were, but he treasured the pail particularly because it had on it a picture of a boy and a dog. The dog was a perky little animal and he called it Dribbler, after one which had visited his neighborhood occasionally. The boy he called Albert Edward, having heard that name mentioned in adult conversation and thinking it a very fine one indeed. In addition there was a paper-covered book about Canada which might interest him conceivably in another ten years, and a small wooden caboose which had once formed part of a toy train.

"Am I to keep these, sir?" he asked.

"They were give to you." The conductor nodded his head. "They're your property now, Ludar, to do with as you like. Not that the people who handed 'em to you had any great rush of generosity to the heart, as you might say." He dropped the ticket punch into a vest pocket which was lined with tape to keep it from fraying. "We'll be getting in very soon now. To the city of Toronto. Where I live myself, Ludar, and a fine city it is."

People were beginning to open parcels of food and eat breakfast. Across the aisle a family of five had a large basket, and the mother started to hand out salmon sandwiches to the three children until the father, who had a beard like a broom, said, "Margaret!" in a shocked voice and then proceeded to say a long grace in a reasonant tone.

The conductor began, "After you've had your breakfast——"

"Oh, please, sir, might I go without breakfast today?" asked the English boy in a pleading voice.

The conductor had risen to his feet and was on the point of leaving. He stopped and frowned uncertainly. "No one goes without breakfast, sonny. It's the best meal of the day. Starts you off with a good solid foundation. You'll never amount to a row of beans if you don't eat big breakfasts. And then there's this eleven dollars and sixty-five cents. I'm responsible for using it to see that you never miss a meal." He sniffed and gave the end of his nose a pinch with finger and thumb. "On the other hand, there's no sense having you sick when you get on the train for Balfour and—and sort of giving the passengers a bad impression of you. I tell you what, Ludar. You drink a glass of milk at the station and we'll call that a breakfast. Does milk come up on you like solid food?"

"I don't think so, sir."

"Well, then, you drink a glass of milk and then you say out good and loud, so everyone will hear you, that you don't want anything more. That will relieve me of responsibility for letting you go hungry with your own funds in my pocket. Then I'll hand you over to Pete Handy, who'll be in charge of your train. *He* will take you in hand and see you get there safe and sound."

"Yes, sir." The boy reached for the toys on the opposite seat. "I better be getting ready, sir."

"No hurry about that. We won't be in for another twenty, twenty-five minutes. I'll come back for you myself."

3

This nightmare of greasy meals and of sleeping on the hard carpet of train seats with the satchel as a pillow, this continual parading of people past him who stopped and stared and discussed him loudly as though he did not exist or at any rate had no feelings to be hurt, had been going on so long that everything else was getting hazy. There had been little enough which was pleasant to remember. Most of his life had been lived with a stern woman called Aunt Callie, in a single room of a house filled with adults who liked to give her advice about him. Aunt Callie went out during the day to a library where she was employed in the repair of books, and so he had spent most of his waking hours alone or penned up in a dreary yard at the back of the house. The lady who owned the house, Mrs. Griffen, gave him his lunch, which consisted of a sandwich, a cup of tea, and a sound rap on the top of his head with a thimble-covered finger.

There had been plenty of things to make his life trying. No one would answer questions, and of course boys of this particular age are little better than animated interrogation points. Everyone was particularly reticent when his queries had to do with himself. Aunt Callie would frown and put a callused finger to her lips or she would say sharply, "That will do, Ludar!" He had ceased, it seemed, to have any name but Ludar because his aunt became quite ferocious if he made use of the rest of it.

This phase of things had started, he believed, on the day when they came to this dismal boardinghouse. Aunt Callie had taken him first to an office, a dark room filled with huge leather books and presses for the storing of letters. Ludar had been instructed to seat himself in a deep chair in a corner. His legs sticking out straight in front of him, he sat there without moving a muscle and watched an old man talk to Aunt Callie. This old man had eyes so dark and penetrating that the boy was frightened when they turned in his direction. The conversation was carried on in low tones, but Ludar was sure it had to do with him, and this added to his discomfort. He had heard someone speak of lawyers and the bags they carried over their backs, so he knew the old man was a lawyer. *His* bag, which was of black velvet, was hanging beside him on the arm of his chair.

Finally the adult talk ceased and the lawyer swung around in his seat and addressed the boy.

"And now, young man," he said, "your name is Ludar Prentice."

"Why, no, sir. My mother told me——"

The old man cut him off with a wave of his hand. "Ludar," he declared, "we are not going to talk about your mother right now. Instead we're going to speak of your father, whom you've never seen. He's a long way off in another country, but someday you are going out there too. Your father's name is Prentice, and if you want to go and live with him, your name must be Prentice. Ludar Prentice. Is that clear?"

"Yes, sir," answered Ludar. But he did not think it was clear at all.

The lawyer rested his elbows on his knees, which brought his terrifying eyes on a level with Ludar's own. "You must forget about that other name. You see, if you were to use it, there might be a great deal of trouble. There would be trouble for many people, but most especially for your father. You don't want unpleasant things happening to him, do you? Of course not. So, your name *is* Prentice, and that other name is never to be spoken or even thought of again. It must be forgotten. You must keep on saying to yourself, 'My name is Ludar Prentice, Ludar Prentice, Ludar Prentice.' If you do that long enough you won't make any mistakes and get people into trouble with the police. Have you been listening carefully?"

"Yes, sir."

"And you'll do what I've told you?"

"Oh yes, sir."

"Then what is it you are to repeat in your mind?"

"My name is Ludar Prentice. My name is——"

The lawyer smiled for the first time since they had entered his office. He handed Ludar a shilling and said, "You're a good boy."

They went to the boardinghouse then and, as soon as they were ensconced in their room, Aunt Callie took possession of the coin. "I'm going to need every penny I can get my hands on to look after you," she said.

Long months passed and nothing happened. Ludar followed his instructions and said many times a day, "My name is Ludar Prentice." Whenever his real name came into his mind he felt guilty and would start to think of something else. Soon, as the old lawyer had said, it came into his mind less frequently and, finally, not at all. Aunt Callie's temper, in the meantime, was getting shorter and she seemed to spend all her evenings writing letters. Ludar would sit in a low chair which barely brought his eyes on a level with the table and he would watch her rapidly flying fingers. He was convinced she always wrote about him. Sometimes her glance would rest on him, and then she would draw in her lips and the pen would travel faster than ever. She would lick the flap of the envelope and the back of a stamp and then weld the whole thing together with a vigorous pound of the hand. She would say: "There! I didn't mince matters," or "I certainly let his family know what I think of them."

Whenever she talked to him about himself it was in a critical vein. She would say: "You take after your father, that's clear. He was a weakling. Why my sister, who could have had any one of a dozen, was foolish enough to marry him is beyond *me.*" Once she said: "Your mother was weak and silly herself. Even though she was half sister to me, I must say it. What right had she to give you a romantic name like Ludar? She must have got it out of a book. Oh, you've no reason to be proud of it. Nor of anything, for that matter."

Sometimes he was able to remember that he had not always lived with this unsympathetic aunt. He would have brief flashes of remembrance of an earlier and happier existence. He could recall that the people had liked him and had been kind to him. There had been one lady with a pretty face who must have been his mother because she had looked after him. It was a large place. What he recalled most often was the wandering he did in shaded gardens and between rows of tall trees, and of catching occasional glimpses of mysterious dark rooms inside the house. The clearest recollection of all was of a clock. It was on the front of a wall covered with ivy and it seemed so high up to him that he thought it must be close to the sun and stars. He had seen the people in the house wind clocks to keep them going and he wondered how this big clock got itself wound. Perhaps a giant came by in the night and reached up with a huge key to attend to it.

His departure from the house with the clock and the beautiful gardens, to go to live with Aunt Callie, had been a sudden one. All he remembered about it was that there was a hush over the place and that the lady with the pretty face, who must have been his mother, was not about. He had never seen her since. There had never been any doubt in his mind that the change had been a most unfortunate one.

After what seemed an endless stretch of time the lawyer came one evening to the house. He talked to Aunt Callie in whispers and seemed in a much more amiable mood. When he was leaving he paused at the door and shot a question at the boy, "And what is your name?"

The answer came without hesitation, "Ludar Prentice, sir."

"Good!" The dark face achieved a semblance of a smile before the door closed on it.

A week later the lawyer returned. He handed over a lot of crinkly papers to Aunt Callie. "Everything is arranged," he said. "His transportation has been secured right through to this town where his father is staying. In this envelope are his steamship ticket and the railway tickets as well. You'll find everything in order."

"And this money?" asked Aunt Callie.

"It's for the boy. Your statement of claim against the family for the expense of keeping him is being gone over. You'll receive a check in settlement very soon. The head of the family is striving to be fair about the whole matter."

After saying this the old lawyer got to his feet. He patted Ludar's head with a hand as bony as Mrs. Griffen's. It seemed, however, to express a hint of kindness.

"My boy," he said, "you're going out to join your father in Canada. Are you glad?"

Ludar nodded his head soberly. "Yes, sir. But I would rather go back to my mother, sir."

The lawyer cleared his throat hastily. "That, unfortunately, is not possible, my boy. Not at just this time. In fact, not for—for a very long time."

Ludar wondered if this important man, who had always been treated with such respect by Aunt Callie, could settle the doubt which had been so long in his mind, the doubt which frightened him when he happened to waken in the night and found himself alone. "Please, sir, is my mother dead?"

"Well, as to that——" The man of the law looked flustered and seemed for a moment at a loss for words. "That, my boy, is a matter we'll explain to you when you—when you're a little older. You mustn't let yourself think about it right now. What you must think about is this trip you're to have on a fine large boat right across the ocean. Very few boys have trips like that."

"No, sir." The boy's voice had become low and with no trace of life in it. The doubt had ceased to exist. Listening to what the stranger had said, and watching the expression on his face, Ludar had become certain that his mother was dead and that he would never see her again. So completely was he filled with the frightening knowledge that from now on he would always be alone, with no one he could turn to for kindness and love and understanding, that he paid no attention to what the lawyer was saying. It was not until Aunt Callie said sharply, "Ludar, pay attention to Mr. Collison," that he recovered himself.

"I was telling you, my boy," said the old man with another smile which made him look almost benevolent, "that I remember your father from the time he was born. I was very fond of him and I'm glad to see how much you take after him in your looks. I'm compelled to add that I hope the resemblance won't be too complete in—in all respects."

Aunt Callie drew down the corners of her mouth. She was dressed this evening in a worn working suit of a dark brown cheviot and looked even more grim than usual. "I should hope not, Mr. Collison," she said. "That man as good as killed my sister."

All trace of geniality left the lawyer's face at once. His manner became so glacial that even Ludar noticed the change. "There will be no more talk of that kind, Miss Railey," he said. "Not, at least, if you want this arrangement to be carried through."

He bowed very stiffly then and stalked to the door. With a hand on the knob he turned and asked the same question, "What is your name?"

"Ludar Prentice, sir."

"You at least seem capable of learning a lesson quickly. You might profit by his example, Miss Railey."

When the door closed after him Aunt Callie put all the crinkly papers away in her purse. She remarked, "You're leaving in four days, Ludar." When he made no response she cried out with sudden impatience, "Well, are you too stupid to tell me if you are glad or sorry to leave?"

The truth of the matter was that he could not express his conflicting thoughts. He was glad, desperately glad, to be leaving her and this mean little boardinghouse and the ugly yard. The thought of sailing on a ship was exciting, the most exciting thing that had ever happened to him. But he had always hoped that when he did get away from this dreary life it would be back to the quiet big house and to wander once again in the gardens and chase butterflies through the tall files of hollyhocks. And nothing could compensate for the sad knowledge the evening had brought him, that he would never again see the pretty lady who had been his mother.

At lunch time the next day Mrs. Griffen peered at him sharply and asked, "Did she give you the money the lawyer left for you?"

Ludar shook his head. He had received no money. In fact, he barely understood what the word meant, for never in the span of his few years had as much as a penny found its way into his pockets. The shilling the lawyer had given him had stayed in his hand until Aunt Callie claimed it.

On the day after, Mrs. Griffen asked the same question, "Ludar, has she given you *all* of it?" When he shook his head a second time, the landlady went into a tantrum of angry disapproval. "I know that Callie Railey through and through. She's going to keep it all for herself. That's what this talk of getting oilcloth and paint means. If boarders with steady jobs weren't so few and far between, I'd show that woman the door. Lor lum me, if I wouldn't!"

Aunt Callie never mentioned money to him. When the day came to take the train to the seaport she surprised him by holding his coat for him. This was strange because she had been refusing sternly to help him dress for at least a year. He thought she must be sorry, after all, that he was going away and that her responsibilities, about which she had written so many letters, were at an end. The coat seemed stiff and unfamiliar, but when he asked about it, she snapped, "Ask no questions because you'll get no answers."

It was a cold day, and, as he had no overcoat, she wrapped a shawl around him. It seemed to him there was something unmanly about appearing on the streets in a woman's shawl, but his protests elicited nothing but angry demands that he behave himself. There was a long ride on a train which he found exciting. When they reached the wharves she took him by the hand and marched him quickly through offices and turnstiles and past lines of people and finally up an elevated walk with sides so high that he could not see the water, although he was sure it must be right beneath them. He tried to hold back and to see something of what was going on around him, the pushing, noisy passengers with bags slung over their shoulders, the activities of men in uniform who must be sailors, the great structure of the ship looming up above them, and the puffing and snorting of tugs; but the stern pressure of her fingers led him on inexorably. Finally they reached a long room which was filled with struggling and indignant passengers, and here they stopped.

"This is the purser's office, and I'm leaving you here," said Aunt Callie.

"You go right up to the desk and tell them your name, and they'll see that you go where you belong."

"But, Aunt Callie," he protested, frightened at the prospect ahead of him, "why don't you take me?"

"I have my reasons," she said, drawing in her lips. She was looking more prim and stern than usual under a gray hat which perched precariously on the top of her graying hair and seemed certain to be swept away by the the first breeze. "I couldn't afford to buy you a Bible or a prayer book, but you must go to the services on the ship and you must listen and take things to heart. You must always be a good boy and not let any of the wickedness in you come out. You must remember what the gentleman told you about your name——"

"Yes, Aunt Callie."

"And you mustn't talk. If you do, you'll say something you shouldn't. Never ask any questions."

Then she suddenly whisked the shawl from around his shoulders and turned toward the door. He was surprised to see that she had started to run and was wrapping the shawl into a bundle as she went.

He became aware with a sinking heart that he was being left in a strange new world, a world of grownups, all of whom looked hurried and angry and unsympathetic. Panic swept over him. He shrieked, "Aunt Callie! Wait for me! Wait for me!" and started to dodge in and around the milling passengers in a frantic effort to overtake the scurrying figure wearing the absurd gray hat. As a result he charged head-on into the round and soft stomach of a large man with bags in both hands. The man had been wheezing with his exertions, and it was some seconds before he could gather enough breath to speak.

"Gor blast it!" he exclaimed. "Here's a nipper gone stark mad!"

An officer in a braided uniform caught Ludar as he ran by him and, in doing so, obtained a full view of the back of the boy's coat. "Well, dance me the tickle-toe!" said the officer, holding him with both hands. "Here's a stumper! This is a new one, this is."

4

As the train began to slow down Ludar heard the conductor in the aisle saying to someone, "Yes, ma'am, this is the boy."

The train official had returned with a whalebone frigate of a woman in tow and, teetering along behind, a small man with white whiskers as fluffy as tumbleweed. The trio stared down at him, and the lady, who had a deep voice, said, "It's an extraordinary thing for people to send such a small child out into the world with no one to take care of him."

"Don't worry about that, ma'am," said the conductor. "He's been well looked after, this boy has. We've done everything we could to make him comfortable."

"You haven't provided him with berths," declared the lady. "I understand he has had to sleep in his clothes. Becoming, no doubt, thoroughly —er—unsanitary in the process."

"Well——" The official seemed at a loss. "I suppose there's never been any berths available. You couldn't expect people who had booked in advance and paid their money down to step aside and sleep in the day coaches, now could you? If I had asked you last night to give up your reservations, ma'am——"

The little man burst into a high-pitched laugh. "He has you there, my dear."

His wife turned on him angrily. "You will please not drag me into this." Then she transferred her attention back to the conductor and went into a lengthy tirade on the sins of railroads. The little man pawed his whiskers into place and winked at Ludar, as though saying, "You and me, sonny, we should be seen and not heard."

Ludar was prepared for her questions when they came. All these curious adults who had stopped by his seat in such a continuous succession had wanted to know the same things.

"How old are you, little boy?"

"I'm going on seven, ma'am."

"And what is your name?"

"My name is Ludar Prentice, ma'am," he answered.

The conductor nodded down at him and smiled encouragingly. "Stand up, Ludar," he said, "and let the lady see the sign on your back."

He had been compelled to do this so often that obedience had become automatic. For perhaps the hundredth time he rose to his feet and turned around.

A square of oilcloth had been sewed to the back of his coat. On it, in crude red letters, had been printed:

THIS IS LUDAR PRENTICE.
He has no money. He is
going to his father
Vivien Prentice at Bal-
four, Ontario, Canada.
BE KIND TO HIM.

"He stepped off the boat at Halifax just as you see him," explained the conductor. "His railroad tickets had been turned over by the purser, so that part of it was all clear. The passengers on the boat had chipped in to make a fund to buy him meals. He had no changes of clothing except one shirt and one pair of stockings. No overcoat or mittens. They say he nearly froze coming off the boat. Just the same, ma'am, he's been brought through and he's been looked after like he was the Prince of Wales hisself."

Other people had been gathering in the aisles and making the same remarks he had heard so often, such as "Well, if this doesn't beat the Dutch!", "What *were* his people thinking of?", and "He's a sickly-looking boy."

The conductor, feeling the responsibility which rested on him, began to make explanations. "The officers of the ship said he was no trouble, except he had as bad a dose of seasickness as they ever remembered. He's been a little ill ever since, in fact. How are you feeling now, Ludar?"

"Not very well, sir," answered the boy.

"I know what's wrong with him," declared the lady. "He needs his breakfast. A good, hot, solid breakfast."

"He isn't much of an eater, ma'am," said the conductor.

The lady snorted. "He's a boy, isn't he? Put the right food in front of him and his appetite will come back." She took her husband by the arm. "Edwin, why don't you do something? Surely you must see what our duty is. We must *not* think of ourselves. We must look after this unfortunate little boy."

The husband pulled at his ineffectual whiskers and nodded. "Yes, I agree with you, my dear."

"Then," declared the lady, "we'll take him right over to the Walker House, or even the Queen's, and put him down in front of a good meal. There will be just time enough for it before the morning train leaves for Balfour."

Having thus made up her mind as to what her duty was, the lady could not be dissuaded. As soon as the train pulled into the station she saw to it that Ludar collected his belongings, including his small collection of gifts. "Come along, little boy," she commanded. "I'm going to do my full duty today. I'm going to treat you as though you were my own grandson."

Ludar looked about him helplessly, trying to find a friend to whom he could appeal. The conductor had disappeared and the faces left about him were strange and, he realized, not really friendly.

Her manner became brusque when he held back. "Give me your hand," she demanded. "What's the matter? Do you think I'm going to harm it?" The thought had entered Ludar's mind that she had some such idea in insisting that he turn it over to her, that perhaps she would have it cut off and pickled and put in a bottle in a museum for other little boys to look at. "Boy, where *are* your manners? And where's your tongue? Has the cat got it? I'm going to take you to have a good breakfast and I want you to seem more willing and even a little grateful."

Ludar did not dare say anything but a meek "Yes, ma'am."

Chapter II

I

With a disdainful smile the butler made his way to the kitchen where the two women servants were preparing furiously for the reception the Tanner Cravens were giving, an annual affair to which all the best people in town were invited and to which, moreover, all the best people came; with a few exceptions, as will be seen later. He disliked his employers and he felt a contempt for the comparative inadequacies of the household. The kitchen, for instance, was very old-fashioned, cavernous and dark and cockroachy, with a huge coal range and wooden cabinets on the walls, from which the green and purple paint was peeling. The kitchen at Ransley had been old-fashioned also, but there had been something impressive about its massive discomfort.

The butler, whose name was Twiller, allowed his eyes to rove over the pleasantly plump figure of Mary Cayne, the cook, and the more slender charms of Delia Connor, the housemaid, who had a dimpled face under her frilled white cap. He was inclined to approve of them both. "Master Norman has been h'extinguishing himself," he announced. "The miserable little blighter has engaged in fisticuffs with his cousin, Master Joseph Craven."

Mary Cayne, who was making chicken salad and being very careful about the proportion of meat to the other ingredients, said in an indignant tone: "Stuff and nonsense, Mr. Twiller! Master Norman was at the party the minister's daughter was giving for her Sunday-school class. They wouldn't fight there."

The butler elevated his nose triumphantly. "But they *did* fight, Cook. Miss Daisy Grange, the pastor's daughter, has just brought Master Norman home. He was in quite a state, the loathsome little nuisance! I heard Madame Still-Waters say, 'Now, Daisy, I must know everything. Everything. I must get to the bottom of this.' And I'm willing to wager, ladies, that she will. Then she'll patter down to the office and tell Sniffle-and-Roar about it." His eyes lit up with a gleam of sheer delight. "I'd give a lot more than a cold muffin to have an ear at the keyhole."

"You have no proper respect for Mr. and Mrs. Craven," admonished Mary Cayne.

The butler answered easily, "Oh, I have some respect for Lady Easy-Does-It, but as for the nooker himself——"

"It's fear of him you have," charged the cook.

"Fear I may have," conceded Twiller, "but respect, none."

Delia Connor, who had the telephone to answer (it had been ringing steadily, although there were no more than a handful of them in town), had fallen behind with the sandwiches. She went right on working as she asked, "And did Mummy's little pet get his darling little face bashed in?"

"Unfortunately," replied the butler, "the conflict, if I may call it such, was stopped before our young hopeful could get his just deserts." He looked over the preparations with a critical eye. "Ladies, you'll have to get a ruddy hurry on if you expect to be ready in time."

He turned with slow dignity and left the room.

"English bloke!" said Delia Connor viciously. "Why don't he help us, then, 'stead of walking round and giving orders? Someday—and it won't be long, Mary—I'll smash his silly nose with a broom handle!"

The cook had begun on the mixing of a dressing for the salad in a huge earthenware bowl. She stopped, with the spoon suspended over the dish. "*Whirrah-whirrah!* How people will be laughing! They've started fighting already, the spalpeens, just like their fathers. Mother of all of us, and them not seven years old yet!"

"What's all this talk about the Cravens?" asked the maid. "I've heard a word dropped here and there."

"You can't expect to get all the gossip when you're so new." The cook had started to beat the eggs with a fast and sure hand. "You see, Delia, it's this way. Mr. Langley Craven is the oldest son and he was president of all the companies, and he lived in the Homestead—he still does, to be sure—and he had the say in everything. But the two brothers, they didn't agree about a single thing. So Mr. Tanner Craven, he went to work and he picked up more stock and he got people to giving him little bits of paper—I don't know rightly about such things, Delia, my girl, but I think it meant he got votes that way—and when it came to the meeting, why, he was able to get his brother's place away from him. Mr. Langley Craven is a very—oh, a very great gentleman, and his manners are always perfect, even if he is running a furniture repair shop now. But he was so taken by surprise that he forgot all his manners and he walked over to his brother, right in meeting with everyone looking on, and ladies present, I'll have you know, and he smashed Tanner Craven right in the face. There was a battle *then,* and they broke a chair and a lamp and some other things not so easy fixed."

The housemaid was so interested that she had suspended work entirely. "Mary," she asked in a whisper, "which one got the best of it?"

The cook lowered her voice also. "Langley Craven got the best of it. He gave his brother a black eye and he knocked out a tooth. You know the gold tooth *he* has in front? That was the one. And it's my opinion that *he*" —motioning in the direction of the front of the house—"took it very hard. I'm telling you this in confidence, my girl, but it's my opinion the master has been brooding over it ever since."

The housemaid nodded her head. "I bet it was just like something I heard a teacher say at school. About—well, about something repeating itself."

"Could it be the newspapers?"

"No, that wasn't it, Mary. But what I mean is, I bet it's been the same this time. I bet Master Norman's cousin had the best of it."

In the meantime Mr. Twiller had returned to his post in the butler's pantry, where he proceeded to polish the silver with the slowest of movements. As he did so he allowed his eyes to rest with repugnance on the snow outside the window. It was piled up over the rose trellis and had even blotted out all traces of the rear terrace. Beyond what had been the terrace, and just over the line of the next property, were two ponds, one small and one large. On the smaller some girls from the neighborhood—which meant they belonged to good families—were sliding and laughing shrilly. On the larger one the boys of the same families were playing shinny with an old baseball.

"One might as well be at the North Pole!" groaned the butler. "And it's all your own ruddy fault, Aubrey Twiller. There you were, butler to the Baron Birdedge of Ransley, with three footmen and twenty-two servants under you. But you wanted to be your own man, and nothing would do but to come out to this ruddy country! You thought they'd take one look at you and make you president of a bank or a railroad." He groaned again with disgust over the magnitude of his mistake. "And here you are, a butler again, for a maker of tables and stools and high chairs—

an ill-bred boor who hasn't a drop of wine in his cellar and pinches the pennies. It's a comedown for you, Aubrey Twiller!"

He was not being fair in one respect at least to his employer. Mr. Tanner Craven was a penny pincher and had never spent a ten-cent piece on the purchase of proper wines, and he was president of the Craven Carriage and Furniture Company, where the family fortune had been made in the first place. The company had become, however, one of his lesser interests. He owned a bank and a block of stores on Holbrook Street and he had, in fact, a finger in every pie in town. It was quite possible that his resources would equal those of the Baron Birdedge of Ransley.

The reflections of the despondent butler were cut short by the ringing of a bell behind him. He glanced at the indicator on the wall, said in a voice of the most intense irritation, "It's the nooker himself!" and set off to answer the summons.

2

The Craven house, which was on Fife Avenue, had begun as a white brick structure with good lines and some pretensions to architectural style but had been undergoing enlargement from the moment Tanner Craven bought it. There had been so much adding of wings and running up of pillars, so much raising of roofs and building of tall chimneys, that it was now a sprawling clutter which did not fall into any known category, although its new owner spoke of it as Georgian. In one of the most extensive of the renovations the architect had found himself with a bit of unexpected space for which there did not seem to be any use. Finally he had given it a window and a door and had turned it into a room just large enough to hold a desk, a chair, and a wastebasket. It had become Mr. Craven's office.

The head of the household was seated in the chair and at the desk when the butler appeared in the doorway with the customary "Yes, Mr. Craven, sir?"

"Twiller," said his employer in a grumbling tone, handing him a list of names, "my secretary had to go and select today for a visit out of town. I'm calling a meeting of the board of directors of the Craven Carriage and Furniture Company for eleven o'clock tomorrow, and the notification must reach them without delay. Three will be here this afternoon and I'll take care of them. The others must be spoken to on the telephone at once."

"Yes, sir," said the butler, taking the list.

Tanner Craven rose slowly to his feet. In spite of his quite spectacular success in business, he had not passed his middle thirties. He was rather liverish in appearance, with a neatly waxed mustache, and he was wearing thick glasses which pinched into the sides of a rather bulbous nose. The condition in which he kept his desk was an index to one phase, at least, of his character. Everything was tidy in the extreme, the papers in neat piles, the inkstand in the exact center, not a speck of dust; and in getting to his feet he had meticulously straightened the pen which had been misplaced by as much as an inch from its usual position. It was quite in line with this that his clothes should be modish, the coat beautifully tailored, the waistcoat of fawn cloth with braided pockets, the trousers showing no trace whatever of bagginess.

"Twiller," he said, "I—I'm in need of a drink. Is there anything in the house?"

His manner had changed. He showed a disinclination to meet the butler's eye and was, obviously, uncomfortable in the need to expose his weakness. He had always been somewhat in awe of his English servant and often regretted that the desire to stand in the social forefront (there was no other household in town which boasted a butler) had led him into this expensive and troublesome form of ostentation.

"There's nothing in the house, sir." The butler decided that the circumstances warranted his taking a bold course. "Some time ago, Mr. Craven, I pointed out the advisability of having a cellar. Permit me, sir, to say now that you should reconsider the emphatic negative with which you responded. I've been in a number of fine homes, sir, and I am compelled to say that it's absurd——"

The eyeglasses of the head of the household had slipped down on the bridge of his nose. The glance he gave over them was an indication that his momentary hesitation had ended. He was angry.

"That matter was settled!" he snapped. Then he relented to the extent of making an explanation. "Mrs. Craven is a sincere leader in the temperance movement and she would be very much opposed. In any case, I can't afford such a luxury. As you must be aware, Mrs. Craven gives a great deal to charity, especially at this time of year. She couldn't be so generous if I didn't exercise economy." He nodded his head. "There are many things I can't afford. People don't understand what it means to give away so much."

The butler's face did not show any expression, but he was saying to himself, "The penny-pinching hog, he makes as much in a week as they give away in a year!"

There was a long pause. Tanner Craven straightened his bow tie. He nudged the pen with a careful forefinger into more perfect alignment.

"Are you sure there's nothing in the house?" he asked.

"Well," answered the butler, "there's a finger or two of brandy left over from the Christmas cooking." He shook his head. "I most definitely do not recommend it. It's for cooking and not for drinking. You should know, sir, because you purchased it yourself."

Tanner Craven gave a deep sigh of relief. "Bring it in," he ordered.

In the renovating of the house a new entrance had been contrived at the cost of many interior changes. What had been the parlor had been converted into the hall, and the ceiling had been ripped out to give a commanding height. The stairway was at the back, branching out on each side from a central landing and curving up to a balcony which extended around three sides. The great expanse of wall space thus provided had been filled with some rather mediocre paintings, a gigantic moose head which had been bought at an auction (Tanner Craven had never had a gun in his hand), and one really fine appointment, a Mortlake tapestry.

Flicking a perfumed handkerchief across his mouth to remove the hint of alcohol from his breath, Mr. Craven crossed this imposing entrance and climbed the stairs to a room which was in reality no more than an alcove off the open space of the upper hall. With the installation of a sewing machine, a frame for yarn, and a small cabinet for spools of thread, it had been converted into a nook for the mistress of the house. It was designed for utility and not for charm, the intention being, clearly, that here needle and thread would be kept busy and little of the usual afternoon relaxation and gossip would be indulged in. The geraniums in the window stood up as straight and stiff as grenadiers. The window itself, instead of looking out over the gardens, was blocked off by one of the great round pillars of the front porch, leaving the suggestion that any idle staring at the outside would go unrewarded.

Mrs. Craven had seated herself in a chair beside the sewing machine in readiness for the summons when the first guests arrived. She was almost tiny, with small hands and feet, and she had a small and inconspicuous nose and, behind steel-rimmed glasses, eyes of a mild light blue.

She was dressed, not too happily, in a heliotrope gown of crépon (heliotrope was the favorite color of the season, but it tended to make her look dowdy) with a long Watteau pleat and very full sleeves. It had the usual high tucked collar.

"Well, Tanner, what is it?"

"I've called a meeting of the directors for eleven o'clock tomorrow," he said. "My mind is made up, Effie. I'm going ahead with the plan to enlarge the plant and get into the manufacture of bicycles."

She did not say, as most wives would under the circumstances, "I'm glad you're going on with *my* plan," nor remind him that she had been striving for several months to bring him to the point of a decision. The only remark she made, and in a quiet tone, was, "I think you are being very sensible, Tanner."

"It will mean taking in the Homestead as well as the grounds. There's no other way to get the space for expansion. The house can be converted into offices. The new wing will be built across the back yard, and the old offices will be part of it."

"Do you think the directors will approve?"

Tanner Craven snorted. "*I* am the board of directors," he declared. "What I want, the others want. When I say no, they can't wait to back me up. Oh yes, they'll approve. Some of them realize the chance there is for big profits in this bicycle craze. Some of them are already talking about a name for our machine. The Balfour Beauty. The Balfour Breeze. The Craven Comet." He seemed to be talking to cover up what was behind the plan. "Do you care for any of them?"

His wife did not answer at once. She was gazing out of the window and turning the plain gold wedding ring on her finger. "Tanner," she asked finally, "don't you think it might be wiser to call the meeting later? Say in a week or so?"

"Why?"

"Because people are likely to think that your decision has something to do with—with what happened today."

The mottled sallowness of Tanner Craven's face turned to an angry red. "Let them think what they like! It's the truth, Effie. When you told me our son got the worst of it, I didn't say anything, but I knew I wouldn't be able to rest until I had forced Langley and his family to leave town. I would have called the meeting for this afternoon if we hadn't had this reception on our hands."

His wife turned and gave him a steady look. There was a tinge of color in her cheeks, but her manner had not changed and she spoke as quietly as before.

"I want to be rid of them as much as you do, Tanner," she said. "Perhaps more. That woman——" For a moment it seemed that her composure would desert her and that her real feelings would be allowed to show. She controlled herself, however, and went on speaking in the same low tone. "You must do what you think best. After all, it concerns you and your family, and you must decide for yourself. I've always felt some reluctance because it's your own old home and you are naturally not anxious to put it to this use. But, Tanner, I always knew that in a matter like this, which promises to do so much good and make money for so many people, you would never let personal considerations stand in the way." She paused. "You've looked into all the legal sides?"

He nodded. "We're on safe ground," he said. "Even if Father did assume that Langley would always be the head and stipulated that he was to live there, the Homestead belongs to the company. We can do what we like with it. Langley's no longer connected with the company and has no right to be there."

"There's a still stronger reason," said his wife. "Public welfare. We'll give employment to at least twenty more men by doing this. If it comes to a fight, *that* is what will win for us."

"I fully expect it will come to a fight." Tanner Craven nodded with sudden gloom, realizing that contests of the kind he anticipated were certain to be costly. "Whenever Langley gets angry he rushes to a lawyer. He's dribbling away his share of the estate fighting me. And it's cost me a pretty penny too." He nodded with a quick lifting of mood. "This time we have a clear case, and it's the company will have to pay the costs."

A few moments of silence followed. Then Mrs. Tanner began to speak with a deprecatory gesture of her hands. "I'm going to be very inconsistent. After saying it's no concern of mine, I still can't stop myself from asking—Tanner, *what* about all the furnishings?"

Craven sucked in his lips. "Something must be done, Effie. I don't care what people will say. I won't stand by any longer and let Langley have everything." He had become excited and angry. "It was all very well for Father to want his collection kept intact, but I have some rights. And so has Lish. Do you realize, Effie, that there's more really fine stuff in the Homestead than in all the rest of the town put together? I know people who would give their eyeteeth for some of the Georgian silver and the French hangings and the oriental rugs." His cheeks had become a mottled red with the intensity of his feelings. "I would myself. I swear, Effie, there are some things I want so much that I would give almost everything else for them. There's the Barbudo-Sanchez. I like it better than any painting I've ever laid eyes on. There's those beautiful Waterford glass decanters. And the old drum table! When Father first got it I climbed under it one day and found that other boys had been scratching their initials on the inside. With dates. Some of them went back one hundred and fifty years!

I scratched mine, too, though I knew Father would whale me if he ever found out. I think I crave that drum table more than anything else."

His wife said in a low tone: "When you want things as much as that, Tanner, I believe you're certain to get them. Especially when it's your right."

Husband and wife were completely at one on this point. They sat and stared at each other in silence for several moments, the husband's eyes openly covetous, the wife's suddenly very intent. Then Tanner Craven said decisively, "I'll go to any lengths, Effie!" And she nodded her head in agreement. "I think you should."

He glanced abruptly at his wrist. In spite of his fastidiousness about dress he wore muffetees during the winter and, attached to the one on the right, a small wrist watch. "They'll be starting to arrive any minute now."

His wife patted her hair anxiously. "How do I look, Tanner?"

He glanced at her briefly over the top of his glasses. "You look very well."

It would have been evident to anyone else that she was disappointed at the casual tone in which he had spoken. Her eyes clouded. "I know I'm not beautiful," she said, "like—like some other women I could mention. But I've been most particular about my hair and I thought I was looking rather well. I'm sure this is a becoming dress. You think so, don't you, Tanner?"

He nodded absently. "Yes, it's becoming." Then he seemed to become conscious for the first time of what she was wearing. He frowned. "It looks expensive."

"It is. It's *very* expensive. I bought it in Toronto. But I feel I should look right today and I'm still within the limits I set myself for my clothes." She gave a sudden turn to the conversation. "Have you seen Norman?"

"No!" He sat up angrily in his chair and pounded one knee with his fist. "Effie, he's going to need a different kind of bringing up. This fight has opened my eyes. We were doing our best to make a mollycoddle out of the boy. From now on there's to be no more of that. He's not going to private school, for one thing. He's going to public school with the rest of the boys."

There was a sound of carriage wheels turning into the circular drive and biting sharply into the icy surface. "Here they come," said Mrs. Craven, getting to her feet with a relieved air. "We'll talk about this tonight, Tanner."

"There's nothing more to talk about. It's settled. Norman joins the Y.M.C.A. as soon as he's old enough. I may even hire a physical instructor to get him into good trim." His eyes, staring at her over the top of his glasses, were filled with angry determination. "Someday Norman is going to fight Langley's cub again. And he's going to *win!* I'm not taking any chances."

Mrs. Craven stared as though unwilling to believe in such a possibility. "Tanner, you don't mean it! You mustn't let yourself be so upset by this. Norman isn't a strong boy. I'm afraid he'll always be a little delicate."

"I can't see anything delicate about him. And another thing!" The head of the household leveled an accusing forefinger at her. "That monkey suit you bought him with its silly lace collar. I'm not going to have him wear it."

"Why, Tanner, it's a Fauntleroy suit. All little boys whose parents can afford it are wearing them."

"It goes back to the McGregor store in the morning. As he hasn't worn it, I can just turn it back and get credit for the full amount."

"But—but he has worn it. He wore it to the party today."

Tanner Craven looked apprehensive at once. "Did he get it damaged in the scrap?"

His wife dabbed at her eyes with a small handkerchief. "I must go down at once. Tanner, this is all very upsetting. Why must you keep on about it now? The Fauntleroy suit was slightly damaged. They broke a mucilage bottle, you know."

A red flush spread over the face of the richest man in Balfour. "Mucilage! Then it's badly damaged. We'll have to see that it's thoroughly sponged, if it takes all night to do it."

The butler appeared at the head of the stairs. "Mr. and Mrs. Beeding are here." His mind, as he bowed and spoke, was filled with acid disapproval. He was thinking: "Why are they skulking up here when they should be downstairs to receive their guests? There's been a to-do, I can tell by her eyes. Has Madame Easy-Does-It been getting the worst of it for once? Pinch-and-Save is ruddy well worked up over something. And what friends they have! These Beedings!"

3

The reception was at its height when Twiller approached his master with the information that Miss Adelicia Craven desired to see him. He had shown her into the dining room as it was the only apartment in the house not occupied.

In the process of enlarging the property it had been found impossible to do anything about the dining room, and it had remained as it was before, too small for comfort and with an ugly fireplace which ruined one wall without accomplishing anything by way of compensation (the chimney refused to draw) and with a single window. The result was that when the Cravens entertained at the evening meal they had to give buffet suppers. A boiled ham would be placed on the refectory table in the drawing room, and the butler, in white tie and tails, would carve it there.

Miss Adelicia Craven was standing in front of the useless fireplace and surveying the room when her brother joined her.

"A poky little hole, Tan," she remarked. "I'll never understand why you let yourself get balled up this way. This house is just like everything you do. Fine on the whole, but with one great flaw."

She was the eldest of the three children of Joseph Norman Craven, who had been so completely the financial and social mogul of the town, and everyone said she was her father all over again. The comparison, however, had no reference to size. She was quite small. She liked to dress in the brightest colors and with plenty of bows and ruffles and spangles and, as her face was broad and rather plain, she looked exactly like a large specimen of pansy. As usual she was wearing her father's watch in her belt and carrying a swagger stick.

"So you've come to one of our receptions at last," said Craven, passing over the uncomplimentary implication in her speech.

"Good heavens, no!" answered his sister in a hearty voice. "Do you think, Tan, that I'd eat your soggy sandwiches and your chicken salad, which is always more celery than chicken, when I can stay at home, or go anywhere else, and have a sound meal and an honest drink or two—or three? You and Effie get all the social climbers on New Year's, and all the little shrinking souls who work for you, or hope to work for you, or owe you money. No, you'll never see *me* at one of them."

Her brother showed his annoyance. "You must have had a drink already, Lish, or you wouldn't talk this way," he said stiffly.

"Certainly I've had a drink." The little spinster's voice had achieved its heartiest note. "I had a scotch and soda before lunch. By the time I'm through this afternoon I'll be quite comfortably lit up, as our father always was on New Year's Day. I'm not like you, little Tan Craven. I don't have to sneak away into a side room and close the door tight before I hoist one. I do my drinking in the open."

Craven was thoroughly angry by this time and was twirling the ends of his mustache with unsteady fingers. "Did you come here to insult me?" he demanded.

She became slightly contrite. "No, Tan. I know you have a tough time of it. Effie puts it on well enough to fool a lot of people. But not me—not for one split second. She looks gentle and easy, but I know she's as hard as a beetle's back inside. I feel sorry for you, living in this big house, and the richest man in Balfour to boot, and yet getting so little out of life. Come to think of it, I've always been a little sorry for you, ever since you were born. You were such an ornery little coot, not strong and handsome and gifted like Langley. . . . Well, I didn't come to talk this way." Her manner changed abruptly, becoming direct and businesslike. "I came about this meeting you've called for tomorrow. I had just heard about the scrap between the two boys when that flunky of yours got me on the telephone. I knew right away there was some connection between the two. I came to give you a piece of advice, Tan. Whatever it is you're planning to do, don't do it."

"I don't know what you're talking about," said her brother in a sulky tone.

"You know damn well what I'm talking about. This feud with Langley. I feel that I'm standing in my father's shoes and that I must do what he would have done. He would have stepped in right away, but I felt my hands were tied. Still, better late than never."

"Are you putting all the blame on me?"

"No, Tan, not all the blame. Langley paid no attention to the business and made a pretty hash of things. I had enough pride in my father's achievements to want what he had started carried on. So I stood aside and let you boot Langley out of the presidency, and I don't suppose he'll ever forgive me. I say hard things about Effie, but I can see her side of it too. It was tough on her, coming to this town as your wife and finding Pauline so lovely and popular. She must have discovered right away that you had been in love with Pauline and that everyone said you had married *her* for her money. It was natural enough for her to start hating Pauline." There was a pause, and then she raised her swagger stick and leveled it at him. "But you've carried the feud too far. Look what you've done to poor Langley. What kind of life do you think he lives on the profits of his little furniture repair shop? He hasn't one share left in any of the businesses Father started. You took them all! Do you suppose I don't know all the shoddy schemes you thought up, or your wife did for you, to get his shares away from him?"

Tanner Craven's face had gone white. "That's all I'll hear from you. Get out of my house. It will suit me if you never come back!"

"I won't go!" she said. "Not until I've had my say. I'm here to deliver an ultimatum. The past shapes the future, as we've seen today. Here we have the two boys taking up the feud. I love both my nephews, and in a way I was glad they had the gumption to fight. It shows they have something of their grandfather in them. But just the same, it mustn't go on. Do you hear me, Tan? This fighting must stop right here."

Craven indulged in a laugh. "Go on, Lish. I'm curious to hear how you propose to put an end to it."

"I'm sure you're planning ways of getting at them because Langley's boy seems to have had the better of it. Just you try it, Tan Craven! Just you make another move against that unfortunate family and I'll take a hand in the fight myself. People say I'm like my father. Well, I am, and I'm proud of it. When I get angry I can fight as well as he could. And let me tell you this, Tanner Craven, my father could have taken you apart like the works of a watch!"

"Has it occurred to you that I take after Father also? What do you propose to fight me with?"

"With the brains I inherited from him!" she answered fiercely. "And I have resources of my own. Oh, I'm not in a class with you when it comes to money, but I've done rather well in a small way. I'd be rich if I didn't get interested in so many damn charities."

Without giving him a chance to say anything more, she turned and walked into the hall. On a table near the door there was, among other garish objects, a large glass globe filled with all the colors of the rainbow. She paused beside the table, lifted her swagger stick, and sighted it at the globe. The stroke which followed was more accurate than she had intended. The point of the stick struck the glass squarely and shattered it to pieces with a loud crash.

She turned around with a look of dismay, then made for the door without waiting for the butler to show her out.

4

Adelicia Craven's next call was at the Homestead. This was a square house of whitewashed brick, of such a solidity of line that everyone who remembered Joseph Norman Craven said it was a complete reflection of him. Another respect in which it suggested the founder of the Craven fortunes was that it stood in a corner of the factory; enclosed by an ornamental iron fence, it is true, and shaded graciously by tall maples and oaks, but with the smoke-stained walls of the foundry looking down on it from the rear, and the offices, which were lower but equally dingy, encroaching on the fence from the south. The all-powerful Mr. Craven had been one of the old-fashioned school and liked to have his work under his eyes all the time. This arrangement, however, had never seemed strange to the people of the town; for Balfour had been growing rapidly, and the business and manufacturing sections had intruded into the residential streets, so that some of the best houses were chockablock with stores and lodges and undertaking parlors, and the smokestacks of foundries shared the same patches of sky with church spires.

Her ring was not answered until she gave the bell another pull, and it was not a maid who responded finally but a gentleman of middle years, wearing a rather shabby velvet smoking jacket and carrying under his arm a blue-bound quarterly. He was handsome in a courtly and dignified way, with a fine forehead, a long and straight nose, and a set of well-tended brown whiskers.

"Are you all alone today, Langley?" asked the visitor, stepping into the hall.

"For the moment," answered Langley Craven.

"Good!" said Lish. "We have a lot of things to talk over."

He led the way down the hall, which was high and wide and quite dark, running a hand lovingly, as he passed, over the satiny finish of a maple Queen Anne lowboy. There was the kind of silence about the house which bespeaks affluence, a compound of carpets so deep that feet make no sound, and of the roominess which keeps all domestic activities at a far distance. The library, into which they turned, was an imposing room. There was a stone fireplace facing the entrance, and all four walls were covered with books behind glass doors. In front of the fireplace was a large English drum table covered with brown leather which was magnificently old and worn. Ivory chessmen had been set up on an oriental taboret with blue-tiled panels, and scattered about the room were bronze figurines and bibelots of many kinds. Dominating everything was a painting over the mantel. It showed the inner courtyard of an ancient Norman castle (Langley Craven, who was a student of the past, said it was more correct to speak of it as the inner bailey), done with great boldness of execution and a rich profusion of color.

Miss Craven walked over to the fireplace and spent several moments in silent contemplation of the painting. "As usual," she remarked, "I pay my respects to the Barbudo-Sanchez as soon as I come into the house. It still gives me as much of a thrill as when I first saw it."

"You haven't paid your respects to it lately, Lish," said her brother. He, too, was studying the picture. "It *is* magnificent. Father was not happy in all his purchases—I've relegated a few of the paintings to the attic, where they belong—but the Barbudo-Sanchez is one of the very finest examples of this particular Spanish school."

After another lingering look at the rich medieval blue of the chatelaine's cloak, the russet tones of the tunics, and the warmth of the sun pouring in around the barbican, Adelicia turned about and faced her brother.

"Langley, where's Amelia?" she asked.

He hesitated for a moment. "We had to let her go." It was apparent to the visitor now that he was in a subdued mood. Usually he was able to maintain his high spirits through all ups and downs (there had been many more downs than ups), but he was making no effort at the moment to conceal the fact that he was depressed.

"When did this happen?"

"Oh, ten, perhaps twelve, days ago."

"And Pauline is doing all the work? She'll wear herself out, Langley. The Homestead isn't a one-woman house."

He gave vent to a deep sigh. "The truth of the matter is—well, we can't afford a maid any longer. Everything has been falling out of the sky for me. I had a deal on, the biggest chance I've had since I ceased to be head of the Craven interests. A new factory was to be located here and I was to be president. Household articles, washing machines, brooms, clothespins. A few local men were coming in to bolster up the outside capital. Then something happened. All the local men, for one reason or another, backed out. Big-Ears had gotten wind of it, and he had gone to each one of my men and laid down the law." Langley Craven's face became red with the intensity of his feelings. "A tooth for a tooth isn't good enough for my beloved brother. Because I knocked out one of his, he intends to ruin me, body and soul. . . . The worst of it was that I had financed the preliminary stages out of my last bit of liquid capital. Pauline and I talked it over and, among other things, Amelia was given her notice. Poor, devoted Amelia! Actually we lacked the money to pay her wages."

There was a long silence. Adelicia seated herself in front of the fireplace—which had not been lighted, although there was a chill in the air of the house—and seemed to be studying the Barbudo-Sanchez. Her mind, however, was on the problem of the family feud, and it was habit which drew her eyes to the vivid canvas.

"You've never forgiven me, Langley," she stated.

"Because you didn't stand back of me when Tanner made his attack?"

Adelicia nodded.

Langley, after finishing his explanation, had begun to pace about. He stopped now beside the taboret, picked up the white queen, considered it intently for a moment, and then set it down again.

"I don't bear you any grudge now," he said. "I confess that I was bitter for a time. Then I began to understand that you had acted out of loyalty to Father's memory. Always your first loyalty, Lish, and I—I like you for it." He drew a small meerschaum pipe from the pocket of his smoking jacket and made a pretense of filling it: an instinctive movement, because he abandoned it immediately and returned the pipe to his pocket. "Perhaps it will relieve your mind if I make a confession. I had become lax. There's no doubt of that. Easy office hours. Two months in Europe each summer. More interest in books and music and art than in dividends. But"—the red flush deepened in his cheeks—"it wasn't necessary to throw me out bag and baggage. I would have stepped aside for Tanner. I could have been made chairman of the board or even an inactive director. But nothing would suit him but to have me discarded publicly like an old shoe——"

"That," said his sister, "was not Tanner's doing—not to begin with, I mean. You can thank someone else for that."

Langley nodded. "How that woman hates us!" He was silent for several moments while his mind went back over the events which had led to his downfall. "If it hadn't been so unexpected, so—so piratical! That was what caused me to explode. And because I did explode, the feud has gone on and on and will never end. I've fought back wildly——"

"Langley," interrupted his sister, "do you mean you *have* forgiven me? That we can talk once more as we used to—freely and openly and lovingly?"

Langley Craven's manner relaxed. He seated himself in a modern red leather chair and smiled at her. "That," he said, "is the impression I intended to convey before I let my feelings get the better of me. I hope that from now on we'll be able to speak freely and openly and, most certainly, lovingly."

"Then I've something to say," declared Adelicia briskly. "I'm convinced that Tanner's cup of iniquity is just about ready to overflow. You aren't the only one he's treated badly. He has victims all over the place. And yet he has had a continuous success. As sure as there's a God in heaven—and a law of averages—this brother of ours is due to meet some setbacks. I'm so sure of it that I'm worried because I have so much of the stock still. I haven't breathed this to a soul, but I—I'm going to unload! In fact, I disposed of a small block last week." She nodded her head several times, a habit of her father's when talking business which she had sedulously cultivated. "As a result I have some loose capital. This is what I'd like to do with it. I'd like to form a little partnership, just you and me—and perhaps we should cut Pauline in for a slice—and then begin to buy into a few small businesses in town where the owners need extra capital and some assistance as well. The kind of thing where you can keep an eye on what's happening to your money and sort of give it a push when necessary. Candy stores, men's clothing, a bicycle business.

"That," she went on, "was what Father did, and he was so shrewd about it that he nearly always made money. What's more, he made friends of the people he went in with. That was when the factory was just beginning to make money. You weren't born yet and I was a very small girl. But," proudly, "he used to talk to me about it, and I remember everything that happened."

"May I interrupt to say," asked Langley, "that the idea pleases me beyond measure?"

"Some of the money, of course, should be put into your furniture repair shop. I've a slight suspicion that it's as wobbly on its pins as a newborn calf."

"It keeps going." Langley felt he had not been completely frank and so hastened to add, "I must tell you, however, that I don't seem able to get to the point where a dollar in profits can be taken out."

"We'll see to that the first thing," said Adelicia firmly. "You won't mind if I start to take an interest in it, will you?"

Langley indulged in a hearty laugh. He had a way of throwing back his handsome head and giving full vent to his mirth, one of the few physical traits he had inherited from Joseph Norman Craven. "Did Wellington complain when Blücher arrived at Waterloo and began to take an interest in things?" He got to his feet again. His mood had changed to one verging on exuberance. "Lish, the thought of having you as a partner has changed the whole face of life. I feel—well, I should like to walk out to the front door and shout, 'Watch out, Balfour, the failure is going to have another chance! Now see how hard he'll work and what prodigies of industry and wisdom he'll perform!'"

Having thus blown off steam in a manner completely characteristic, he left the room and returned in a few minutes with a bottle of sherry and some glasses on a tray. "We should drink to this, Lish. The burying of the hatchet. The formation of a new Craven partnership. The first step in the making of new business history. Here's to us, my dear sister, and may we not only do justice to the memory of a great businessman, but may *our* cup of iniquity be as empty always as is humanly possible."

"I'll drink to that." Lish took a sip of the sherry and could not keep the muscles of her face from wrinkling up in a grimace of dislike. "Damn it, Langley, I just can't get myself to like this sweetish pap. Don't you remember that Father always used to call these wines 'wind and belly wash'? Why not a small slug of whisky?"

Langley looked horrified. "Whisky in the afternoon? Great Scott, Lish, that's a habit for—for barbarians."

"Very well," said his sister resignedly. "Here she goes then. To us, Langley." She put the slender sherry glass, with its elaborately encrusted brim, on the table beside her with an air of haste. "I'm making one stipulation, Langley. You get Amelia back at once before anyone else snaps her up."

A few minutes later they heard the front door open, and Langley started for the hall eagerly. "That's Pauline," he said. Then he paused and a look of disappointment crossed his face as he noted the brief space of time between the opening and closing of the door. "She's alone. I wonder what she's done with the children?"

The French clock on the mantel struck four as Pauline Craven came into the room. The chimes were melodious and high and brittle, like the age of the Sun King which had produced them, and they seemed intended as a welcoming salute. She was a beautiful woman in her early thirties, with fluffy blond hair clustering on her forehead under an impudent toque of gray velvet. Her whole personality matched in vividness her eyes, which were an electric blue. She was tall and wore her clothes smartly, and there was about her manner and carriage, and even in her gait, a suggestion of impulsiveness.

She had already stripped off her coat with its very fashionable small cape and she dropped it on her husband's arm as she swept into the room. Then she paused, patted him on the cheek with one gloved hand, and said, "Hello, Langley dear, and did you miss me?"

"I've been miserable without you." He smiled and indulged in a favorite quotation from *Nicholas Nickleby*. " 'Horrid miserable—demnition miserable.' "

Now that she had taken off the coat it could be seen that her dress was of Niagara blue, a rich shade between peacock and turquoise. It was draped in front and arranged in fan pleats behind, with sleeves and epaulets of silk, and the skirt swept the carpet and swayed and rustled as she walked. She still carried her muff, attached to her right wrist with a silken cord.

Langley thrust a glass of sherry into her hand and said: "Drink a toast, my love. Drink every drop because it's an important one. We'll explain all about it when you've settled down in comfort."

His wife drained the glass. "Thanks, Langley dear," she said. "I hope the toast has to do with Lish being here after such a very long time."

"It has. It has indeed."

Mrs. Craven sighed and relinquished the glass. "I needed it." She walked to a small round mirror of Spanish-American make and took off her hat in front of it, being careful to wind the veil of spotted chiffon about it. The removal of the toque revealed that she had a great deal of hair and wore it in a Psyche knot on the top of her head. "It was necessary, Lish," she said, "to take my little Joe to Mr. Saunders, the barber at the Cameo House. He's closed for the day, of course, but he very kindly agreed to cut my small son's hair if it could be done at his own house. Ah, how terribly I felt as I watched those lovely curls being clipped off so casually and unfeelingly! It was like having a knife driven into my heart."

"He'll look a lot better, Pauline," declared Lish, who believed in speaking out. "Joe is a manly little fellow, but you were making him look like a mollycoddle by keeping his hair long. I suppose this painful operation had to do with the fracas at the minister's house today."

Langley nodded in response. "I agree with you, Lish," he said. "Joe will look a lot better without them. But, Pauline, where is our man-child? And where is little Meg?"

"I stopped in at the Ludyards' on the way back. The Ludyard tribe were having such a good time that I left our two with them. Basil will bring them back himself in a little while."

She seated herself in a not-too-substantial-looking fauteuil at one side of the dark and empty fireplace, spreading out her full skirts fanwise to save them from wrinkling. Her cheeks were pink with the cold and her lips were a healthy red. She made, in fact, a picture of triumphant young wifehood. Her mood seemed to change, however, as soon as she seated herself. She ceased to smile; she became pensive and paid small attention to what was being said.

"Pauline!" said her husband, who had not taken his eyes from her. "Something's gone wrong. I know your moods inside out, my dear. You've been offended or hurt or disappointed. What is it? You know I can't bear it when you're not 'blooming like a demnition flower-pot.'"

Mrs. Craven sighed. "Oh, it's nothing much, really; I suppose I'm silly to think anything about it. It's the annual drive for the hospital. Last year

I was a vice-president, Lish. There were seven of us, it's true, and so it wasn't such a very great honor. Still, I was proud and I worked very hard. I certainly expected to be one of the officers again this year. Today I found out that I'm not. Langley"—she seemed suddenly on the point of tears—"I'm not even on any of the committees. I'm not going to be asked to have any hand in it at all."

Langley raised his hands in the air and expelled his breath in an angry puff. "Isn't there to be any end? It's easy to see in this the hand, neither fine nor Italian, which has so often wielded the stiletto."

"Yes, quite easy," affirmed his wife, dabbing at her nose with a handkerchief. "It was expected that Mrs. Alvin Power would be president this year, but the nominating committee brought in the name instead of—of our kind and ever-loving sister-in-law. That's why I'm not to have any part."

The telephone rang in some far-distant part of the house. Without making any move to answer it Langley wrinkled his nose in an expression of disgust. "What now? What creditor is starting off the new year with a vigorous personal dunning of Langley Craven?" He rose slowly with a deep frown. "My dislike of that thing is so great that I take it as a personal insult whenever it rings. That buzzing, strident, gossip-spreading instrument is going to usher in a new kind of life. One, moreover, which will be neither fine nor gracious." The telephone rang again, more insistent and shrill, it seemed, than before. "I won't hurry. If they really want me, they may keep right on ringing."

As soon as he left the room his wife leaned forward eagerly toward her sister-in-law. "Lish," she said, "I can't tell you what it means to see you again. In the house, I mean, and talking to Langley in such a friendly way. I've missed you so much!"

"I was afraid Langley didn't want me. That he still held a grudge."

"He gave it up long ago. But he thought you should take the first step."

Lish nodded. "He was right. And so, as you see, I have taken the first step."

Pauline began to speak in a low tone. "I did something today that Langley may not approve. I was going through some of my things and I found an album of old photographs. There was one of Tanner and me taken on a picnic at Tuscarora Park. He had on one of those collars which reached his ears." She shuddered. "He looked cheap and sly. Even if he is your brother, I can't understand why I ever encouraged him. Of course it *did* lead to my meeting Langley. And there was one of his wedding. We were in it, although I'm sure the bride didn't want us. To see it again, and to think of all the dreadful things which started there!" She had been speaking with deep feeling, but at this point she allowed a sense of amusement to temper her descriptions. "There was one of *her* in a bathing suit! Lish, will you think me catty because I can't refrain from remarking that her lower limbs are somewhat spidery?"

Although no one of good breeding would be guilty of using the word openly, the interest in the subject of the feminine leg was immense and all-pervading. There was a general belief throughout the town that Mrs. Langley Craven herself was quite fortunate in this respect. At any rate, there was the testimony of Sloppy Bates. The old *Star* reporter had been standing at the corner of Holbrook and Bourse streets one damp day when she came pedaling along on a new bicycle and looking very chic in an epangline skirt, with a straw sailor perched on the top of her head. The front wheel had become stuck in the mud and the rider had taken a fall, and for a brief second her elastic-sided riding shoes had been turned straight up to the skies. A few moments later a dazed Sloppy Bates had made his way into the nearest bar and had addressed the company excitedly. "Great gods and unhappy mortals!" he exclaimed. "What a privilege has been vouchsafed me this day!"

The heroine of this episode went on speaking of the various family pictures she had found. Finally she looked at the guest and, with eyes snapping angrily, she added: "I've destroyed them all! Everything that contains as much as an elbow of one of them or their scrawny little beast of a son, I tore into a thousand pieces. Then I took the pieces down-cellar and threw them into the furnace."

Now that she had started on the subject, Pauline continued to speak with the utmost frankness of her feeling for the other branch of the family. "Langley says we must be philosophic," she declared. "I don't agree with him. He let himself go the time he punched Tanner in the face, and I was never so proud of him in my life. I don't try to be philosophic. I hate them!"

The sound of a footstep on a bare part of the hall caused her to stop abruptly and place a forefinger on her lips. "You're the only person I've ever said this to, Lish. Please don't repeat it, not even to Langley."

She fumbled hurriedly in her muff and produced a tiny piece of looking glass, set with seed pearls, and a small book of paper sheets. Tearing one of the sheets out, she began to rub her eyes with it while watching the operation carefully in the mirror. Lish had heard of rice paper which took the place of powder (no self-respecting woman would confess to the use of powder or rouge), but she had never seen it applied. Her round eyes, set wide apart in her large plain face, watched with the most intense interest.

"I'm—I'm quite flabbergasted to see you using cosmetics, Pauline," she said. "You've got such a fine, high complexion. Surely you don't need those paper things."

"Well," said Pauline, finishing the operation and replacing the toilet articles in her muff, "there's always the tip of the nose. Haven't you noticed how shiny it can get on even the best-regulated faces? And when you've been foolish enough to indulge in your feelings—as I've just done—your eyes require a little attention."

Langley was frowning when he returned. He stood in the doorway and fumbled again with his pipe. "A man named Alfred Hull was on the telephone," he said. "He's at the Balfour Realty Company, which Tanner owns, of course. You both must have seen him around. He has a long black beard and he fixes you with his eye like the Ancient Mariner and asks if you're prepared for the end of the world. Everyone calls him Calamity Hull."

"I heard him preach once," said Pauline. "I was just a young girl and I went with some others to a tent on a vacant lot in the East Ward. He talked so long that we got up and left. And what did Mr. Calamity Hull want?"

"An appointment. He's coming here to see me at twelve o'clock tomorrow. He was very mysterious about his errand, but I knew what it was without being told. He's another of Tanner's emissaries. He'll try to bribe me into leaving town and going out west."

His sister sat up straight at this. "Well," she said, "this is something new."

Langley went on in a half-amused, half-angry tone. "Persuading me to leave town with all my worldly goods and chattels, my wife and children, my menservants and maidservants, seems to have been Tanner's favorite occupation for the past year or so. His intermediaries have offered as high as five thousand dollars. My answer has always been that I was born in Balfour and intend to die here, even if I have to fill a pauper's grave. I cut them off sharply, and so we never get down to terms. Except that I know I would be expected to sell everything in the house to Tanner at a figure to be agreed upon."

"Well!" said Lish, her cheeks coloring indignantly. "I must say that was very considerate and generous of Tanner. What about me? Was I to be brushed aside?"

"Oh, he'd have offered you some small sum by way of compensation."

Langley seemed to be able now to regard the situation as amusing. "The last one to approach me was old Shearstone. He spoke to me in the most fatherly way, combing his beard into two neat and equal parts while he talked. Have you ever noticed what a fruity voice he has? He talks as though he has his mouth full of plum pudding. He called me 'my boy' and fairly oozed advice. The old hypocrite and liar!"

Langley seated himself beside his wife. Her face had assumed a worried look, and he patted her hand reassuringly. "You mustn't take things so much to heart. Now that Lish is on our side, we'll show Big-Ears and his gentle little adder of a wife something they won't like." He turned toward his sister. "Now that we've mentioned the furnishings, Lish, what was your understanding of what Father said?"

His sister did not find it necessary to pause. Her memory needed no refreshing on the point in question. "He called us in, the three of us, and said he hadn't mentioned the contents of the house in his will because he was sure we would all agree to what he wanted. I've always felt, Langley, that what he meant was that a will becomes a public matter, and he felt that what was to be done with his collection was a family concern and should be settled privately between us." Her voice sank to a low pitch, and she found it necessary now to dab at her eyes with a large red silk handkerchief. "He was so far gone that he couldn't say much. Just that he had great pride in the collection and he wanted it kept together. He wanted you, as the elder son, to keep the things here, and he stipulated that you were to make it up to Tanner and me by a financial settlement. We were to arrange it among ourselves. I'm sure he had spoken to the lawyers because he used a legal phrase, and that wasn't his usual way. He was so outspoken and forthright, Father, in everything. He said, 'in lieu of their reasonable expectations of participation.' And that was all he had the strength left to say."

Langley nodded his head. "That's exactly as I remember it. At first, as you'll recall, I made repeated efforts to reach an understanding with Tanner as to the amount of the compensation. Father said *reasonable expectations*. Was Tanner ever reasonable?"

Lish considered the point. "About as reasonable as a street robber who demands your money or your life."

"It became my belief that Tanner was deliberately standing in the way of a settlement because he didn't want Father's wish carried out about keeping everything together. He wanted some of the things for himself. The best things, of course."

"That same suspicion has been in my mind."

"And then," went on Langley, "the feud started and I became unable to meet the most moderate terms. It has drifted such a long time that I've no expectation of ever getting the situation straightened out. If Father had only said what I was to pay you and Tanner, I would have made the settlement at once."

Lish took fire instantly at this hint of criticism. "He preferred to leave it to our common sense and fairness." She indulged in a rueful laugh. "How was he to know we would cease to be a happy family so soon after he died? He was certain the three of us could get together and come to an agreement without any difficulty. We flew at each other's throats instead. Then you went and lost your money. One thing never entered Father's head, I'm sure—that you would find yourself in a position where you couldn't carry out the terms of the will."

Langley rose to his feet and walked to the fireplace. Leaning an arm on the marble top of the mantel, he stared down into the empty grate. "What a mess I've made of things. How blind and stupid I've been! Here I am, surrounded by all the beautiful things Father collected with such care. And I can't afford a fire in the library on New Year's Day!"

5

Adelicia Craven was a fiery radical in her views, but when it came to anything concerning herself she was the stoutest of Tories. There was, for instance, the matter of the front steps on her house at the foot of Wilson Street. Serving the joint purpose of an approach to the main entrance and an arch over the kitchen door, the steps were undeniably ugly and should have been done away with long before. But the owner hated change. "They're homely as all get out," she would concede. "Still, I'm used to them and they give me plenty of training for that long and uncertain climb I'll have to try sometime." The undignified front was left as it had been when she acquired the property.

It was a tall and thin house, looking a little like a six-eyed Chinese giant, with the above-mentioned steps providing a drooping pretense of mustache. It was a completely unsuccessful house, a gangling, rawboned affair. For reasons quite apart from its outward guise, it had become known throughout town as The Waifery.

On returning from her two family visits, the owner walked first to the stables, which, by way of contrast, were in the very best tradition, with beautifully proportioned gables and an iron weather vane in the form of a truly classic eagle. Here she spent a pleasant quarter hour with what she called, quite accurately, her best friends, talking in her deep voice, slapping rumps, and smoothing manes. "Ha, Marcantonio!" (to the Dalmatian dog who was swishing a rawhide tail in the ecstasy of her presence). "Well, Boney, and how are you and the Iron Duke hitting it off these days?" (to the two old horses who went on munching but whose ears twitched at her approach). "Grouch," (the long superannuated pony) "do you hate everyone as much as ever?" "I see, Burley," (the goat, named after a former mayor of the town) "you haven't grown a new tail yet. That old one is the mangiest thing I've ever laid eyes on." The barn cats, many generations of them, poking their indifferent faces out from the yawning haymow or around the iron posts of the stalls, were affectionately greeted. "Ha, Bouncer!" "Well, the Yellow Kid himself!" "What have you done with that last batch of kittens, Virgin?" The chickens she shooed away indignantly, muttering that they were the least layingest hens in captivity and never had pulled their weight.

She might profess a need for the exercise provided by the front steps, but she had been careful to make progress easy inside the tall coffin of a house. Facing the kitchen door was an elevator, the only one in Balfour, a cranky, rickety affair which achieved its way slowly from floor to floor under the control of an undersized Englishman named Al Hanley. Al had been added to the household staff when Miss Craven's protests failed to get him reinstated at the carriage factory, from which he had been discharged for protracted alcoholic absences. He had become expert with the cables which provided the motive power, and only at his touch would the elevator, with much groaning and grunting and rattling, condescend to move. Miss Craven had tried on several occasions to work it and had emerged with red-faced threats to rip the thing out of her house as she would a malignant growth.

Al was standing at the entrance to the unfriendly lift when she came in from the yard, and he grinned and bobbed his head with a laconic "Mum!"

"Shorty," she said, using a nickname which she alone employed, "what's going on upstairs?"

"Mrs. Wasson's at it again," said the elevator man. "A-entertaining at euchre in the drawring room."

"The same gang?"

" 'S, mum. They've been a-pounding down the cards with one 'and and eating sangwiches with the other, mum, for 'ours."

"I don't want to greet Mrs. Wasson's guests today, Shorty. Would it be safe to stop on the next floor?"

"Lor', no, mum! Cousin 'Ubert's at the piano again. I tell you, mum, a boiler factory's nothing to the third floor when Cousin 'Ubert goes to work. And the biby's a howling fit to kill."

"Then take me right up to my own rooms, and try to make no noise so we can get by without being seen."

While the creaky hoist went up to the top floor, which Adelicia Craven kept for her own exclusive occupancy, it might be advisable to pause for a few words of explanation. The owner's goodness of heart, which she tried to cloak with a brusque manner in imitation of her father, made it impossible for her to resist anyone in distress, and gradually the tall house had filled up and had begun to deserve its name, The Waifery. The first staying, but not paying, guest to be ensconced on the third floor had been a cousin on her mother's side, one Hubert Mark, who taught music when he could get pupils, which was seldom, and was slowly starving to death. He had arrived without a cent in his pockets, but shortly afterward his grand piano had been delivered and had been hoisted up to his room by block and tackle, the whole operation costing Adelicia a pretty penny. The next to come was Cousin Mary Ann Wasson. Her husband, a plumber, had taken all his tools with him—for the first time, perhaps—when he died; at any rate, he had left nothing. There were others: Velva, a pouty girl of sixteen who had been in the orphanage, a once industrious little seamstress named Lizzie Bain whose trade had left her when she supplied the world with a puny potential seamstress, and a mysterious female whose story was known only to Miss Craven and who was called Mrs. Twain.

"Shorty," said the mistress as the elevator rose through the waves of Wagnerian din which filled the house, "any chance of a poker game tonight?"

The diminutive elevator man nodded. "We're starting off the new year right, mum. A little game of draw."

Miss Craven drew some bills from her purse. "I'll stake you to the usual, Shorty. Three dollars. And if you lose this time, I'll never put up another cent."

The operator nodded without enthusiasm. "Sometimes, mum, I think the cards is possessed," he said. "I gets on the wrong end of every bumping bee. When I 'as two pair, they beat my brains out. When I catch a full, they won't invest as much as a nickel."

"Sometimes, Shorty," said his mistress, "I wonder if you understand the game. Are you aware of rules like threes beating two pair? The game's full of little things like that, you know. My father was the best poker player in town, and sometimes the two of us would have little games of stud between us. I tell you, Shorty, I could clean up these friends of yours if I had a chance to show them what my father taught me."

"Mum, I never lost a shilling at poker in England."

"That's what is wrong. You play like an Englishman—— Blast! Cousin Hubert's at that chorus again. . . . Shorty, if you win tonight, which seems to me highly improbable, don't bring this noisy contraption up to my floor before I'm awake. I can wait for my cut until I've had a cup of coffee. Hold the news until I come down. If there are people around, just nod your head if you've taken a little tallow out of your friends. A nod for each dollar. If you've lost, don't make the mistake of catching my eye, because what you see there will give you a start."

"I'll do my level best, mum."

"Play 'em close to the chest, and don't bet when you see the whites of their eyes. That means they've got something."

As they reached the top the bell on the ground floor rang. It was a peremptory ring and made it clear that the ringee would stand for no delay whatever.

"That will be Miss Fitch," said the mistress, stepping out of the cage.

"Yes, mum. Miss Fitch it is. I'd know her ring, mum, if all the bells in 'ell was buzzing at the same time. She can't wait a second as soon as she 'ears you come in. She'll be up 'ere laying of complaints about me."

Miss Craven tried to look stern. "What have you been up to, Shorty?"

"It's nothing as I've done, mum. It's what I 'aven't done. I didn't get the ashes out and the slops emptied. 'Er Ladyship keeps an eye on things."

"If it's no worse than that, Shorty, I may be able to forgive you. This once only, of course."

"Mum," he acknowledged, "it may be as there's a few other matters. But nothing you might call serious, mum, not 'ardly."

The main room on the top floor combined the functions of bedroom, boudoir, and sitting room. It was enormous and had both north and west exposures, with the result that the rays of the setting sun were pouring through the windows and laying bright streamers across the gray carpet. There was a large fireplace, and over it a portrait of Joseph Norman Craven. The artist had pictured him as a rather beefy man with high color and stern eye. He looked, it must be confessed, the typical captain of industry and with considerably more than a hint of pomposity. To the affectionate daughter, however, he represented everything that was fine and worth while—Honesty, Integrity, Judgment, Generosity, Magnanimity.

Apart from the portrait, Miss Craven's chief pride was in a bed which she had picked up herself in a little French town on a sly hint from an antique dealer that it had belonged to Napoleon and had been in Malmaison. This might easily have been true. The bed was Bonapartist in material and design, with an arched headboard of laminated wood, the edge elaborately carved in a gilded pattern with the three bees of Elba, and a low footboard from which projected a beautifully carved Napoleonic eagle with a wingspread of three feet. The cover was of blue velvet like the cloak of Marengo. Once a friend had remarked that it was impossible to look at the couch without seeing in it the form of the great conqueror.

She seated herself near one of the windows and began to reflect on what had happened during the first visit of the day. Having thrown down the gauntlet, she was committed to go on, but she was becoming a trifle apprehensive. "Tanner is as sharp as a bear trap," she said to herself. "You don't stand to get any good out of going to law with him." Certainly Langley had come out badly in all the legal moves he had resorted to in efforts to recover the leadership of the Craven enterprises. Most of his inheritance had been swept away in the long legal battle. The rest had gone in the abortive efforts he had made to set himself up in opposition to the factories. "Poor Langley!" she thought. "He's so fine and intelligent—and without any more business gumption than a louse!"

She believed that Tanner would take her warning to heart and desist from any further aggression. But if she had accomplished nothing more by her ultimatum than to stir him to renewed hostility, what was he likely to try? She puzzled over this for some time and finally had to acknowledge to herself that the direction of his next move was beyond her comprehension. She was certain of one thing only, that she would have to consult Ted Laird in the morning. Ted Laird was the dominant partner in the law firm of Thompson, Laird and Fine, and she spent a great deal of her time, and quite a bit of her money, in his office.

"If Tanner turns ugly," was her final rejection, "he'll find, at any rate, that he has a fight on his hands this time. He won't be against a mere babe in the woods like Langley."

The elevator came to a stop at the door of the shaft, and a brisk middle-aged woman stepped into the room. Miss Minnie Fitch had been secretary to the founder of the Craven fortunes and, on his death, had been given notice and a very small pension. Feeling sorry for her, Adelicia had taken her in as her own secretary and house manager, with the result that ever since the highly capable Miss Fitch had been ruling things with stern exactitude. She had acquired the name of Chief. She was seldom seen without a pencil in her hair and paper cuffs over her sleeves, and she was so addicted to accuracy and literal transcription that she kept notes on everything that occurred.

"Well, Chief, what's on your mind?"

Miss Fitch drew some sheets of paper from under her belt, where she always carried letters and memoranda. "Many things," she said. "First, a caller. A man looking for work. That Englishman who hangs about town. Vivien Prentice is the name. How can the English think of such names? He drinks. Heavily."

"So I've heard." Miss Craven had become thoughtful. "I always feel sorry for men like that."

"Don't feel sorry for *this* one," said the secretary sharply. "The house is full to overflowing now. I can't think of anything we need less in this house than a drunken Englishman—unless it might be another piano player. Have you ever stopped to think what He"—*He* was, of course, the late Joseph Norman Craven—"would say if he saw the state of affairs you've gone and got yourself into?"

"Of course I've thought of it," declared the head of the household. "My father was a charitable man, and he would approve of what I've done."

"He would *not!* I was his right hand for twenty-seven years and I knew the way his mind worked. Charitable, yes, but a well-ordered philanthropy. I'm sure He would turn in his grave if He saw the hit-and-miss way you have of taking in all these weaklings."

The daughter made no response. In her silence lay the clue to the control which Miss Fitch had assumed. She claimed to know what Joseph Norman Craven would have said and done under any conceivable circumstance. Adelicia, believing her father perfect in all things, had not been able to find any way of substituting her own conception of him for the stern theories of Miss Fitch.

Miss Fitch was consulting her notes. "I put down everything he said, even if it didn't seem to make sense. He asked if we had a piano to tune, and I said no at once. Then he asked—and I'm reading you his exact words, 'Have you any Greek or Latin translating I can do for you? Anyone to be coached on Flemish primitives? That's the sort of thing I seem best fitted to do.' I was in no mood for joking. I said, 'We'll have no jokes, please.' He said, 'In that case, I'll come straight to the point and ask if you have any menial work for me, such as shoveling snow or taking out ashes or splitting wood. I don't seem able to make a living at anything else.' I was suspicious of him by this time, but I said you had a fund for needy cases and that you might be willing to let him have some money. Now listen to this." She studied her notes. "This is what he said. 'I don't know to what depths I'll sink before I'm through, but I'm sure I'll never sink low enough to accept charity.' I asked him if he was hungry and he said, 'Yes, I am always hungry, but that is a matter which concerns only my stomach and myself.' Then he bowed and left. You would have thought I had insulted him."

"You had, Chief."

"How was I to know he had so much silly pride?"

Lish Craven shook her head slowly. "Not silly pride. It seems to me that he has the kind of pride you can admire." She sighed. "Was he drunk?"

"I didn't notice anything on his breath."

The owner seated herself in a large Morris chair. It was clear that she had suddenly become depressed. "Chief," she said, "I don't like the looks of this. That poor man! He must have been in a desperate mood."

"He's just a tramp. Why bother your head about him?"

"Because I think it's time some head was bothered about him. It's quite clear that he's a gentleman. Why must there be so much trouble and unhappiness in the world?" She got to her feet again with an air of sudden resolution. "I'm going to find him. Right away."

"No," said Miss Fitch sternly. "Not right away. We're having dinner early tonight. Some people are coming in for the evening, and we've got to have everything out of the way. It was to be a kind of surprise for you. George Tapweed is bringing his banjo, and Sam Davids is coming and he has a lot of new songs and jokes. Miss Magna Readlong will be back from T'ronto and she wants to talk over plans for a big charity concert. For the Mission to Lepers, of course. It's going to be a full evening, I can tell you."

Adelicia sank back into the chair a second time. These were all old school friends, and so it would indeed be a full evening. She did not want a moment of it to be lost. She asked weakly, "What have you planned for dinner, Chief?"

"The perfect dinner," was the answer. "A roast of beef. The first ribs, of course. Yorkshire pudding with the roast. Parsnips. Creamed carrots. Baked potatoes. Hot biscuits. Mince pie."

It was the perfect dinner. The owner of The Waifery agreed so completely that she would have been content to sit down to it every evening of her life. When Miss Fitch said, "If you must see this Englishman, I guess it can wait for the morning," she nodded and acquiesced. "It will be better to wait," she said. "That will give me a chance to think it over and see what plans I can make to get him work. Father always believed in having something definite to go on, didn't he? And make sure, Chief, that Cook doesn't get the beef overdone again."

Chapter III

I

It was such a wonderful day that anyone who did not know William Pitt Milner to be the most irritable man in Balfour would have been puzzled to account for his air of ill nature as the express train from Toronto rolled and jolted and puffed along. He was sitting alone, his neck encompassed by a well-starched collar and a maroon tie with a black pearl stickpin. He was also the most influential man in Balfour and one of the richest. For years he had been investing in things—stocks, bonds, factories, stores, and real estate—with invariable success. Recently he had purchased the city newspaper, swearing that he would make something of it. He was rather stout and somber of mien, a man born to say no. His brow had one outward function only, and that was to frown; his voice rumbled with indignation when it was not going off in petulant explosions. It was generally supposed that the state of his liver dictated his moods, a verdict in which his doctor concurred.

The conductor came down the aisle and stopped at the seat occupied by the great Mr. Milner. "The boy's still sick," he announced.

The publisher laid aside the pamphlet on printing presses, which had been engaging his attention, and frowned. "Do you suppose," he inquired, "that he's brought one of these infectious diseases into the country? Cholera or something like that?"

The conductor shook his head. "No, Mr. Milner, it's just his stummick. I'll tell you exactly what happened. He came in on the night train from Montreal and should have gone on to Balfour this morning. But some old fuss-button of a woman took him to the Walker House for breakfast and stuffed him full of food. The boy has no overcoat, and on the way back to the station he got chilled, so they squatted him too close to a fire in the grate in the stationmaster's office. When they took him down to put him on the train there was a big party of settlers from the old country starting for the west, and this Alderman Appleton, who never misses a chance to make a speech, was there to give them a send-off. He's a regular wheeze-gut, this Alderman Appleton, and nothing would do him but to get the boy up in front of him and make him an example of how well newcomers are being treated in this country. The boy had his satchel in one hand and some toys in the other, and he was feeling so ill he could hardly stand. 'Look at this fine little young man,' says the alderman, taking him by the lapel of the coat. 'He comes into this country without a cent of money—this sign of his back says so right out—and I want to tell you that generosity has been shown him at every stage of his trip. This new citizen of ours has money in his pocket now, donated by kind people. This morning it was felt that a hasty bite was not good enough for him, and he was taken instead to a local hostelry and was filled with sausage, applesauce, fine thick pancakes with rich maple syrup——' At that point it happened, Mr. Milner. No, he didn't lose the sausage and pancakes and maple syrup all over Alderman Appleton, though I wish it had happened that way. He just crumpled up like an a-cordeen and went out in a dead faint.

"That," he went on, "was how I came to get him. I went to him and found him looking at a tin pail he had fallen on when he went down. It was flattened out. 'I broke my pail,' he says, and for a second I thought he was going to cry. But he didn't. I said to him, 'Boy, you made the best criticism of the oratory of Alderman Appleton that was ever known. What's more,' I said, 'I hear you got more throwing power in that stummick of yours than any National League pitcher has in his arm——' "

The publisher interrupted to ask, "Did you get that wire off for me to my newspaper?"

"Yes, Mr. Milner. Your editor has it by this time, and he's probably rushing to the station this very minute." He started to move down the aisle but changed his mind and came back. "I won't be seeing this editor of yours, so I'd better tell you what I've heard about this young fellow. He's the most patient and polite boy you've ever seen. During the whole trip by train there's never been a word of complaint out of him. And it was the same on the boat. Most boys would have been scared to death at being sent off alone like this. I'm told he hasn't cried once, unless perhaps at night when no one was around. He said to me just now, 'Do you think, sir, my pail can be fixed? Dribbler and Albert Edward were friends of mine.' Now what do you suppose he meant by that?"

"I don't know, I'm sure. What kind of a family does he belong to?"

"A poor family, if you go by his clothes. But he's got the manners of a young gentleman."

The publisher frowned. "From what I know of this father of his, the boy isn't going to have an easy time of it. I guess some organized effort will have to be made to get him looked after properly." He added in a voice which declared the conversation at an end, "Thanks, Alling."

But he was not through with the incident. A loud feminine voice in the aisle said, "How lucky for you, Mr. Milner, that you're on this train." The publisher realized that one of the worst possible calamities had happened to him, that he had been trapped by Miss Magna Readlong. He looked up into a face like a batch of rising dough, in the midst of it a pair of eyes like small black currants. He extricated himself slowly from the parcels surrounding him and rose to his feet.

Miss Readlong spoke as though she found it difficult to be polite. "I don't know why I'm doing this for *you*. I suppose I've always been over-sensitive to duty, and it's my duty to tell you there's news for you on this train. There's a boy——"

"I know all about the boy, thank you."

"Oh!" It was clear the lady was disappointed. She did not, however, give up. "Are you quite sure you have all the details? I can give them to you if you haven't."

Mr. Milner made no effort to suppress his annoyance. "Great Scott! Do you think I'm a reporter, running around with a pencil and taking down items? I pay people to do that for me——"

"But you don't pay them much, Mr. Milner."

"One of my reporters will meet the train, and he'll get the news about this boy. Don't bother your head about it any more."

"*Very well!* I was just trying to be helpful. You needn't snap at me as though I worked for you." Miss Readlong was breathing hard. "Mr. Milner, I believe in speaking my mind. It's my opinion that you, sir, are ruining the fine, human paper Daniel Cape published for us. You seem to be trying to make the *Star* an imitation of the newspapers in T'ronto. You've had it nearly a year, and it's going downhill fast."

William Pitt Milner's face had become white with anger. "I appreciate your tact in airing your opinions where so many people can hear you. When I'm convinced you're a good judge of newspapers—or of anything, Miss Readlong—I'll give some thought to what you've said."

The indignant lady turned, adjusted her neckpiece of black bear fur, and swept down the aisle with a rustle of skirts like the blowing of wind through a cornfield.

The train started to buckle and buck and snort and then to slow down, advertising the fact that it was coming into the junction. There was no need for people to hurry, because the shuttle train for Balfour would not start until all of them were ensconced in its two poorly heated coaches. Nevertheless, they began to pick up their parcels with uneasy haste and to crowd out into the aisles, where they fought their way forward as though every foot was going to count. William Pitt Milner, getting his packages into his arms and raising himself to his feet, found himself face to face with a man as heavily encumbered as himself.

"Hello, Billy Christian," he said in a tone which would have surprised most people by its amiability.

"Hello, William."

Billy Christian was a man of medium height and of a thinness which manifested itself in the looseness of his collar and the bad fit of his suit. He was dressed in the cheapest of clothes. His mildness of disposition showed in his light blue eyes which had deep wrinkles at the corners, in his tiptilted and far from aggressive nose, in his limp, sandy mustache which had never made the acquaintance of wax. He was a man born to be liked but destined to be disregarded. It was apparent at the first casual glance that he would always be trampled on where money was concerned.

The publisher looked at him suspiciously. "Where have *you* been?"

The eyes of the man in the aisle twinkled with excitement. "T'ronto. I took advantage of the holiday to run down. I went to the library to read up on some things. On"—he hesitated, as though in apology for mentioning his interest in such important matters—"on some scientific principles. And I went to see a man who manufactures machines."

"You're working on another invention." The publisher's voice had an accusatory ring, as though the other man had been guilty of a misdemeanor.

Christian nodded. "Of course. I've always got something on the fire. This time"—his eyes began to sparkle—"I've got a real money-maker."

"You never make anything out of your inventions. Do you remember, Billy, when we were boys together back in Minefield and you made a slingshot which carried a pebble twice as far as any other? All you got out of that was a licking from the teacher. It's always been the same. What trinket are you working on now?"

The inventor tapped the top of the largest parcel. "I've the model of it here. You know, of course, that most wood used in making furniture is veneered. This machine will do away with veneering. It will stamp the surface to make it look like any wood you want. What do you think of that?"

Milner pursed his lips. "I must say, Billy, it sounds more practical than anything you've done so far." Then, not being able constitutionally to stay long on the affirmative side of the fence, he began to scowl. "Who was this man you went to see? Someone you know well?"

"No. Never saw him before in my life. But I was told he was interested in taking on ideas."

"Have you put in for a patent?"

An unconcerned shake of the head. "Not yet. I'll need a more complete model before I can get one. Besides, patents cost a lot of money, more than I'll ever be able to scrape together. I figure on having the man I sell the idea to put up the cash for the patent."

Angry protest showed in the eyes of the most influential man in Balfour. "Billy, you haven't the sense of a two-year-old!" he said. "If you told this man all about your idea—I see you did, by the guilty look on your face—he can have a model made and get the application in before you can turn around. Don't you know enough to protect your own interests?"

William Christian was a carpenter by trade. He worked every day from seven o'clock in the morning until six at night and, as though ten hours of steady labor were not enough, he never failed of an evening to get to work on the ideas which filled his mind; drawing and measuring, hammering and sawing, and gluing pieces of wood together to form his beautifully concise little models. It was when thus engaged on something for which he need not account to anyone (except to Matilda, his wife, who was an eminently practical woman) that he was happy. To be creating something was all that mattered. He had never acknowledged to himself that the possibility of making money out of his ideas did not seem very important, and certainly he had never confessed as much to anyone else. He would talk to Tilly of the fine house they would have when something "clunked," his usual expression, and of the servant girl they would hire; but, when he finished one thing, he was so full of what he was going to do next that it did not matter much if the completed gadget (generally it was not quite finished) had no better fate than to rust unwanted thereafter on a shelf in the summer kitchen. Only distant fields looked green to William Christian. No failure could quench his zeal for the great things he was going to do in the future.

In spite of his easy optimism, however, he found himself disturbed by what the publisher had said.

"The man seemed honest and—and straightforward——"

"Great Scott, Billy!" Milner spoke with the deepest resentment, as though the unworldliness of this old friend who had happened to be born in the same village as himself was in some way a reflection on him. "When a man looks honest and straightforward, keep a hand on your purse! Never take it for granted that a stranger is going to be a friend. Start by thinking he's out to do you, and you'll be a lot better off in the long run."

William Christian smiled at this and unexpectedly nodded his head twice in agreement.

"That's right, William," he said. "That's quite right. You must always look out for Number One. I believe that and I try to act on it."

The train jolted to a stop. The people in the aisle lurched forward, then began to push and shoulder their way ahead. The two parcel-encumbered men moved along with them.

"Billy," declared the publisher, "the last time I heard you say you looked out for Number One was when we were in school together. You had some trading on with a couple of other fellows." He shook his head. "I tried to warn you, but you said you could manage things—that you were looking out for Number One. At the end of the trading you had nothing left but a busted guitar string." He frowned with a half-amused irritation. Then he thought of something else and frowned even harder. "And let me tell you this, Billy. You work too hard. I don't believe I've ever looked over the back fence at night, damn it, without seeing a light in your workshop. You're going to kill yourself if you keep this kind of thing up."

2

The sun had disappeared when the shuttle train came rocking and belching into Balfour, and it was already so dark that the street lighter was busy with the lampposts at the ends of the platform. The illumination thus provided gave the low station of blackened red brick and green wooden trim an atmosphere of cheerful activity.

A man with intense black eyes, who wore neither overcoat, muffler, nor gloves, darted out from the waiting room and came across the tracks to the train. He walked with bent head, muttering to himself. "Nothing from nothing," he was saying. "Nothing from nothing. Nothing remains." His manner changed when he spotted Conductor Cobbler alighting from the front coach. He ran up to him and laid a hand on the lapel of his coat.

"You're late again, Cobbler," he said accusingly. "Seven minutes."

The conductor laughed. "Only seven minutes? Why, that's the same as being on time. We can generally manage to be much later than that." He stopped laughing and went on to say in an impatient tone: "Now look, Jerry Fowley, you're a busy man, being president of two railways like this. You can't spare the time to bother with the likes of me. Why don't you go over to the express offices and make them show you the books?"

Jerry Fowley, whose shoulders were bent with the heavy responsibilities he carried, was not willing to be put off. He drew out a dollar watch of the kind known as a turnip. "Seven minutes *and* fifty seconds, Joe Cobbler," he said. "I won't have this sort of thing. If you're late again, I'll throw you off the pay roll."

For the first time in his life, perhaps, Cobbler had an inspiration. He turned and faced his critic. "Listen, Jerry. There's been a wreck on the other line. Why don't you go over there?"

An excited gleam came into Fowley's eyes. "A wreck?" he cried. "Why wasn't it reported to me, eh? Were they scared to tell me? That was it. They were scared of me. Scared I'd throw the whole kit and biling of 'em off the pay roll." He clutched at Cobbler's arm. "Is the wrecking crew on the job? Has the ambulance been called?"

"Certainly not. They don't dare do anything until you get there. They're standing around and waiting for orders from you. I hear the groans of the dead and dying can be heard all the way to Eagle Nest. You better get over there as fast as you can."

The ruse was successful. Jerry Fowley ran down the platform, shouting over his shoulder as he went, "Call the hospital, Joe!" In something under a minute his bent back and long thin legs had vanished around the corner of the dingy station.

This was the start of Balfour's longest running joke. Every day from that time on, poor Jerry Fowley was sent back and forth from station to station with reports of wrecks. Whereas he had spent his time previously in brisk visits to the freight yards and roundhouses, staring through the windows at busy clerks and issuing absurd orders in his deep voice, he now kept on the run, shouting the news of death, suffering, and destruction, and babbling orders for the relief of the victims.

Publisher Milner materialized out of the dusk at Cobbler's shoulder and asked, "Where's the boy?"

"Right here, Mr. Milner."

Ludar was standing behind the conductor of the shuttle train and looking about him in astonishment at the huge quantities of snow. The branches of the trees were weighted down with it, and it was piled up in globular whiteness on the roof of the station. The wind was rising. The boy, in his overcoatless condition, was shivering with the cold.

"Well, my boy," said the publisher in a voice which he strove to make kind and sympathetic, "here we are. We must see about getting you out of this at once. I don't suppose you're accustomed to this kind of weather."

"N-no, s-sir," answered the boy, his teeth chattering.

The publisher frowned in the direction of the station platform. "Where is the confounded fellow! I'm expecting one of my staff, Cobbler. Bates, I imagine."

"And here he comes." The conductor pointed. "Never knew him to miss this train. Not Sloppy Bates. He gets too many good items out of me. About people who've been traveling. I guess your paper couldn't get along, Mr. Milner, without me and the six-two train."

The stooped figure of a man approached them, peering about him as he walked, and staggering slightly. Resentment had been piling up in the mind of the irascible publisher, and he now proceeded to vent it all on his inebriated employee.

"You drunken fool!" he exclaimed. "What do you mean, being in this condition when you represent the *Star?"*

"Of course I'm drunk." Bates said this in an aggrieved tone. He nodded his head, on which he wore a dilapidated tam-o'-shanter. "Why, damn everybody's eyes, I'm always drunk at this time of day. You know that, Mr. Milner. Everybody knows it. As this is New Year's, I'm just a little drunker than usual."

"I wish there was some way of getting rid of you!" said Milner, literally gritting his teeth.

"But there isn't." The old reporter performed a dance step with his stiffened legs, something between a buck and wing and the first movement of a Highland fling. "There isn't, Mr. Milner! You'll never be able to get rid of me."

This was true. Daniel Cape, who had founded the *Star* and had run it for thirty-five years by methods and ideas all his own, had been very reluctant to sell the newspaper and had resisted the importunities of Milner for many years. When poor health made it impossible for him to continue he had given in, but he had insisted on inserting a list of conditions into the agreement of sale. This, the new proprietor must always do; this, he must never do. One of the things he must never do was to discharge John Kinchley Bates, who had been a faithful reporter almost from the birth of the paper. Unfortunately for Bates himself, the exacting founder had added a further condition. "The aforesaid Bates is never to receive an increase of pay until he has gone three months without alcoholic stimulants passing his lips, his drinking habits having always militated against the proper execution of his duties and having been a constant bone of contention between him and the party of the first part." Thus the party of the first part, Mr. Daniel Cape, had succeeded in having the last word in the long bickering. As Sloppy Bates could not conceivably go three hours without a drink, not to speak of months, he would never get the raise the stern founder had denied him.

Milner restrained the capering of the aforesaid Bates with both hands. "Now listen to me, you clown," he said. "I may not be able to fire you, but I can put you at sweeping out the office and cleaning the toilets if I've a mind to. I've been thinking of doing it——"

Bates cackled. "I wouldn't be good at it, Mr. Milner. I promise you, the office will get into a state if you do."

"Forget about that now. You've got to pull yourself together and handle a story. I hope you're not too muddled to get the facts straight."

"I write my best when I'm drunkest," declared Bates with an air of offended dignity. "Everybody knows that. I saw six moons in the sky the night I wrote the murder of Fanny Barley. It was after that when Mr. Cape said he would fix things so I would never have to leave the *Star*."

"If I didn't know that Daniel Cape was a teetotaler I would say he was drunk himself when he made that promise. Because you were so intoxicated that you wrote a good account of the murder of an Indian girl, I've got to put up with you and your filthy ways as long as you live!"

"Perhaps not. Perhaps I'll decide to buy a newspaper of my own." This idea struck Bates as so funny that he cackled still louder. Then, regaining some degree of sobriety, he went on. "What is this story? Let me at it and I'll write a report which will march through the columns of the *Star* like a regiment with colors flying."

Milner drew him aside and proceeded to tell him what the story was about.

"Him?" asked Bates, pointing a finger in the direction of the shivering boy.

"Yes. I want you to go along with him and his father and get everything. It's human-interest stuff, Bates, and I'm sure it's worth a lot of space. Get the boy talking about what happened to him on the way over. And find out all you can about this man Prentice."

Bates let out a sound like a cork being drawn from a bottle of wine. "What did you say, Mr. Milner?"

"I said to question the boy's father, this Englishman Prentice."

There was a moment's silence while Bates teetered on his unsteady legs and gazed at his employer with astonishment and dismay.

"No!" he said finally. "It can't be. I tell you, it's impossible. It's against all the rules for a thing like this to happen. Damn everybody's eyes! It can't be so!"

"You're so drunk you don't know what you're saying," declared the publisher in a tone of final disgust.

Bates drew himself up as straight as the permanent crick in his back would allow.

"Mr. Milner," he said, "your story has turned into a tragedy. The boy's father isn't here to meet him. A telegram arrived for him this afternoon. Being a holiday, they didn't deliver it right away. In fact, it was just an hour ago that the messenger went to Creek and Stapley's stables, where he's been sleeping. It was just too late. The poor fellow had taken a shotgun and blown the top of his head off."

Chapter IV

I

The publisher of the *Star* had buttonholed William Christian as the latter was leaving the train and had suggested he accompany the boy to the police station. "I'm in a great hurry and can't go myself, Billy," he said. "But I don't like the idea of leaving him in the care of this drunken reporter of mine. Just go along with them like a good fellow and sort of keep an eye on things." William Christian was in a great hurry too. If he didn't get home promptly, Tilly would be pacing the floor when he arrived or standing at the door of their cottage at 138 Wilson Street, her arms grimly akimbo, her expression unforgiving. But he had nodded his head and said he would go along.

There was no mistaking the police station. It was a three-story building which seemed as unsteady and teetering as the drunks it harbored, its brick walls splashed over with a dark red paint, and a gaslight, blazing away like the all-seeing eye of Justice, above the front entrance. The boy drew back at the worn stone steps and said, "I don't think I like this place, sir."

"No one likes this place, sonny," said William Christian. "But you and me, we don't have to care. We're just going in to see about getting some things done for you."

On the way from the station in a hired hack (paid for by the *Star,* of course), the carpenter had realized that the boy could not be kept in the dark and had told him that his father had gone away for a long time and that he, Ludar, would not see him after all. The boy had looked frightened and had asked in a tense whisper where he would go now. The explanation that he would probably be sent back to his relatives in England had thrown him into an immediate panic. He had clutched at Christian's arm and had said no, he would not go back. When asked if he didn't like the lady he had been living with he had cried and said that she was a very mean lady and didn't like little boys. Realizing then that he was in deep water, Christian had said nothing more.

"You'll laugh yourself sick, boy, at the things which go on here," said Sloppy Bates. Now that a good look could be had at the reporter, it was apparent that he richly deserved his nickname. Two buttons were missing from his threadbare overcoat, which came almost to his heels, and the black ribbon of his tam-o'-shanter was dangling down his back. "Yes, boy, they're out-and-out comedians in here."

The boy still held back. "Come, Ludar," said Christian. "I'll stay right with you. All you'll have to do is to sit down for a few minutes in a nice warm room. You won't have to say a word."

They found themselves, after entering, in a room which opened to the left of the main door. It was heavy with what is called the institutional smell. A big man in uniform was sitting behind a high desk and growling at two disheveled individuals standing in front of him in the care of another officer.

"Eliven of them," said the big man in a pleasant brogue. "And all in one afternoon."

"There'll be more before this night's over with," declared the other officer.

"Do they think all the whisky in town must be lapped up in one day? It makes me ashamed of the human race, it does." The sergeant turned to one of the delinquents. "So, it's you again, is it, Connie Stockarrah? You'll have to tell me how to spell your name. It fools me every time. How can people have such names?"

"You don't have to spell it," said the drunk indignantly. "You just write it down."

Bates went to a corner of the desk between the two drunks and said to the sergeant, "If it pleases the guard who holds the flaming sword at the entrance, I'll have a word with the Lord of Law and Order."

"Be getting along to the chief's office, then," said Sergeant Feeny, treating the reporter to an unfriendly stare. "And don't be giving me any more of this highfalutin talk."

Bates crossed the hall and vanished through a door to the right of the entrance. For ten minutes or more Christian and the frightened boy waited. When the reporter returned he nodded to them and pointed a finger at the ceiling. "We're to go up. The Lord High Mucky-muck's holding special court tonight, and we'll have a consultation as soon as it's through." Then, in a sudden burst of drunken laughter, he asked the boy, "Well, Masther Ludar, and is it the good laugh ye've been gettin' out o' Sergeant Feeny?"

The face of the man at the desk, with its long upper lip and wide mouth, flushed with the slow anger of a good-natured man. "And it's a laugh I'll be havin' at you, Mister Bates, when ye're booked with the rest of the drunks as you deserve."

2

A stairway had faced them as they entered the building, and they proceeded now to climb it, taking great care in doing so, for it was easily the steepest stair ever designed by an architect faced with lack of space. At the head of it was an open door leading into a high room which made Ludar think of a church because it had long narrow windows and an elevated chair, which looked not unlike a pulpit, on the south wall behind a railing of yellow wood. The resemblance did not go farther, for the place had no suggestion of the peace and stillness of a church; and the atmosphere, in spite of a round ventilator in the center of the ceiling, was as heavy as in the booking room below.

There was an old man sitting on the elevated chair, directly under a large crown on the wall. The old man was leaning his elbows on the yellow desk. He looked very tired and his eyes were closed.

To the left of the magistrate's chair was a small square space closed in by a solid wooden partition, and in this stood a skinny little individual with a sly face and a strip of court plaster on his nose. An officer was reciting the details of the prisoner's misdoing, which had taken the form of a brutal beating of the woman who stood to him in the relationship of common-law wife. When the sordid story was completed it became clear that the hunched figure on the bench had not been sleeping. The eyes opened at once.

"It is an indictment of our civilization, surely," the magistrate began in a low voice, "that a man who has spent his whole life in the service of the community must sit with aching bones on a hard chair on a holiday, which is a day of rest for everyone else, and listen to the revolting details of the wickedness of nasty little specimens of the human race!"

William Christian and the boy were in the front row of seats reserved for the public. The former had watched Ludar closely at first. As soon as the magistrate began to speak, however, he became so intent that he paid no attention to his charge. It was not until he heard the boy's feet touch the floor that he looked down again. He was just in time to reach out a restraining hand as Ludar began a stealthy exit.

"What's the matter, young fellow?" he asked in a whisper.

Ludar struggled to get away. "Please, sir," he pleaded frantically, "let me go!"

"But you can't go yet. Some men are going to talk to us in a few minutes. It won't be long now."

The boy continued to struggle. "Please, sir, I'm afraid of that old man."

William reached down and lifted the boy back into the seat beside him. He lowered his head and whispered, "You mustn't be afraid of Mr. Jenkinson, because he's really a very kind old man."

Ludar was keeping his head down, as though he did not dare risk another glance at the stern figure above him. "He doesn't look kind, sir. He looks like the giants who eat little boys."

The head of the magistrate turned in their direction. "Must I remind spectators, who are here only at the discretion of the court," he demanded, "that I do all the talking in this room and that I am not to be interrupted?"

He proceeded then to demonstrate how well qualified he was to do the talking. He excoriated the prisoner and depicted the crime of wife-beating as one of the most heinous on the calendar. While doing so he often paused in the delivery of his well-rounded sentences to utter asides in a low tone. They were not intended for ears other than his own, but long practice of the habit had made him careless and now all his interpolated remarks could be heard. Christian covered his mouth with his hand to conceal his smiles, and Sloppy Bates sniggered aloud several times as the peppery old magistrate interlarded the delivery of sentence with such fragments of unofficial opinion as "Sneaking, crawling rat!" "Lecherous ghoul!" and "Pindling little Bill Sikes!"

For five full minutes the magistrate kept his eyes on the pad in front of him while he indulged in a verbal lashing which cut like the cat-o'-nine-tails. Then he raised his glance to the prisoner.

"I sentence you, Abel Padner, to three months at hard labor in the county prison."

"Well," whispered Christian to the reporter, who sat on his other side, "Mr. Jenkinson certainly was going it."

"He's always going it," answered Bates. He glared up at the bench.

"The other day he was going it at me. Said he didn't want reporters in his court who were drunk." The reddish-topped head of the newspaperman gave an angry snap. "And I couldn't say a word back. I *was* drunk. . . . I know something about that croaking old raven that would cause trouble for him if I let it out. He's going blind. From where he sits he can't see the witnesses or the prisoners in the dock. Parsons has to whisper things to him. If the public knew about it he might get bumped off the bench."

The police officer led the prisoner out of the wooden box and toward a stairway beside it which descended to the cells below. Abel Padner, however, was not yet content to leave court. He drew back and, looking up at the judge, indulged in a bow.

"Thanks for all your kindness, Judge, Your Honor," he said in a thin, high voice. "You'll never need to beat your wife, Judge, Your Honor. All *you* will have to do is talk to her like you talk to helpless prisoners. She'll suffer more than if you gave her a good flick in the face with your fist."

The magistrate turned in his chair and stared down at the man. Then he smiled and bowed back.

"Was it not enough, Abel Padner," he said, "that I sat for an hour and listened to the recital of your bestial conduct while the white-hot pincers of rheumatism racked my frame? No, seemingly it was not enough. You have had the effrontery"—*You cowardly muckworm!*—"to display insolence in this court of the sovereign people and to express contempt of the appointed instrument of the King's justice.

"There was a time," went on the indignant voice, "when I would have given you the punishment you deserve, a round dozen lashes as well as a term of detention. But there were people who thought this too brutal"—*The weaklings, the soft puddings, the driveling psalm singers!*—"and so it became necessary for me to bow to their will. But what has just happened in this court cannot be passed over, and I shall revive in a certain form the sound punishment in which I still believe. It is the further sentence of this court that on a day to be determined by your jailers, you, Abel Padner, shall be led out to the courtyard of the county prison. The woman you have so cruelly assaulted shall be present, if she so desires, and she shall be allowed to apply five lashes to your bare back, if again she so desires, with a whip provided by the authorities for the purpose."

Fear had driven all insolence from the prisoner. "Judge, Your Honor!" he cried. "You're not in earnest. You can't be. This is a joke, Judge, Your Honor. You don't know the woman or you wouldn't do this. She'll use the whip, Judge. She'll lace it to me so hard she may kill me!"

The eyes of the old magistrate shone with the purest pleasure. "I am glad to hear it. I repeat the words of the great poet who wrote *John Gilpin,* 'May I be there to see.' "

When the protesting prisoner had been forced to descend the stairs the magistrate straightened his bent back and asked the court clerk, who sat at a flat-topped table beneath him, "Well, Parsons, is there anything to delay further a very weary and very sick man?"

"There's the matter of this boy, Your Honor. The chief will be up in a moment to explain."

Ludar grasped the arm beside him. He whispered: "I'm afraid, sir. What will they do to me?"

3

The head of Chief Jarvis appeared with the suddenness of a jack-in-the-box in the stairway down which Abel Padner had just disappeared. He was sandy of complexion and rather amiable of expression. His uniform was a quiet one and absolutely devoid of brass buttons, a black serge with strips of black ribbon by way of decoration.

He stepped up beside the magistrate, and for several minutes the two chief custodians of the law in Balfour conferred in low tones. Then Mr. Jenkinson nodded his head with an air which suggested a reluctance for the task ahead of him.

"The present problem, then," he said, "is to make arrangements for the temporary keep of this boy."

"Yes," said Chief Jarvis. "I've already taken steps to locate the relatives in England, but there may be some delay about getting the information."

"There shouldn't be any delay," protested the old man. "The cable was laid across the Atlantic a great many years ago. Have you heard of it?"

The chief of police smiled as though accustomed to the irascible humors of the old man. "I've already cabled the steamship offices in London," he said. "But there *is* a difficulty. A very real one. I have reason to believe that Prentice is an assumed name. It's certain from some letters we've found in the man's things that he belonged to a good family, and it's probable he assumed the name of Prentice when he came out to Canada. Perhaps he was in some kind of trouble and was running away. Or it may have been through some quirk of pride. In either case, it may take the shipping people a while to get to the bottom of things."

Magistrate Jenkinson tapped his knuckles thoughtfully on the cork top of his desk. "What, then, do you want me to do, Chief? Issue authorization for the boy to be kept at the Orphans' Home until he can be sent back home?"

There was reluctance also in the nod given by the police head. "I'm afraid that's the solution, Your Honor."

William Christian was listening to every word, but he was not watching the two figures on the raised platform. His eyes were fixed instead on the small boy snuggled up close beside him as though seeking protection. He was a nice-looking boy, he said to himself, just the kind of son he would have liked if Tilly had been able to have one. The little English boy had fair hair with the slightest touch of gold in it and a natural wave across the forehead. His features were rather delicate, and his eyes were a light blue. "He's going to make a fine man," thought the carpenter. "And he's smart. As smart as they make 'em at that age."

The formidable pair on the platform had turned their eyes in the direction of the three watchers in the front row, although the magistrate would not be able to see anything of them, if what Bates had said of him was true. It may have crossed the mind of the chief that there was a similarity between the man and the boy, the kindly, diffident man and the small waif whose immediate fate rested in their hands. It was clear, at any rate, that both were very much disturbed over what would come out of this conference.

Without looking up Ludar whispered, "Are they talking about me, sir?"

"Yes, Ludar. They're trying to decide where you're to stay."

"Couldn't I—couldn't I stay with you, sir?"

William Christian had considered this solution but had put it away because of the opposition which could be anticipated from Tilly. It would be pleasant, he thought wistfully, to have a youngster about the house for a while. He found himself marshaling arguments in his mind which could be used to convince his strong-minded spouse. The Reverend George Grantly would approve, and Tilly was one of his most devoted supporters. The dead Vivien Prentice had belonged to a good English family; and Tilly, being a Hull of Coldwater, would be impressed by that. It would not be very expensive.

"You!" said the magistrate suddenly, waving a hand in the general direction of the group of three. "Sitting beside the boy. Come up here and tell me what you have to do with the case."

William Christian looked more diffident than usual, if such a thing were possible, as he pushed his way through the swinging gate in the railing and walked up beside the table where the clerk sat.

"I've nothing to do with the case, Your Honor," he said. "I was on the train from T'ronto, and Mr. Milner asked me to—to sort of look after the little fellow until it had been decided what was to be done."

"And what's your name?"

"William Christian. I live at 138 Wilson Street."

"I've heard of you, Mr. Christian, but I'm sure you couldn't have been born in Balfour. I fail to recall any family of that name."

"I was born on the Isle of Man, and my father brought me to Canada when I was two years old. He settled on a farm near Woodstock, and I never saw Balfour until I came here to learn my trade."

"A Manxman, eh?" Magistrate Jenkinson placed the palm of one hand over his eyes and, opening a space between two fingers, stared down at Christian through this triangular opening, an aid to vision which he employed often. "There was a great Manxman once named William Christian. Are you a descendant of his?"

William's face flushed with pride. "I often heard my father say he was sure there was a connection."

The magistrate had fallen into an amiable mood. "A descendant of William Christian should be of some help to us in a matter like this. What do you advise we do with the boy?"

And then William Christian found himself taking a stand without any fear of consequences at all. In a lifetime of striving to please other people there had been few occasions when he had been so rash. "Let him stay with me," he heard himself say. "I'm sure my wife would be glad to take care of him."

But would she be glad? Immediately he had spoken he began to have doubts. Suppose he took the boy home with him and Tilly put her foot down with that unshakable determination of hers? What could he do then? At the very least he would have made a fool of himself publicly.

"Now this is a solution I confess to liking much more than the other," declared the magistrate. "Whenever I inspect the orphanage I come away depressed with the ills of the world. It would be with a reluctant hand that I would sign an order sending this boy to that den of hopeless childhood." He closed his eyes and gave the matter some thought, snuffling as he did so, for he had a bad cold. "It's certain I could get the other police commissioners to agree that we should pay for the boy's keep until he can be sent back: for, say, a space of time not to exceed two months. Would a dollar and a half a week suffice? Perhaps two dollars would be a fairer amount."

The chief of police had been watching the carpenter while this conversation went on. He now asked a question. "Mr. Christian, didn't you marry a Miss Matilda Hull?"

William bobbed his head in response. "Yes, Chief."

"I remember her. A tall, dark girl. In fact, I knew several of the Hulls of Coldwater." Perhaps an association of ideas prompted his next remark, which was addressed to the man on the bench. "I'm inclined to think, Your Honor, that the weekly payment should be fixed at two dollars."

William felt a wave of intense relief sweep over him. Now he could see his way clearly. The chance to have this additional income would appeal to Tilly so much that it would be a relatively easy matter to reconcile her to this step he had taken without consulting her. She was a Hull of Coldwater through and through when it came to money. How fortunate that Chief Jarvis had known some of the Hulls. Well, he didn't know a tenth of what he, Billy Christian, knew about these hardheaded, grasping people who had developed such a reputation for themselves in the country above Balfour and who, moreover, had such a belief in their own importance. Did the chief know Uncle James Hull, for instance, who saved nine hundred dollars out of the first thousand he earned? Had he heard that Cousin Abimelech had been so outraged by his mother's will that he had gone to her grave on the night after the funeral and destroyed all the flowers, pounding them into the earth with his heels?

The old magistrate was speaking. "Then we may consider that we have settled this matter on the best possible basis." He waved a hand in dismissal of the matter and then blinked his eyes in the direction of the cause of all this discussion. "Perhaps now I may be allowed a closer look at this most unfortunate young gentleman."

William walked back to the front row to bring Ludar up for inspection. Although tongue-tied and self-conscious himself in all his dealings with people, he had a way with children, and in a few moments he had succeeded in calming the fears of the boy. "Now, my fine big fellow," he whispered, "it has all been arranged that you're to come home with me, just as you wanted. Yes siree, you're going to live with me and we'll have lots of fun. I've a fine gray cat——"

"I love cats," said Ludar. "But I've never had one of my own."

"This Old Cat of mine has had seventy-nine kittens in her day. What do you think of that?"

The boy's interest was now completely won away from the ordeal of going up to be inspected by the fierce old man. "Have you all of the kittens still, sir?" he asked.

William patted his head. "No, there isn't a single one left. They've all gone out into the world, those grays and blacks and tortoise-shells of Old Cat's, and most of them are grandparents themselves by this time. But we're making the kind judge wait. See, you take my hand and come up with me to get everything settled."

Ludar walked to the front and was lifted up to a standing position on the table of the clerk. Mr. Jenkinson repeated his habit of squinting down between his fingers in order to get a clearer view. Christian saw that Ludar's knees trembled while he underwent this trying inspection, but otherwise the boy made no sign. "There's a little soldier for you," thought the carpenter proudly.

"A regular young Saxon, this boy," said the magistrate. "I can see he's of the right stock, descended straight from the men who marched with Alfred and died with Harold at Hastings. I think he may grow into a useful citizen, Mr. Christian."

William made an attempt at a joke. "I don't think, Your Honor, that he'll grow up to be a hod carrier."

The attempt was a success as far as the magistrate was concerned. He indulged in a rumble of laughter. "No, this boy won't be a hod carrier. With a head shaped like that, he might become a schoolteacher, or even an artist or musician. But not a man who carries bricks for other men. And now, turn the boy around if you please."

The sight of the placard gave the old man another moment of amusement. "What an idea! This sign brought the boy safely from England, across the ocean, and to my court. But now, I think, the time has come to relieve him of it."

Chief Jarvis took a jackknife from his pocket and ripped out the threads holding the square of oilcloth to the boy's coat. "Here we have Exhibit A," he said.

"What are you going to do with Exhibit A?" asked the magistrate.

"If I might make a suggestion," said William, "I think it should be kept for the boy. When he grows up he's going to be proud of what's happened to him. No other boy's been sent out all alone to go halfway around the world. He'll want to talk about it and he should always have the sign to show."

The magistrate nodded his head in complete approval. "An excellent idea. You are a man of discernment, Mr. Christian. Our young man here will grow up, and he'll want to show this sign to his own children and grandchildren. And now, gentlemen, I believe I may take it on myself to seek the comfort of my fireside. Will you permit me, Mr. Christian, to send you and your tired charge home in a hack at my personal expense? You see—ha, ha!—we are all under direct orders in this matter." He pointed at the sign. "We must be kind to him."

4

"Come to my room, Mr. Christian," said the chief. "I've something to show you."

The office of the head of the force was small, and the chairs it provided for visitors were in such bad repair that the stuffing was sticking out in many places through the horsehair coverings. There were a number of photographs of women on the walls, and one of these showed a plump female, with an enormous feather in her hat and a sash draped over her hips, standing between two heavy-looking men. The men were in plain clothes, but there was no mistaking the fact that they were police. This was not surprising, for the woman was Mrs. Rosa Rushon, who had been tried for the poisoning of her husband. She had been acquitted and had immediately turned ugly, bringing suits against everyone concerned, the judge, the lawyers, the witnesses, the press, even one of the jurors who did some talking afterward. Nothing had come of the suits, and eventually the broad-beamed Rosa had gone out west and married a minister. Chief Jarvis kept her picture in full view as a warning to the force on the need to be careful about the nature of evidence.

To lend respectability to this gallery Mr. Jarvis had a framed picture of a pleasant-faced woman in a black hat, Mrs. Jarvis, no less, and a steel engraving of Queen Victoria.

Seating himself at his desk, the chief drew out an envelope from a drawer, and from the envelope produced a watch. This he placed in Christian's hand.

"Look at it carefully and tell me if you see anything curious about it."

After an inspection of several moments the carpenter said, holding the watch on the palm of his hand as though weighing it: "It's heavy. A lot of gold in it."

"Don't you see anything else?"

"N-no. I'm afraid I don't."

"Look at the initials on the back of the case."

William turned over the watch and then said: "Now I see. The initials are V. R. and not V. P."

"Exactly." The chief nodded his head with satisfaction. "The man's real name was not Prentice but something beginning with *R*. What's more, the people in England who sent the boy out were aware of the fact. They used the assumed name for the boy."

He replaced the watch in the envelope, and the envelope in the drawer. Then he produced a small and somewhat faded picture of a group of children standing by a dogcart to which a fat little pony was hitched. This he also handed across the desk to his visitor.

"There's something about this."

In a childish hand at the bottom of the photograph appeared five names: Nancy, Poppy, Cyril, Vivien, and Goonhilly.

"I suppose," said William, "that this curious word——"

"Goonhilly?"

"Yes. Does it refer to the pony?"

"Naturally. The goonhilly is a type of pony, raised, I think, in Cornwall. Or Devonshire."

Chief Jarvis retrieved the picture and dropped it back into the drawer. "With these two bits of evidence we may safely take it for granted that the name Prentice was assumed by the man. It's clear also from the value of the watch—he seems to have preferred to starve rather than part with it—and the fact that the children had a pony—that the family had position and wealth. The last piece of evidence we found in his belongings was this letter from one of his sisters. At any rate, it's signed Poppy, which would be the pudding-faced little girl second from the left, and it's dated nearly two years ago. All names are carefully scratched out. I'm not going to read you the letter, but it offers more proofs of the importance of the family. She tells him his wife is dead—that would be the boy's mother, without a doubt—and that she wishes she could send him more money than the small postal order enclosed but that she and her husband, who is referred to only as Dear Old Fishface, have bought a place of their own with a trout stream and are strapped. She says she has named her second son after him and that she wishes 'dear old Viv' would be sensible and come home. She's sure that all would be forgiven. Now there we have a clear picture of this man's background."

Christian nodded his head. "What a sad end for him!"

"The coroner says he was pretty far gone with consumption. No doubt that decided him to make an end of it."

Sergeant Feeny put his head in at the door. "The hack's here," he announced. "He's stewed, Barney Grim is, and it's quite a ride you'll be having, mister. If someone yells at him, 'Barney Grim, who's your father?' and he gets his whip out to be lashing at them, you just be taking it away from him. It will be trouble you'll save by doing it."

William nodded his head, although it was highly unlikely that he would ever assert himself to the extent of taking a whip out of the hands of a drunken driver. He was, as a matter of fact, paying little attention, his mind being fixed on the need to get home as soon as possible.

"And speaking of fathers," went on the sergeant, "it would be only right and fitting to take the boy to the undertaker's on the way. It's the one chance he'll be having to see his father."

"I don't think that would be wise," said William, becoming attentive at once. "Children shouldn't be shown dead people. It would frighten a small boy like this."

The policeman looked at them blankly. "He'd be scared, you say? Why, mister, there's been all manner of people asking to see the body and bringing in children with eyes on them as round as saucers. And why should the sight of a good clean corpse frighten them?"

"Mr. Christian's right," said the chief.

5

Barney Grim had enjoyed a good New Year's Eve drunk, but now he was getting over it. He was muttering to himself when they started out, but there were no boys on the street to jeer at him and allude to his clouded paternity, and so the whip was left in its socket. The hack proceeded at a rate which might be called morose.

It was a night of supreme beauty, cold and clear and still. The stars twinkled and the snow was packed down so hard on the streets that the steady clippety-clop of Barney Grim's old horse had almost the effect of gunnery. The pleasant light of oil lamps, which lacked so completely and so happily the all-pervading insistence of electricity, was seen in the windows of the quiet houses. It was suppertime, and the streets were deserted save for one tall young man in a form-fitting gray overcoat and a pair of brown fur gauntlets who was singing in a low voice, "Juanita . . . Oh, oh, Juanita . . ." as he strode along. An icy crossing sent him sprawling and brought the music to an abrupt end. Barney Grim said, "Blasted dude!" and, as though revived by the incident, took out his whip for the first time and lashed at the horse.

They swung at a much sharper gait around St. Paul's Church, where the organ was booming (Sterling Rennie, the new organist, was playing there for his own pleasure); and a sleigh, heaped high with rugs and with sleigh bells jingling at a furious rate, came around the corner from Grand Avenue. Christian saw that it contained Mr. and Mrs. Tanner Craven and knew that they were following their usual custom of leaving the house, after the last reception guest had departed, for a solitary dinner at the Cameo House. He knew they would have exactly the same dinner as on all previous New Year visits: celery and olives (William had never seen an olive but had heard it was a most unpleasant-tasting thing), a thick vegetable soup, roast turkey, cold ham, cold tongue, mashed potatoes, canned peas, and plum pudding. "The nobs!" he thought with a shade of bitterness.

He dismissed them at once from his mind and began to think of a more important matter. Tilly would be looking through the colored glass panel at the front door and would be surprised to see him come dashing up like this. "She'll think I've taken leave of my senses," he said to himself anxiously.

He began then to rehearse the case in his mind, wondering what would be the best way to break the news. By the time they came to the corner of Wilson and Bently, which was only a block from the house, he had decided. He would jump right into the heart of the matter, saying to her, "Well, Tilly, here's a real stroke of luck, a sum of money every week. You'll be able to add every cent of it to your principal." No, on final thoughts he decided to leave out the reference to principal and let Tilly come around to that herself. The money she had laid away, most of it being a legacy she had come into, was a favorite topic with her. "I added to my principal today," she would often remark, folding her hands in her lap and looking important and proud, even though the addition could not have been more than a dollar or two. William had no idea what the bank account contained. On occasions when he worked up his courage to speak of the need he had for money, for some special wood, perhaps, or for castings, she would be quite furious with him. "Do you think, William Christian, I would rob my principal for you and your inventions!"

He was getting nervous now that the testing time was at hand. Perhaps it would be wise to coach the boy as to how he should act.

"It's this way, young fellow," he began in an apologetic tone. "You're a well-mannered boy, and I want you to be the politest you've ever been in your whole life when you meet your new aunt. You say, 'Thank you, ma'am,' and 'I won't be any trouble, ma'am,' and don't speak until you're spoken to. I wonder now if we could think up something complimentary for you to say, something that would please her right off. Let me see now. . . ."

The boy's only response was a deep sigh, followed by steady breathing. He had fallen asleep.

Chapter V

I

Ludar awakened to a sense of deep comfort. He was free at last of the clothes in which he had spent so many weeks. His body had the relaxed feeling which follows a steaming hot bath (he had roused only partly in the tub), and he was wrapped from chin to toe in a garment which had served William once as an overcoat (how old and worn it must have been for him to discard it!) many years before.

He was sitting in a Morris chair in which pillows, collected from the beds, had been piled up to serve as cushions. The room was not large and must have been in the center of the cottage, for there were five doors opening off it, and two windows. It was plainly furnished: a much-nickeled coal stove which was throwing out a wonderful heat and had at least twelve feet of pipe suspended on wires from the ceiling, a tall and ornate secretary of reddish wood (William had made it in his spare moments) with six or seven books in the space provided for fifty, a table with a chenille cover on the middle of which stood a lamp as round as the belly of a Buddha and with a shade of the same proportions, and a carpet-covered stool with a wobbly leg.

Through one of the doors he caught a glimpse of a larger room which must be the kitchen, because it had a coal range and a wood box and a string of pots and pans on the wall. It was serving also as a dining room, for there was a table with a red cloth at which two people were plying knives and forks in the total silence possible only to couples who have been married a long time. William was one of them. He was facing the door, and at the other end of the table was a severe-looking lady in an apron which covered her from head to foot.

William perceived at once that the boy was awake. He put down his teacup, brushed off such moisture as might be clinging to the tips of his mustache, and came into the room where Ludar sat. He said in a hearty tone, "Well, young fellow!" at the same time nodding with an expression which hinted at a secret between them. When he came close enough he said in an urgent whisper: "Remember what I told you, sonny. Put your best foot forward."

As Ludar had been asleep when the matter of his conduct had been mentioned, he was at a loss to understand why he should put his best foot forward. It added to his perplexity that he did not know which was the better of the two and that both were bound up so securely in the old coat that any movement at all was practically impossible.

The lady in the apron came into the room with a look on her face which said, as even a small boy could tell, "Just try to make me like you!" She would have been a rather good-looking woman if it were not for a habit of compressing her lips and drawing them down at the corners. Her eyes were a light blue and her hair was dead black, and she had, it must be confessed, a rather large and formidable nose.

"Well, Ludar," said this forbidding lady. She added after a moment, in most proper tones, "I am informed that such is your name."

"Yes, ma'am." The boy had been put on the defensive immediately. "Ludar Prentice, ma'am."

"I was informed of the whole name."

William, standing behind his wife, was forming words with his mouth: "Aunt Tilly. Call her Aunt Tilly." The boy showed his nimbleness by grasping what was meant quickly.

"Yes, Aunt Tilly," he said.

Mrs. Christian turned and said accusingly to her husband: "You put him up to that. Don't tell me you didn't, Billy Christian."

William became apologetic. "I said it would be best if he considered us his uncle and aunt while he lived with us. He has to call us *something,* doesn't he?"

His wife continued to face him as though determined to stare him down. "I won't be shoved into liking him. I won't be rushed, Billy Christian. You should know that by this time." As indeed he should. She even began each day with such a determination not to be rushed that he would prepare and eat his breakfast before she as much as turned over in bed.

"Don't you think he should have some supper, Tilly?" suggested William.

Mrs. Christian turned back and looked at the occupant of the chair in which ordinarily she spent her evenings. "Are you hungry?" she asked.

"Not very, ma'am."

William felt an explanation was in order. "He has a weak stomach. He was sick on the train coming from T'ronto."

His wife turned stonily in his direction. "And why do you think it necessary to tell me he was sick, Billy Christian? Didn't I see the state his clothes were in? Are you reminding me that *I* was not on the train from Toronto but here at home alone all day?"

William smiled at the boy as though telling him that none of this meant anything at all. "Would you like some soup, Ludar?"

The boy nodded. "Yes, sir."

"What kind do you like best?"

"There you go!" exclaimed his wife. "Spoiling him already. Does it matter what kind he likes best when there's a pot of one kind on the stove this very minute?"

William looked abashed. "It's barley soup, sonny. I hope you like it."

As it happened, barley soup was a special favorite with the boy. He answered, "I like it very much, sir."

"Good. I'll get you a plate right away. It's a fine and filling dish."

He was back in a few minutes with a tray on which reposed a steaming plate of the soup, a slice of ginger cake as piping hot as the soup, and a large spoon. Ludar took a bite of the cake, and his face lighted.

"Um-mh!" he said. "That's good, ma'am."

It was clear at once that some progress was being made. The severity of Mrs. Christian's expression seemed to diminish slightly. She did not smile, but she said in a voice which told that she had been pleased: "Eat it up, boy. There's more in the kitchen."

A knock sounded on the front door. William said, "That's your uncle Alfred." He could never be mistaken about Uncle Alfred's knock and often said that, when the Lord arrived at the front door of the world to announce that the end of things had come and the judging would begin at once, He would knock in exactly the same way. He moved with reluctant steps to admit the visitor.

2

Alfred Hull deserved the name of Calamity by which he was known in Balfour. He was an elderly man with a long black beard and an air of deep gloom, wearing a pea jacket, a variety of overcoat much in use because of its relatively small cost (it came only a few inches below the waist), and with a limp leather Bible under his arm.

His greeting, as he came into the room with a shuffling step, consisted of one word, uttered in a tone of reproof, " 'Tilda!"

"Yes, Uncle Alfred."

"I've heard what you've done." His voice implied that the Christians,

William and Matilda, had murdered someone and buried the body in the cellar, or had been guilty of something equally terrible. His eyes roved about the room and came to rest on Ludar, who had finished his meal and, for the first time since his pilgrimage began, was feeling no ill effects. "So! This is the boy. This is the boy, is it?"

William asked in a tone which was almost sharp, "How did you hear about it, Uncle Alfred?"

"Sergeant Feeny's sister lives next door to us. She came over before she had finished the supper dishes and told us the news. I lost no time. I put on my hat and coat—*and* my muffler *and* overshoes—and here I am!"

Calamity Hull helped himself to a chair in the kitchen and carried it in. Placing it directly in front of the boy, he sat down, crossed one knee over the other, and set the Bible on top. Then he fixed his deep-set eyes on the newcomer.

"Now," he said. "Here we have this boy. He looks like a healthy enough boy. There are no outward signs, at least, of depravity. There's nothing about him to suggest that some morning you may wake up to find your throats cut. Observe his hands. They are ordinary hands, cleaner, if anything, than most boys'. But can you be sure they're not itching, as we sit here, for matches? How can we be sure he isn't waiting the first chance to set the house on fire?" He held one of his own hands in front of him. "You know nothing about this boy. You have no way of telling what thoughts fill the mind behind that face. He may be full to the bursting point of jealousy, ingratitude, gluttony, filthiness. How do you know what designs he's hatching?" He shook his head and then turned to his niece. "I put it to you, 'Tilda. Is this wise? Is it the kind of risk you should take? Is it the kind of thing a Hull of Coldwater should do?"

While this speech was being delivered William looked first distressed and then resentful. He was so incensed that he kept his eyes lowered as he replied to the visitor.

"This—this is uncalled for, Uncle Alfred. If you wanted to say such things you might at least—well, you might have waited to say them to us and not—not before the boy."

"I have no such foolish scruples," said Uncle Alfred. "When I think a thing should be said, I say it. If feelings are hurt, let them be hurt."

"Do you really think, Uncle Alfred," asked Tilly, whose satisfaction with the arrangement had been oozing from her visibly as she listened, "that there might be—that there actually might be—danger?"

Calamity Hull took the Bible in his hands and held it up. "There must be something here to cover the case. I'll search for it and, when I find it,

I'll come back and read it to you. It will be something to make you see sense." Releasing one hand from its hold, he raised it to his head with one finger cocked over an imaginary trigger. "Have you ever heard the word heredity? Do you know what it means? Well, 'Tilda, think it over before you get yourselves committed."

Ludar had listened with the fatalistic acceptance forced on him by a childhood in which he had been continually repressed and subjected to criticism. It had been too good to be true then, this quiet home where kindness had been shown him. This old man with the dark beard and eyes which could chill a boy's blood at one glance had come to put an end to things. What, he wondered with a sense of rising panic, would happen to him now? Would he be turned out of the house before he had a chance to set it on fire with matches (but he was afraid of matches and never touched them) or—or cut someone's throat? Would they even wait to let him get dressed?

He looked in his distress toward William Christian and could tell that the latter was very angry. Was he angry because he had been cheated? Because he had not known what a bad character he, Ludar, was before he agreed to bring him home?

William was so angry that his shoulders twitched and his face became red around the eyes. He began to pace about the room. "I won't have it!" he said. "Not in my own house. Such talk! I guess I know a good boy when I see one." He came to a stop beside the visitor and glared down at him. "Do you know what I think? I think the Hull family went to Coldwater when they came out to this country because they knew the name of the place fitted them. They're always throwing cold water! Every last one."

"William Christian!" cried his wife. "How dare you say such things!"

"Now that I'm started, I may say a lot more things. Things I've been storing up to say."

The differences between the three might have developed into a real quarrel if someone had not chosen this moment to rap on the side door. William went out to the kitchen, and they could hear him say as he opened the door, "Well, Catherine," in a voice from which all ill feeling had departed. He called in to them, "It's Catherine, Tilly."

Catherine McGregor, who lived in the two-story white brick house next door (which was comparatively grand, having a veranda and a bathroom), came into the room like a dove of peace. She was rather small and very pretty and as friendly as a young puppy.

"So this is the little boy," she said. She walked over to the Morris chair and went down on one knee to bring her face closer to his. "Why, what a *very* nice little boy! What is your name?"

"Ludar, miss."

"Ludar? And a *very* nice name it is. It's a most unusual name and it seems to suit you." She glanced up over her shoulder at Mrs. Christian. "Don't you think it's just perfect for him?"

Tilly's mind was still full of the doubts which Calamity Hull had instilled. She said in a far from convinced voice, "Well, perhaps it is, Catherine."

"I'm so glad, Ludar, that you've come here to live," went on the pretty neighbor, transferring all her attention back to the boy. "I'll get out the sled I had when I was a little girl and I'll take you out for rides. And we'll throw snowballs and have a grand time, won't we?"

Ludar's spirits rose a little. He thought to himself, "Perhaps this nice lady will tell them I'm not a bad boy and they needn't send me away." He sat up in the chair so suddenly that the coat slipped from around his thin bare shoulders. "Yes, miss," he said. "But I've never played in snow. Won't it be very cold, miss?"

William had been moving in chairs from the kitchen, plain wooden chairs with spindle backs which he had made himself. He put one down in front of the girl. "How did *you* hear about Ludar?" he asked.

"In the usual way." Catherine laughed. "Mrs. Chief Jarvis's servant girl is a sister of Margaret. Margaret was through for the day, but she came back to tell us. So of course I had to come in and see my new neighbor."

She got to her feet and walked back to the kitchen, motioning Tilly to follow her. When they were alone the girl whispered: "I have another reason for coming, Mrs. Christian. I hope you won't mind."

"I'm glad you came when you did," the older woman whispered back. "The two men were having a quarrel. William was in a great rage."

Catherine could not restrain herself from laughing. She had straight fair hair and an oval face with a few freckles on the bridge of her nose, and when she laughed she brought one dimple into play. "Uncle Billy in a rage? I won't believe it."

"But he was." Tilly shook her head solemnly. "I thought—I really thought he was going to strike Uncle Alfred."

"Well, of course . . ." The girl's tone left the impression that this made everything understandable. "But what I want to tell you is this, Mrs. Christian. Clyde's coming too. Any minute now. Clyde Carson, you know. He doesn't go to our house because Father gets into a rage when he does.

So I told Clyde—it was yesterday afternoon and I just happened to be in W. and M. Appleby's store—that I would be here tonight and that he might drop in and see me for a minute or two. You *don't* mind, do you?"

"N-no." Tilly spoke as though the explanation had poured out too rapidly for her to grasp it all. "But why, Catherine, doesn't your father want him in the house?"

The girl's brow clouded. "Oh, that father of mine! He thinks Clyde isn't *worthy*. That's the word he always uses in our arguments. You see, Aunt Tilly—— Do you mind if I call you that?"

Mrs. Christian's face flushed with gratification. It had quite apparently been a sore point with her that up to this moment their lively young neighbor had called her husband Uncle Billy but had never progressed beyond Mrs. Christian with her.

"Mind? No, no, child. I'd like you to."

"Well, Aunt Tilly, Clyde's people aren't much. They live away down in the East Ward, and Clyde's father has a little grocery store. Clyde has always wanted to be a doctor, but there wasn't enough money to send him to college. So he took a business course instead and now he's keeping the books at W. and M. Appleby's. Oh, he'll get on in the world. But Father disapproves. He says I must make an important match."

"He has your best interests at heart, Catherine. Your father is a fine man."

The girl frowned and laughed at the same time. "He's more than a fine man, Aunt Tilly. He's a remarkable man. Oh, how I've suffered over that sign! All my friends tease me about it. But when I speak to Father he frowns majestically and says, 'I am a remarkable man, child, and I believe in letting the world know.' He says it's a real trade getter." She sighed. "I suppose it is. But that doesn't make it any easier for me."

"Your father knows best, Catherine."

"About some things. But not about what I want to do with my life. You see, I've a mind of my own. Just as my remarkable father has. . . . Oh, there's someone coming to the front door. I guess it will be Clyde. May I let him in?"

"Of course, child."

Catherine fairly ran through the sitting room and from there into the cold front hall. They could hear her say, after opening the door, "Well, you *are* a prompt young man, Mr. Clyde Carson."

The young man's face, when he came into the sitting room, was red with the cold. He was pleasant-looking, with yellowish hair which could not be controlled by the most systematic use of brush and comb and bay rum, and his eyes were so full of Catherine that he stumbled at the place where the hall carpet joined the straw matting.

"Excuse me," he said. "I always was clumsy on my feet. Good evening, Mr. Christian. I hear you've a new member of your family."

William's face registered surprise. "Now how did *you* find out?"

Young Mr. Carson spread his hands in front of the stove door. "Barney Grim lives down in our part of the town," he explained. "It's not a tony part, you know. If Mr. McGregor was here I wouldn't need to say that. He would say it for me."

"Clyde, please!"

"You know he would, Cathy. I was only stating the truth." The boy had broad shoulders, and there was ample room on them for chips. "Well, Barney came into my father's store—it's a very small store, Mr. Christian, nothing remarkable about it—to get a plug of tobacco, and he said he had driven you and the boy home."

Uncle Alfred thought it time to inject himself into the conversation, having been silent for fully five minutes. "It amazes me how news spreads," he said, shaking his head. He raised the Bible and sighed deeply. "But the messages here are neglected. I often ask myself, 'What is the world coming to?' There's no reason for me to ask, because I know the answer. The world is coming to destruction. To brimstone and fire. To the End, in a great blast of smoke and flame."

The young man squared around and faced the speaker. "I know what the world isn't coming to, Mr. Hull. To a sense of justice, to a realization of the need for all people to have equal rights and chances. This perfect world!" His face became even redder. "How can men think of it as perfect? It's a world of cruelty and unfairness and hypocrisy——"

"Now, Clyde!" interjected the girl hastily. "People will begin to think you're a socialist if you talk that way."

"Perhaps I'm not so far off from being a socialist, if the truth were known. I don't think this world perfect. Although"—his voice rose to a much higher pitch—"it would be a lot closer to perfection if I could put a stick of gunpowder under a certain sign——"

"Uncle Billy," said the girl, "could Clyde and I have the crokinole board and take it out to the kitchen? I won't promise that we'll play. We—we always have so much to talk about, Clyde and I."

William produced the crokinole board from a clothes closet and held it out proudly for their inspection. "This is a very special one. I made it myself. The wood's curly maple and the leather bands on the pegs are a lot thicker. I thought we might put some out like this; at a higher price, of course. But Mr. Alloway"—the owner of the planing mill where he worked—"figured the cost would be prohibitive." He sighed. "It was too bad. I'd hoped to collect a small royalty. Mr. Alloway never likes to change his way of doing things."

The young couple took the board and disappeared into the kitchen. Immediately the click of the disks could be heard over a steady undertone of eager conversation. Anyone who understood the game would have known the shots to be random ones.

Uncle Alfred glanced suspiciously toward the door leading into the kitchen and then motioned to William and his niece to draw closer.

"I'm going to tell you something which must never be repeated," he said in a voice just above a whisper. "This afternoon—I was summoned. I was summoned to appear at the home of a certain wealthy man who happens to be the silent owner of the Balfour Realty Company and so is my employer. There was a social gathering at his home of the best people of the town—in their own eyes, may I say, and not in the eyes of the Lord, who someday will cut them down in their pride like corn before the blade of the reaper. Was I brought in among them like an honored guest? I was not. I was escorted from the front door by a butler who seemed to look at me with the end of his nose, and I was led to a mean little room with one desk and one chair. Pretty soon in came this certain wealthy man and seated himself in the one chair. He was in his carks and moods, which was not surprising because, as far as my experience of the man goes, he always is in them. He began to give me orders. I must talk to people with property adjoining the furniture works and get evidence that there would be concerted opposition to any extension across the streets. Then at twelve o'clock I must go to Mr. Langley Craven and inform him that the board of directors have decided to take over the Homestead and use it for an extension, and that the matter of his vacating the property has been put in the hands of the Balfour Realty Company." Uncle Alfred rolled his eyes from one to the other. " 'The rich ruleth over the poor, and the borrower is servant to the lender.' That's what the Word says. Langley Craven has been the borrower, and now he will become the servant."

William showed the indignation he felt. "It's all a move to get Langley Craven out," he said. "And mark my words, it's the doing of that woman."

"It is not!" cried Tilly, whose views were always narrow and always violent. "I won't hear a word against Mrs. Tanner Craven. She's a good Christian woman. I know because I've worked with her in the church and I've helped her with the poor baskets at Christmas and the tea meetings."

Calamity Hull shook his head in dissent. "The woman is a hypocrite," he declared. "All who are high up in the churches are hypocrites, and this woman the worst of all. Right here in the Word it's said that a hypocrite with his mouth destroys his neighbor. That's Mrs. Craven for you. She won't rest until she's destroyed the other family. As for Tanner Craven"—his eyes flashed with indignation—"that man is a whited sepulcher, a Midas, a Judas, a Jacob stealing the blessing from Esau!"

Tilly fell into a reminiscent mood. "When I came to Balfour to learn the millinery"—this had been an unfruitful interlude because she had shown no knack for the making of hats—"I stayed with my cousin Hetty German on Grace Street, just opposite the Terrys'. Many's the beaux that Pauline Terry had! I used to sit in the front window, behind the lace curtains, and watch. I used to see Langley Craven arriving, so grand with his team of horses and his whip with a ribbon on it, and wearing a high collar and carrying flowers for Pauline and boxes of candy. But when Tanner Craven came to see her he came on foot, and I couldn't see that he ever brought any presents. He was close even then. It was easy to see who would win. When Langley arrived Pauline would come out, her skirts sweeping along the brick walk—she always wore the longest skirts!—and she would begin to laugh and talk." Tilly sighed, half in envy, half in nostalgic pleasure. "She always looked wonderful, that Pauline. I saw her in old clothes in the mornings and before she had done anything to herself, and she looked just as lovely then as she did in the evenings after she had been primping. I saw her at *that* too."

"Do you remember," asked Uncle Alfred, allowing his voice to rise, "when the Langley Cravens were married? It was in the winter and there was a snowstorm. Tanner went to T'ronto on business that day so he wouldn't have to be there. He came home that night, and Barney Grim drove him from the station. Barney says he was as drunk as a lord and that he began to sniffle when they passed St. Paul's Church."

"But a lucky man he was in the end to get as fine a woman as his own wife," asserted Tilly.

It suddenly occurred to Uncle Alfred that the company in the kitchen might be listening to this frank talk. He raised a finger in warning. It was clear at once, however, that they had not been overheard. The young couple were too concerned with their own problems to be interested in those of others. Catherine's highly excited and very happy voice could be heard above the deeper rumble of the young man's talk.

The footsteps of two people in the street came clearly to the occupants of the room. William said, "Must be getting colder when you can hear like this." The footsteps turned in, and a loud and demanding knock was heard on the front door. The humble home of William Christian was proving the most visited house in all Balfour on this cold New Year's night.

Catherine came to the door of the sitting room on tiptoe. "That's Father," she whispered.

William got to his feet and walked in the direction of the front hall, but the girl gestured to him to wait. She went over and whispered in his ear: "He mustn't find us here or he'll be very angry. I think, Uncle Billy, that Clyde and I had better go out through the summer kitchen. We'll wait until you've let Father in, and then we'll climb the fence and get away." She was drawing on her coat as she talked, a red flannel one which reached to her ankles. With excited hands she knotted a handkerchief over the small bonnet she was wearing. "Come on, Clyde. Isn't this fun?"

Clyde did not follow at once. He stayed in the doorway, his hands in the pockets of his coat, his forehead drawn into a frown. "I don't like running away like this, Mr. Christian," he said. "I wouldn't do it except that the remarkable man would be angry with you if he found us here. I don't want to get you involved in this hole-and-corner kind of business, but—but a fellow gets kind of desperate at times." A second knock sounded on the door, a peremptory one this time. "Well, that settles it. The wrath of the Lord is rising. Good evening."

3

Lockie McGregor strode past William in the hall, saying in a brisk voice, "Catherine's here, isn't she?" He was a stout man, with a round head, a thick neck, and a solid body as soundly coopered as a keg of nails. There had never been a moment in Lockie McGregor's life when he had not overflowed with energy and confidence. There had been few moments when he had not demanded the same qualities in those around him.

He said, jerking a thumb over his shoulder to indicate a man standing behind him: "This is Alvin George. Smart young man. I've taken a fancy to him. He travels for a Toronto hat firm."

William said: "Oh yes. Come in, Lockie. And you, Mr. George." There was no light in the front hall, and so it was impossible to get any impression of the smart young man who traveled for a Toronto hat firm. "Catherine was here. She came in to see the boy. But she's gone."

"Gone? Where? Good grief, Billy, that girl is always coming or going.

I never know where to find her." The energy his daughter had inherited from him seemed reprehensible when it interfered in any way with his plans. He strode into the sitting room, saying, "Good evening, Tilly." Then he stopped, his eyes on Uncle Alfred. "Oh, you're here, are you? Spreading good cheer as usual, I suppose. Any more Catholic plots to take over the city and set up the Inquisition again? Any more secret drilling on Strawberry Hill?"

Having come from next door, neither man wore an overcoat. Alvin George followed the merchant into the sitting room. William saw now that the smart young man had a stolid and rather pasty look to him and that he was dressed in the height of fashion, with a fawn waistcoat and a watch fob dangling below it in the form of a galleon with spread sails.

"Tilly," said William, hesitating, for introductions worried him, "this is Mr. George from T'ronto. My wife, Mr. George. And this is her uncle, Mr. Alfred Hull."

The stranger said, "How-da-do," in an indifferent voice and, seating himself in a corner, spread his feet wide apart, placing his hat on one knee. His eyes had yellowish pupils and looked like a pair of poached eggs. The merchant drew William off to one side and began to whisper in his ear: "He sells me a big bill of goods twice a year. Sharp as a trap and knows his business. I brought him up to meet Catherine. Why can't that girl of mine be around when I want her?" He let his voice drop another notch. "What you think of him, Billy?"

"I guess he's smart enough."

"He'll make something of himself, that young man. Did he stay at home because it's New Year's and there were parties to go to? No, sir, he came on from Toronto so he could get an early start on his work in the morning. Called me up from the Cameo House, and I took him with me to the mayor's party this afternoon. I tell you, a young man who passes up all this New Year foolishness is going to make something of himself. He might be"—he hesitated, then continued in the lowest of whispers—*"just the husband for Catherine!* At any rate, I thought I would trot him out tonight and put him through his paces." He gave his head a confidential nod. "I've big ideas for that girl of mine. Her mother thought I named her for an aunt in the family. I'll tell you a secret, Billy. I didn't. I named her after the Queen. You know, Catherine the Great. She was my daughter, and the name of a queen was none too good for her." He fell into a brief moment of reflection. "Sometimes I wonder about her. She's bright and she's good and she's a pretty little chick. But occasionally I think she's a little giddy. What do you think, Billy?"

"I've never seen anything in Cathy that I couldn't approve and admire."

The merchant was pleased. "I guess you're right. She's like her mother in some ways. But," hastily, "not in too many. You never knew Mrs. McGregor. A fine woman. She had to be or I wouldn't have picked her out. But she had her faults. She was stubborn, Billy, stubborn as a mule. She was against the sign, dead set against it. I had the figure made and gilded and all ready to go up in front of the store, but she said no, she would leave me if it was used. That expensive sign lay in the attic for seven years, getting dusty, and all because of her stubbornness. Then she did leave me. She died. The funeral was at two in the afternoon. At four the sign was up. And I want to tell you it caused a lot of talk at first." He nodded his head with deep satisfaction. "And ever since, for that matter."

Realizing that this private conversation had been carried on long enough, the merchant raised his voice to address the hat salesman. "This is the boy I was telling you about, George. What do you think of the little codger?"

The salesman studied Ludar before replying. "He'd be a hard one to fit," he said finally. "Quite a long head he has. He'll have to be fitted out with English caps. You'd never be able to suit him with one of these billycocks, or even one of these straw boaters."

Ludar had been feeling reassured with all the coming and going of people and the friendliness of the pretty lady from next door, but now his spirits took another downward plunge. Was it a bad sign to have a long head? Did it indicate that he was dangerous and likely to cut throats or burn down houses with matches? He watched the salesman with apprehensive eyes, fearing more revelations of his own unworthiness.

The merchant went on talking in loud and confident tones. "It was quite a shindig the mayor gave this afternoon. He greeted each one of us with the usual 'It sure beats hell how heaven keeps going!' Finance Chairman Huffer threw nickels out of the window at some boys sliding on the sidewalk, so you can judge how well lit up he was. Yes, it was quite a party. What did you think of His Honor, George?"

"He's a Turp," answered the hat salesman promptly. "In the trade we divide people up into three classes: Sharps, Flats, and Turps. This boy here is a Sharp. You're a Flat, Mr. McGregor. The mayor's a Turp. Short for Turnip, you know." He shook his head as though he had revealed a municipal scandal. "With the size head he's got, we couldn't fit him with a single hat from our whole line. I'll bet fifteen cents right here and now that he has all his hats made to order."

What further efforts the energetic merchant might have made to draw his companion out were rendered impossible by a sudden outburst of sound down the street. There was a loud tooting on a horn, the jingling of sleigh bells, and a continuous volume of cheering and singing.

"Now what foolishness can this be?" asked Lockie McGregor, getting up and starting for the front door.

"Wickedness and vanity!" exclaimed Uncle Alfred. Nevertheless, he deposited his Bible on the seat of his chair and walked out with the rest.

Every front door on Wilson Street, it seemed, had opened to watch the passing of two large carryalls packed with young people, all of whom were wrapped up in blanket coats of red and white and were wearing toboggan caps. One of the men had elevated himself to the front seat of the first of the two lumbering vehicles and was attempting to play "Little Brown Jug" on a horn, but without much success. The rest were waving to the people in the doors and shouting, "Happy New Year! Happy New Year!" Never before in its history had staid old Wilson Street been treated to so much unrestrained noise and merrymaking.

"A generation of vipers!" said Uncle Alfred bitterly.

As the second carryall passed, a man dropped off the end and walked over to the group in the open door of Number 138. It was clear that he had not belonged to the merrymakers, because he was wearing a bowler (which no one ever called anything but a "christy stiff") and thin gloves.

"You're William Christian, aren't you?" he asked, nodding to the carpenter. "I'm Allen Willing of the *Star*. I stole a ride up. What a time they're going to have! They're on their way to the Carlton farm and they've got three gallons of oysters and fifteen pounds of soda biscuits. Every pretty girl in town must be in that party—and all of them looking their very best. I tell you, it takes a cold night like this to bring out the roses in the cheeks and the sparkle in the eyes. If I hadn't had some business with you, Mr. Christian, I think I'd have found a seat for myself between two of the liveliest of them."

"Young man," said Tilly, looking at him sternly, "aren't you married?"

"Oh yes, I'm married," the newcomer answered airily. "Good gad, how married I am! Thirty years from now they'll take a picture of my wife and me sitting together on a horsehair sofa, with one of these new-fangled phonographs playing 'Home, Sweet Home.' And they'll call it *The New Darby and Joan*."

"I must say," declared Tilly, "that you don't talk like a married man."

"Now that's what I call a compliment. Apparently the fetters haven't cut too deeply into my wrists yet. I know what you're thinking, Mrs.

Christian, that newspapermen are a wild lot. Oh, we are!" He turned to William. "Have you the boy here with you, Mr. Christian?"

"Yes. We're going to keep him until he can be sent back to his relatives in the old country."

"Good. I'd like to ask him a few questions. That is, if Mrs. Christian will allow a moral leper like me in her house."

"Come in," said William. "The boy's in the sitting room, and that's the only place we have to receive company. You'll have to put up with the rest of us, I'm afraid."

"I work my best with people looking on. Brass bands and cheering multitudes excite me to prodigies of achievement. I'm glad Sloppy Bates passed out just as he was starting to come here, because I was fairly itching to do the story myself. I'm going to put some writing into my account of the arrival of this boy and the death of the father. I'm going to pull out all the stops and let the chords of my fancy thunder and roar. It will get the whole front page if necessary."

When they went inside again, it could be seen that the editor of the local newspaper was a rather stocky young man with lively brown eyes, an inquisitive nose, and unruly hair. He walked into the sitting room and promptly took possession of the chair the hat salesman had been occupying. When no one made any move to supply him with another Mr. George draped himself against the side of the secretary and studied the contour of the Willing head, which did not fit into any of the three categories of Sharps, Flats, or Turps.

"Well, Ludar," began the newspaperman. "You've had quite a trip for a boy of your age."

"Yes, sir."

The fullness of the evening and the constant procession of new and strange faces were having the inevitable effect. Ludar could hardly keep his eyes open.

"Past your bedtime, eh?" said the editor. "Well, just fend off the sandman for a few minutes while the Delane of Canada asks you a few pertinent questions. First, do you like brandy balls? Here's a bag of them for you. I got them at Milton Truck's before starting up." He winked at the circle of adult faces. "The Machiavellian touch."

Ludar was sure the large green candy balls in the bag would not be good for him, but he could not resist them. He popped one into his mouth and found it a new and exciting experience.

"Now," said Willing, "where did you live in England?"

"With Aunt Callie."

The editor gave his thigh a resounding smack. "Shades of Betsey Trotwood! Of course there would be a female protector in this story. Now we are on our way! This lady with the Jane Austen name, was she kind to you? Did you love her very much?"

"No, sir. She was cruel to me."

"Better still! An unfeeling guardian. Her last name wasn't Murdstone, by any chance?"

While the voluble newspaperman plied the boy with questions and Ludar sucked the brandy ball and rubbed his eyes while striving to find satisfactory answers, Lockie McGregor moved his chair over beside that of William.

"Sharps! Flats! Turps!" he said in a fiercely indignant whisper. "Can he think of nothing but hats, this suet-head? I take back everything I said about him. He won't do for my Catherine. But mark my words, Billy, I'll get a real man for this little girl of mine."

Remembering the happy light he had seen in the girl's eyes while she talked to Clyde, William said, "Why don't you let her choose one for herself?"

The Remarkable Man snorted. "Never! Do you know the kind of son-in-law I'd get if I let her have her own way? That smart-aleck son of a flyblown grocer in the East Ward! No, Billy, this is too important for her to do the deciding. I'll get her the kind of husband she *ought* to have."

In the meantime Allen Willing had uncovered by his brisk cross-examination the existence prior to his life with Aunt Callie which lingered so uncertainly in Ludar's lagging memory, and was off on the new scent, baying loudly. He began to discover all there was to know with his skillful queries: the large dark rooms in the mysterious house, the extensive gardens, the avenue of tall trees, the hedges cut into the shapes of animals, the cast-iron deer which had startled Ludar every time he saw it because it was so lifelike. The people remained more vague. Some of them had worn white caps and aprons. The boy produced painfully a few names: Bessie, Frank, Mrs. Mearns, Old Baffle. Strangely enough, Ludar did not mention the pretty lady. Perhaps he considered that his memory of her was his own and could not be shared. Finally he mentioned the clock.

"But," said Willing, puzzled at the importance which Ludar seemed to attach to it, "there must have been many clocks in such a large house."

"It wasn't *in* the house, sir."

"Not in the house?" Willing's eyes began to show signs of mounting excitement. "Where was it then?"

"It was way, way up, sir. With leaves around it."

"A clock tower!" The editor turned about and nodded to the listeners. "Now we have a real clue. There can't be many private homes, even in England, with clock towers."

The probing continued, leading to the discovery that the boy retained also some recollection of stables. They were big, and he had seen horses looking at him from behind bars. There had been a pile of manure "as big as a house," and always there had been men working at it with pitchforks. He had found his way once to a distant part of the buildings and had fallen into a pigpen. He had been as black as a pig himself when some of the men had hauled him out. And then he contributed, out of a blue sky, a most important piece of information. "Old Mr. Cyril laughed at me."

Old Mr. Cyril! The editor promptly bombarded the boy with questions, receiving answers which, while both vague and scanty, added much to the picture. Old Mr. Cyril was not around a great deal. He, Ludar, had not seen him often. He always carried a book under his arm. He was quiet mostly, but sometimes he became angry and shouted at Frank and Old Baffle and the rest. He had whiskers.

"Were these whiskers of Old Mr. Cyril's like—well, like those of Mr. Hull here?"

The boy laughed. "Oh *no,* sir!"

"What else can you remember about him?"

Ludar thought for a long time. "He had spectacles, sir. But he'd lost part of them."

"A monocle!" The editor got to his feet and stared with a look of triumph around the circle of intent faces. "We now have some facts to go on. It seems certain that this Old Mr. Cyril was the head of the family, the boy's own grandfather, in fact. One of the sons in the family group was called Cyril. Young Mr. Cyril, he was probably called by all these servants. We have a picture now on which we can build—an extensive estate, a large house with an ivy-covered clock tower, a whole train of servants. The children don't seem to have been about. Married and out on their own, no doubt. All we have to do is find in England an estate which answers this description, owned by a gentleman of bookish habits, wearing a monocle and answering to the name of Old Mr. Cyril Something-or-other. The family undoubtedly had wealth and more than likely a title."

Uncle Alfred asked a question. "Was the boy's father the oldest son in the family?"

"According to the photograph, yes."

"Then," said Mr. Hull, once of Coldwater, with a triumphant flourish of his hand, "the boy is the heir."

"That's the way it looks," answered Willing. "He's another Lord Fauntleroy or I miss my guess."

Lockie McGregor, who had shown unusual restraint in remaining silent while all this was going on, now cleared his throat. "I'm not a lawyer and I'm not a newspaperman," he said, "but I'm not satisfied we've got to the bottom of this yet. Now, if you don't mind, Mr. Willing, I'm going to ask a few questions myself. I think I can make him remember what his real name is."

The opportunity, however, had slipped by. The boy had fallen into a sound sleep, and William refused to let him be wakened. "The poor little fellow is tired out," he said. "Do you realize that he hasn't slept in a real bed for at least a month?"

Tilly had vanished a quarter of an hour before, and so William lifted the sleeping boy and carried him into a bedroom which opened off the sitting room. It had no light, and he left the door open as he turned back the covers and deposited his charge under them. He said to himself that this was a shabby place for one who belonged to a family with a clock tower and stables. It was small, and the lace curtains hanging at the one window had been removed from the sitting room when they started to fall into pieces. There was a bureau which William had made. It was somewhat too fancy in design, and he counted it among his failures.

"I don't believe there's a toy in the house," he thought. "What will the poor little fellow have to amuse himself with?"

There was no use thinking of buying toys out of the money the city would pay each week. Tilly had already declared that every cent of it was to be added to her principal. After a moment's thought he produced from one of the drawers of the bureau a rather tattered copy of *Pilgrim's Progress*. Shaking his head over this doubtfully, he placed it on a chair together with the flattened tin pail and the wooden caboose, and then shoved the chair against the side of the bed to prevent the occupant from falling out. "Little boys waken early," he said to himself. "These will keep him busy if he does." He had never dipped into the volume himself. It belonged to Tilly, and she had been too solicitous about the preservation of the weakened binding. He was not much for reading, anyway; a year before Tilly had decided they could not afford the *Star* and had cut it off, and he had missed it little.

He returned to the sitting room at the same moment that Tilly arrived with a tray containing cups of chocolate and a plate heaped high with buttered gingerbread. Each man in the room sat up straight and smiled expectantly. For several minutes thereafter no sound was heard save the clicking of cups on saucers.

Uncle Alfred had mellowed as the evening progressed. He was now smiling with an expansive generosity which even included William. " 'Tilda," he said, holding his cup of chocolate in front of him and blowing on it, "if this boy you've taken into your home and circle had his rights, there would be a butler to get together a bite of supper like this. And hand it around, too, in velvet knee britches."

Willing nodded his head. "No manner of doubt about *that,*" affirmed the editor. "Mrs. Christian, you may be acting *in loco parentis* to a future earl or duke."

"That will make no difference in how I treat him," declared Tilly. "I intended to do my duty, and that's what I'll still do." She was enjoying herself to the utmost. It made her proud to have company like this and to have them praise her food. It was clear also that she already had visions of dazzling rewards, checks for large amounts, signed with ducal flourishes, which could be added to her precious principal.

William was even more content. He could see how completely his wife and her uncle had changed sides. There would be no further opposition or criticism. When the company had gone he filled the tray with the dishes and carried it out to the sink in the kitchen. "I'll wash them tonight," he called to Tilly, who was pulling down the blinds. "Might as well, because I need hot water. I'm going to give the clothes the poor little fellow was wearing a thorough scrubbing. If they're not ready for him in the morning he'll have nothing to put on."

"I'm going right to bed," answered his wife. "This has been a very trying day."

Chapter VI

I

Ludar wakened early the next morning as his new guardian had expected. It was not early enough, however, to precede the departure of the head of the house. William had helped himself to a bowl of oatmeal with milk, which he preferred to cream, and a slice of toast without butter, for which he had no fondness, and had already hurried off to work in his worn brown overcoat with a patch on one elbow and his round hat of some quite suspicious fur pulled down snugly over his ears.

The boy's eyes lighted first on the book. He lifted an arm from under the covers and took it up. It was full of pictures, but the first one his eyes encountered showed Beelzebub with a pitchfork in his hand, and that destroyed his interest entirely. He put the book back as though it had the power to bite.

His eyes encountered next two points of gray fur above the ridge of comforter created by his feet. He raised himself cautiously and found that the points, as he had hoped, belonged to a round gray head with half-closed green eyes.

"Old Cat!" he cried delightedly.

His pleasure, however, was short-lived. Old Cat, out of the experience of long years, wanted nothing to do with small boys and, at his first move, sprang from the bed and vanished through the door. Ludar called, "Old Cat, I won't hurt you!" The veteran knew all about that kind of promise and did not return.

In sitting up the boy had made a third discovery. His clothes were hanging over the back of another chair and, moreover, they looked fresh and clean. This was hard to believe, and so he got out of bed and went over to investigate. Running a hand along the corduroy, he discovered it to be soft from washing and almost unbelievably spruce and neat. He had an instinct for cleanliness, and this discovery compensated for his disappointment over the departure of the cat.

His fingers moved rapidly in dressing himself, and in a very few minutes he had made his way into the sitting room and from there to the kitchen. The stern woman into whose not too sympathetic hands his immediate fate had been entrusted was sitting at the table, eating a boiled egg and dabbing at her mouth with a table napkin.

"Well!" she said, staring at him as though she could not believe her eyes. "Now haven't I something to tell Cousin Ellie! I'm sure her Leroy was seven years old before he could dress himself like that."

The day had started well. It was lucky he had pleased her, because a serious moment came later when he said that oatmeal made him sick. He could have eaten an egg, but the only offer made was a bowl of corn-meal mush left over from the previous day. This he managed to consume without any bad effects.

The morning passed quietly. Ludar spent some time stalking the cat, but without any success. He opened doors and stared into dark rooms (it was believed necessary in winter to keep the shutters up in the bedrooms day and night) and found them dismal and rather frightening. He slipped out to the summer kitchen, which was almost as cold as the outdoors, and searched for the old soapbox which had served as a bed for a dog named Gyp, a little black-and-tan which the Christian family had once owned. He found it in a corner, partly concealed behind the coal scuttle. Gyp, he had been told, had been dead many years, but the box had not been changed or moved. Ludar's eyes filled with tears. "Poor little Gyp!" he said aloud. "I wish we had him now."

By eleven o'clock the house had been cleaned and the dinner was on the stove. Mrs. Christian put on a severe bonnet which poked out above her forehead like a coil of rope, and muffled herself up in a voluminous coat with some very scraggly fur on the collar.

"I've an errand to do," she explained to the boy. "I won't be long at it as I'm only going to see my cousin Celia. Will you be all right till I come back?"

"Yes, Aunt Tilly."

"Now, don't you get fresh with the cat," she warned. "She'll scratch you if you do."

At first he was glad to be alone, for he was still apprehensive of her moods and of the use she might make of her authority. This feeling dissolved rapidly, however. He began to realize there was no one in the house and that it was strange and silent. He was afraid now to open any of the doors because of what might be behind them. Seating himself by the window in the sitting room, he tried to occupy his mind by looking out. This was not successful, for a high board fence closed off the view, and all he could see was a cold blue sky shot brilliantly with sunlight.

A few moments of intense staring and he could no longer hold his feelings in check. Laying his head down on his arms, he gave way to tears. He wept loudly and desperately for many minutes, his back shaking with the force of his sobs. It was not homesickness which had gripped him, because he had no desire whatever to go back to where he had come from. It was a sense of lacking the two things which children need most and which life so far had withheld from him, security and love. He could not remember a day, except at the first, when fear of the morrow had not filled his mind, a dread of being dragged to some new place or of being subjected to some new indignity. He had heard himself discussed so often and so unfavorably that he had no illusions. He would always be a weight, he knew, on those who were afflicted with his care.

By the time Tilly returned, however, he had recovered, and if his eyes still happened to be red she did not notice it. She carried a bundle in one arm, and her face wore a look of triumph.

"I guess that was a good day's work," she said, placing the bundle on the open lid of the secretary. "You mustn't touch it, Ludar. You'll see what it is in plenty of time."

Dinner was ready at ten minutes after twelve, when William returned from the factory. He was in the best of spirits and even indulged in a skipping gait as he carried his coat and hat to the pegs on the kitchen wall.

"The whole town's talking," he said. "About *us*."

"Is that any excuse for dancing?" demanded his wife. "What kind of an example is this to set the child?"

She was holding the loaf of bread against her apron front and cutting slices from it with an inward stroke of the knife. Putting the slices on a plate, she seated herself with an air which said he would find it difficult to atone for such conduct.

The dinner consisted of a beef stew with potatoes and onions and carrots, and for dessert a steamed suet pudding with raisins popping out from its sides. Ludar was not interested in the stew, but he was entranced with the pudding and not only finished his own helping but downed every one of the raisins which William picked out of his slice and laid on the side of the boy's plate.

"If you don't appreciate the puddings I make for you with my own hands, William Christian," said his wife, "you might at least refrain from making the boy sick."

Ludar was not worried by that possibility. Food that he liked never seemed to give him those sudden pains in his stomach. He liked the pudding so much that he would gladly have gone on eating it until the buttons burst on his newly washed corduroys.

"Everybody was talking about what we've done," said William, leaning back in his chair and waiting for his cup of tea to cool. "They all made excuses to see me, and then they pumped the questions at me. Mr. Alloway called me into his office and said, 'Well, Billy, I hear the *Star* is going to be full of the Christian family tonight.' He said we had done a worthy thing. Archie Parsons—the Englishman on the boilers—said to me, 'I hear ye've got the Duke of Buckingham in disguise at your place.' Never before in my life have I had so much attention paid me."

Tilly forgot her pique over the dancing. "Do you really think the *Star* will print much about it?" she asked. "I've only been mentioned in the *Star* twice. When we were married—and I don't mind telling you, Billy

Christian, I was angry at the short note they gave us—and when my aunt Agatha Hull from Coldwater visited us."

"Didn't you see Mr. Willing taking down notes? He asked me all kinds of questions, and I'm afraid I gave him the impression that I *am* a descendant of William Christian. And I told him you were a Hull of Coldwater."

When the whistle of the Alloway Planing Mills gave warning that only ten minutes were left before the labors of the day would be resumed, William got to his feet, bowed ceremoniously, and said, "As one famous person to another, Mrs. Christian, I bid you farewell."

Her eyes achieved the faintest glimmer of amusement. "You're always full of nonsense," she said.

William said to the boy as he went out the door: "Good-by, Ludar. I hear you've given a good account of yourself this morning. You just keep on being sensible and obedient and you won't be a hod carrier, after all, when you grow up."

The afternoon proved to be eventful and exciting. First of all, Catherine McGregor paid them a visit. When Tilly saw her come climbing over the back fence she shook her head and muttered, "Even if her mother did die when she was six, she ought to know better than that!"

The girl came into the house through the summer kitchen. "I'm so happy!" she announced, throwing off her coat and hat and depositing them on a chair. "Clyde and I went for a long walk after we left here. All up Fife Avenue and clear to the start of Whisky Hollow Hill. We talked about serious things—marriage and children and socialism—and I gave Clyde a good talking to when he complained about the tony houses we were passing. And we threw snowballs at each other, and once I hit an old man who was going by, and he was very much annoyed." She paused for breath. "At breakfast this morning Father was as nice as anything to me. He said I didn't miss much by not meeting that young fellow he had with him last night. He was kind of sheepish about it."

"I thought Mr. George a very proper young man," declared Tilly.

"And now, Young Tinkybobbin," said the visitor to Ludar, "what are we going to do for your amusement? It's too cold to go out, I'm afraid. How would you like to play me a game of crokinole?"

Ludar had been very much interested the evening before in the handsome shiny board and had listened intently to the clicking of the disks. "Oh yes, miss!" he said.

"Would Uncle Billy mind if we used the board?" asked the girl.

"Not if it's going to amuse that boy."

Ludar spent an ecstatic half-hour, therefore, learning how to shoot the smooth disks and to send his opponent's men into the gutter. In fact, he picked off so many that she finally gave up in despair.

"You'll beat me all hollow in no time, young man," she said. "I guess you're going to be a regular shark at games."

She told him a story then, selecting "Jack the Giant Killer." He listened with the greatest attention, although he felt compelled to correct her at one point. "I think, miss, that what Blunderbore said was, 'I'll reserve this fellow for my breakfast tomorrow.'" One of the passengers on the boat had told him the story.

"I believe the giant did say that." She looked a little crestfallen. "I must go now. Good-by, Ludar. I'll be in to see you tomorrow."

Shortly before four o'clock a knock came on the front door. Mrs. Christian was in the summer kitchen getting a fresh scuttle of coal, and so Ludar took it on himself to answer. The visitor proved to be the red-headed reporter.

"Will you come in, sir?" Ludar was doing the honors with the greatest politeness. "It's Mr. Bates, isn't it?"

"Yes, it's Mr. Bates. I haven't been called that in I don't know how many years. The populace, the commonality, the *hoi polloi,* prefer to call me Sloppy. A name, I may say, which I've honestly earned." He ran the back of his hand across his mouth. "I don't feel myself today. Perhaps it's because I'm sober. Haven't had one drink. But give me a chance, Little Lord Fauntleroy. Just give me one more hour and I'll be properly pie-eyed." He stared down the hall into the sitting room. "Now I want to speak to Mrs. Christian."

Tilly came in from the summer kitchen with the scuttle of coal. When she saw who the visitor was she wiped her hands hurriedly on her apron. "Well, Mr. Bates," she said, "and what is it this time?"

The reporter took off his dilapidated bonnet. Then he fumbled in the pocket of his overcoat and produced a copy of the day's paper. "Here, ma'am. I brought you this. You'll find yourself and your husband and the boy here, and that poor fellow with the hole in his head, all plastered together on the front page. *Mister* Willing, who thought he could write it better than I could, got hold of the dictionary and spilled big words all over the paper. There's an editorial too. Mr. Milner wrote that. It's headed *Be Kind to Him.*"

Tilly was almost breathless with suspense and anticipation. "I must read it at once, Mr. Bates. I'm sure this is all very kind of the *Star.*"

"What I really came about was what happened after the paper hit the streets. People started calling up and saying *they* wanted to be kind to the boy. In no time at all twelve dollars had been sent in to provide him with new clothes. Mrs. Tanner Craven, who must have heard that the greatest of these is charity, sent two dollars. The boss chipped in with two. The Remarkable Man sent over a quarter. *Mister* Willing chipped in the same. I pungled in with a dime. It was all I had." He did not have to unbutton his overcoat to get at the inside pocket where he had the money. Two of the buttons were missing, making access easy. "Here, ma'am. Twelve dollars. *That* ought to be enough to outfit one small boy. It's more than I've spent on myself in as many years."

"Ah, how useful this money will be," declared Tilly. "Will you please say in the *Star* that I'm most grateful to everyone who has helped us carry the burden?"

"And now," said Sloppy Bates, "the minion of the daily press has performed the duty which brought him here. He goes immediately on a more pleasing errand than any he's performed today." He winked at Ludar. "Little Lord Fauntleroy here knows what it is."

2

"We're going downtown, Ludar," said Mrs. Christian as soon as the door had closed on the reporter. "First, we'll go to the shop and get Mr. Christian. That Mr. Alloway won't dare dock him if he leaves on *this* errand."

She then proceeded, with many mysterious nods of the head, to untie the parcel she had brought in. It proved to be a garment of some kind, made of a peculiar shade of fawn with hints of both green and pink. When she held it up Ludar saw that it was a boy's overcoat.

"I just happened to remember," she said proudly, "about this fine, warm coat. My cousin Celia's Jackie grew out of it years ago, and I knew how careful he had been with it. I bought it," her dark eyes snapped proudly, "at a very great bargain."

But when the coat was tried on it became apparent that when Cousin Celia's Jackie abandoned it he had been a much older and bigger boy than Ludar. The sleeves came down to the tips of the fingers, and the tail, instead of ending at the knees, went on lower and lower and nearly touched the ground. The neck was so much too large that it stood out an inch from the collar of his coat.

Tilly stepped back and looked him over. There may have been some doubt in her mind, but she promptly cast it out. "Well," she said, "there we are."

Ludar was so disappointed that he was finding it hard to keep back his tears. Most boys are interested in the clothes they wear, secretly, perhaps, and he was no exception to the rule. Never having had any clothes of which he could be proud, he had been immensely excited when he had seen the money handed over and had known it was to be spent on him. The overcoat was a sad awakening.

Even his fear of the formidable lady did not prevent him from voicing a faint protest. "It—it seems pretty large, Aunt Tilly."

"Nonsense!" Her tone was sharp and resentful. "Boys shoot up so fast it's always necessary to buy their clothes large so they can grow into them." The old, familiar, hateful argument! "I can remember when this coat was bought for my cousin Celia's Jackie. It was much too large for him. But did he say a word? He did not. He was a true Hull, that little Jackie. He was able to wear it for more than four years."

But whereas the coat had been large for Jackie, it fairly swallowed Ludar up. He knew that all children who saw him in it would laugh and jeer. The childish mind has not had time to develop a degree of philosophy to accept trials of this kind, and the smallest matter can become a tragedy. To Ludar the overcoat was nothing short of that.

Tilly produced a muffler and stuffed it down inside the neck. "There!" she said. "Now no one can say that it isn't a fine overcoat for a little boy who didn't have one at all. There are mitts in one of the pockets, Ludar. Put them on. It's very cold out, and I don't want you getting the hot-ache in your fingers."

The decision, then, had been made. The overcoat was his, and he must wear it. He would have to appear in it on the streets where other children would see him in it. Nothing that had ever happened to him before had seemed as bad as this.

The boy's worst apprehensions were realized as soon as they crossed over to Caledonia Street and passed by the North Ward school. It was an ugly white brick building with a plank platform which covered all the space in front. As they drew abreast of the school the front door opened and out poured a swarm of children, released from study for the day and screaming like Indians on the warpath. Ludar hoped to get by unnoticed, particularly as one of the bigger boys began at once to organize the others into ranks of threes.

"Come on now, you kids!" the leader bellowed, producing a whip which he had kept wound around his belt and snapping it viciously in the air. "I want to see some good marching today. I won't have you crawling along like the sissies and silly-billies the Dogans say we are. We're going to march right past the Dogans today and show them what we think of *them.*"

It happened that his eyes lighted at this moment on Ludar and his guardian. He was so interested at once that he forgot the squads he was forming and allowed the whip to trail on the ground. "Well, will you look at that!" he said with a loud laugh. "Where's the coat going with the boy?"

Ludar said urgently, "Let's get away from here, Aunt Tilly," and tried to run. This was impossible, partly because of the way the tails of the coat wrapped themselves around his ankles, but even more because of the firm grip she kept on his hand. The remark of the leader, however, had drawn the attention of the whole school to them. Children were now swarming across the street to get a better look. A fat boy shouted delightedly, "Hey, come out of that coat, I know you by your boots," and was rewarded by gales of laughter. Then one of the girls said in an excited voice, "That's Mrs. Christian, and so it must be the boy!" This put a different face on things. Another of the girls asked, "If he's the boy, where's the sign on his back?" A third addressed herself to Mrs. Christian. "Please, ma'am, is he really a lord or something?"

The girls had all been won over, but the attitude of the boys still reflected skepticism. "What's he doing in that coat if he's a lord?" demanded one of them. Suddenly a boy exploded with "Be kind to him!" in a falsetto voice. This seemed very funny to the others, and they all joined in a chorus of "Be kind to him! Be kind to him!"

A fortunate diversion was created at this moment by the leader, who was the much-feared Peter Work, acknowledged boss and bully of the school, as Ludar was to learn later. He had not crossed the street but had been watching proceedings from the edge of the plank walk. He remarked now in a loud voice, "I'm not a lord or anything, but I could knock that kid into the middle of the East Ward with one smack of my fist."

The troop had broken ranks and were clustered around their leader. One of the smaller ones said in a wheedling tone, "You can hit harder than anyone in the world, Pete-O."

Peter Work turned angrily on his follower. "Who asked you for an opinion, Snotty-Cuff?" he demanded. "And how many times must I tell you not to call me Pete-O? How many times, eh?"

"But," said the small boy with a startled look, "but you told me you liked to be called——"

"Never you mind what I told you. Some days I like to be called Pete-O and some days I don't. Do you suppose I'm going to tell you kids which days is which? And if you call me Pete-O on days when I don't—well, you'll just have to take the consequences. And now, Snotty-Cuff, the question comes up about what your punishment's to be."

The great Peter Work regarded his follower with a thoughtful scowl. "I could make you carry my schoolbag home and throw it over the fence and take the boot in the behind my old lady would give you if she caught you at it. I could shake what money you got out of your clothes and buy candy with it. You always have a nickel or a dime or two on you." Then he shook his head. "No, this is what I'm going to do with you. I'm going to make you march in the front squad today when we pass the Dogans' field. If they've got any ice snowballs or rotten potatoes ready for us, why, you'll get the full benefit, see? Now who else's turn is it to march first?" He cast a stern eye around the group. "You, Bulgy, and you, Stink Smith. That makes the three. Now!" He burst into a sudden roar. "Get back into line!"

3

The shop proved to be most exciting. All the windows were so covered with dust that a hazy light filled the place. As soon as they went in through a door marked "No Admittance," a roar of machinery greeted them. The walls were thick with the dust. It was heaped up in corners; it covered the machines and the men who worked at them. In addition there were tall piles of shavings everywhere. Apparently this process of accumulation had been going on for years and would keep on going until the shop would be filled up to the ceilings, and then no one would be able to do any work. Such, at least, was Ludar's impression of the place.

He was so interested in watching the dusty men run planks through huge saws and other kinds of machines, from which the wood emerged in various shapes, that he forgot for the moment the humiliation of wearing the grown-up Jackie's coat. He paid no attention when Mr. Alloway issued out from his cubbyhole of an office and told Mrs. Christian that William would be free to accompany her downtown without any deduction from the—ha, ha!—the little envelope at the end of the week. He kept his back turned when the owner asked if this was the boy and what had been done with the sign?

Then Uncle William arrived, looking very pleased, and Mr. Alloway brought a whisk from his office and helped him get rid of the dust with which he was well coated. Uncle William was in a hurry, apparently, for he got the farewells over and had them out in the street in no time at all. He stopped as soon as they were outside and looked at his wife with a puzzled air.

"My aunt Sally Ann!" he exclaimed. "Tilly, what have you done to the boy? What kind of a getup is this?"

"I got him an overcoat," answered his wife. "From my cousin Celia. I couldn't bring him out in this kind of weather without one, could I?"

"No, of course not. But—did you—uh—get the coat after the money arrived?"

"No, I got it from Cousin Celia this morning."

His relief was immediate and great. "For a moment I was afraid you had—uh—invested some of the funds in it. Well, now we can get him a new one and you can take back this thing you borrowed from Cousin Celia."

Tilly said in an ominously quiet voice: "I didn't borrow the coat from Celia. I bought it. I paid for it out of the money he had left over from the trip."

They were walking down the street now with Ludar, in the center, holding onto a hand of each.

"Bought it?" William looked apprehensive. "My aunt Sally Ann! You must have closed the deal before you tried it on. Well, your cousin Celia is a sensible woman. She'll be willing to take it back now that we've been given the money to buy him a bang-up new overcoat."

"I will *not* take it back to Cousin Celia." Tilly's voice had acquired a battle edge. "What would she think if I did? I don't go back on bargains, William Christian. And this *was* a bargain."

William looked down at the small figure between them and at the over-long coat flapping ludicrously about the boy's legs. "Whatever you paid, Tilly, it was not a bargain."

"William Christian," said his wife, "this is a matter of principle. Money is not to be wasted. I've spent money for this coat, and so the boy must wear it. There's no manner of good in arguing with me. My mind's made up. You ought to know by this time that once it's made up I can't ever change it."

William was seriously disturbed. "But, Tilly," he protested, "what would people say? Here they've gone and contributed this money for new clothes. If we let him appear in this thing they'll say we've kept the money for ourselves. That wouldn't do."

"It's as good as the one you've got on your own back."

"That," stated William, "has no bearing on the matter. What I wear is my own business. What the boy wears is the business of the people who supplied the money." He nodded his head several times by way of emphasis. "Now you listen to me, Tilly. Not one cent of this money is to be added to your principal."

That, he knew only too well, was the crux of the matter. Ordinarily she would not hesitate about taking the coat back to the original owner and demanding the return of the money she had paid for it. Tilly, however, had never been able to let money pass through her hands without saving some of it. This motive was so deep in her that it had become the dominating principle of her life. She would scrimp and leave bills unpaid, she would let food and fuel supplies run low, but each week she kept something out of the pay envelope to be added to the sacred principal. Even if it were no more than a quarter she would feel that she had done her duty. William understood how the possession of twelve dollars—this money which had fallen unexpectedly into their laps—would affect her. He was sure that she had been seeking ways and means of keeping some of it and running to the bank with it, and that she saw in the second-hand coat the chance to do it. "It's a passion with her," he said to himself. "It's like gambling or drinking, with other people."

Nothing more was said at this point, much to Ludar's relief. He had listened with lively dismay, realizing that this dispute over what he was to wear might have very serious results. Perhaps Aunt Tilly would get so angry that she would remember the things the old man with the beard had said and refuse to let him stay in the house. If Uncle Billy gave in and she had her way, then he would have to go on appearing in the coat and be laughed at by everyone.

As they neared the business section of the town the roads became filled with vehicles on runners, the sleigh bells jingling and children stealing rides on the backs. They began to pass by stores and offices and restaurants. The cold was showing a slight tendency to abate, and people had come out to take advantage of it. At every step, it seemed, they were meeting acquaintances and there were nods and smiles to be exchanged, and greetings of "Happy New Year." If Ludar had been old enough to be interested in such things he would have realized that this was a very nice and friendly place in which to live. But the mind of the boy was weighed down with the trouble which had arisen to steal away his newly acquired feeling of security. The sense of unhappiness stayed with him until they stopped in front of what was obviously the very largest store on Holbrook Street. Above the entrance was the life-size figure of a man with a board under it on which words were printed. He became immensely interested at once.

"Uncle Billy!" he exclaimed. "That's a picture up there of Catherine's father!"

"You're a pretty observant little boy," declared Christian. "Yes, Ludar, that's Mr. McGregor, and this is his store." He glanced at the gilded figure in its long-tailed Prince Albert coat and top hat, and indulged in a smile. "So you think it's a good likeness, do you?"

"Oh yes," answered the boy. "What do the words say, Uncle Billy?"

"They say: *Lockie McGregor, The Remarkable Man.*"

"What does 'remarkable' mean, sir?"

"Well, it means what Mr. McGregor is. You see, Ludar, he was a poor boy, and when he came out to this country from Scotland he hadn't one farthing in his pocket to rub against another. In fact, he's very proud of telling that he had borrowed the clothes in which he came out and that it took him a full year to save the money to pay for them. And now he owns this big store, and they say he's becoming a very rich man. That's what remarkable means."

They went into the store. Making their way through crowded aisles and past counters piled up with goods, they came to a corner which was much quieter. Here were clothes in great quantities, hanging on racks and even displayed on dummy figures. There were chairs for customers, and they took possession of three. In less than a minute Mr. McGregor himself came to see them.

"Good day, Tilly, and you, William." He seemed in rare good spirits. "The new year is starting off well. We've done a fine day's business. I suppose you've come in to get the boy outfitted with the money that's been collected. I'm one of the contributors. Did you know?"

Tilly nodded her head. "Mr. Bates of the *Star* told me."

"And here we are," said William. "Trot us out the very best, Lockie."

"You've come to the right spot. The Remarkable Man can dress this young gentleman up to the Queen's taste." He paused and gave the problem some thought. "I could give you a great bargain on something very extra special. A customer just brought it back. It's a Fauntleroy suit, just a little bit damaged. I could knock a full dollar off the price. Perhaps even a dollar and a quarter."

Ludar's heart leaped eagerly. He knew that a Fauntleroy suit was something very special indeed. In fact, he had seen a boy wearing one on the street. It was made of blue velvet and it had fine bright buttons.

If they bought him one he would not mind wearing the secondhand overcoat so much. If he had a velvet suit of his very own he might then hear people say what they often remarked about other boys but never, alas, about him: "There goes a fine fellow." But Aunt Tilly squelched the idea immediately.

"Not practical at all," she declared. "It's to be worn only on Sundays and to parties. *This* boy should have a suit for steady use."

The merchant nodded. "You're quite right, Tilly. But I'll tell you what we can do for this young man. I've a few suits of navy blue with brass buttons. They wear well and yet they look almost as good as the Fauntleroys."

William had noticed how eagerly the boy had responded to the suggestion of a velvet suit and the look of disappointment he had shown when the idea was discarded. He said at once: "Let's see one of these blue serge suits. But remember, the buttons must be very shiny."

The first suit produced was the right size, and it looked so well on him, moreover, that Ludar strutted up and down in it, proudly aware of the size and shininess of the brass buttons and the fact that there was a little braid around the pockets. Never in all his life had he possessed anything which gave him so much pleasure.

"We'll take it," said William before the question of price had been discussed. "And now, what about all the other things he needs? Shirts, underclothing, boots, a hat, an overcoat."

Tilly spoke up firmly. "He has an overcoat."

The merchant had noticed what Ludar was wearing, and he now walked to the chair where the article in question had been deposited. "Well," he said, holding it up for a closer inspection, "yes, he has an overcoat, in a sense. It's an odd color, isn't it? Frankly, I can't put a name to it. But he'll have to grow up to this coat, Tilly. Give him another year on it."

William leaned over to his wife and began to talk in earnest tones. "Listen to me, Tilly. We're not free in this matter. I explained it on the way down. The money has been given for one purpose, and it must all be spent on the boy. We may be expected to make an accounting. There's enough for a new overcoat, and we can't let them see him in a secondhand one." He looked up. "Isn't that so, Lockie?"

"William's right about it, Tilly."

Her face flushed with annoyance. "You men! You always stand together. But you needn't think you can come it over me like this. I tell you, I won't consent to such waste of money. If these people want an accounting, I say, send the money back to them! Toss it back in their faces if they don't trust us!"

The merchant decided they had better settle the issue between them. "I've a telephone call to make in the office," he said. "You two come to an agreement about this by the time I'm back."

After his departure Tilly retreated into the silence to which she resorted whenever their wills clashed. Her lips were drawn in tightly, and she refused to meet his eye. William was so familiar with these symptoms that he fell into a mental panic. What she had said about not being able to change her mind, he knew, was true to this extent: she never had changed it, whether through inability or disinclination, he could not tell. She had always won in the struggles between them, generally by sitting still and saying nothing. "It's taken hold of her," he thought, "the chance to save four or five dollars out of this fund. I can't do anything about it."

The silence continued for several minutes. William was reaching the inevitable stage in all disputes when, as one of them had to give way, he would decide to yield and get it over. As a final effort he whispered to the boy:

"Ludar, what about it? What do you want?"

Any ordinary child who disliked an article of clothing as much as the English boy did the discard of Cousin Celia's Jackie would have protested his dislike loudly and tearfully. But Ludar, through continuous repression, had learned one of the lessons which adults are supposed to understand, the need to step carefully, to avoid giving offense. He had already seen that Aunt Tilly's attitude was likely to prove a danger.

He gulped and said, "Uncle Billy, we better keep the old one."

William had expected a different answer. "Why do you think so, Ludar?"

The same sense of caution warned him of the inadvisability of any explanation. He repeated, "I—I think we better."

"But—but surely you don't prefer the old one?"

There was a long pause. "No-no, Uncle Billy. But I guess it's a—a warm coat, sir."

William sighed. Then he got to his feet and took a few steps to where his wife was sitting. "Ludar thinks we had better keep the old one," he said.

Lockie McGregor had gone to the elevated cage where the cashier sat and made change from the carriers which came buzzing in over the wires. He spent a great deal of time here each day, watching the sales slips and taking advantage of the opportunity to look down upon a large part of the store. Miss Arabella Finch, the cashier, saw to it that a fresh pad of paper and a well-sharpened pencil lay on the tiny desk at which he sat so that he could make notes while there of anything which popped into his head. There was on the desk also a large bell with which he could command the attention of the whole staff. It was not unusual for him to stand up in the cage and address words of admonition or praise to his employees in a voice which carried from front to rear of the emporium. He generally selected moments for these announcements when the store was well filled with customers.

There was no telephone call demanding his attention. He had used it as an excuse to leave the Christians alone with their problem. He sat down at his desk and said to Miss Finch: "There's a man down there who has always been a slave, and he's got a chance right now to knock off the shackles. But is he going to do it? I wonder."

Miss Finch had danced many times with William Christian in the days when he was serving his apprenticeship and she was wrapping parcels in The Remarkable Man and they had both been going out to parties. She answered in a snappish tone, "You mean Billy Christian, I suppose. If he has a chance to put that wife of his in her place, I certainly hope he does it."

The merchant had been watching what transpired in the clothing department. He got up now and remarked: "I might as well go down. It's a lost cause. I can tell by the expression on Billy's face that he isn't going to do anything about the chains today."

When she heard what Ludar had said Tilly brightened up immediately. "If he has decided he wants to keep the old one, then he's a very sensible little boy," she declared.

A detached look came into William's eyes, and he raised one hand to smooth his sandy mustache. "What you just said upsets me, Tilly," he remarked. "Isn't it a sad thing when children of that age have to be sensible about things? It's not natural for them to be sensible, you know."

"What nonsense are you talking?"

"I'm thinking," said William, "of something which happened to me. I was just his age. We were living on the farm near Woodstock and we were very poor. Our mother was dead and we had a stepmother who didn't like us much, me or any of my brothers. I've never spoken of this before because I've always felt—well, kind of deeply about it. On Christmas Eve we hung up our stockings—but in the morning there was nothing in them. We all believed in Santa Claus and so we were—kind of puzzled and hurt. But none of us said a word. We had glasses of milk and pop for breakfast, then we started in to do our chores. As I was carrying in kindling wood I heard my stepmother say to my father, 'They're being sensible about it, just as I said they would.' But, Tilly, we weren't being sensible. Something else kept us from complaining. I'm not sure we quite understood it then, but now I can look back and I know what it was. We were sorry for Father and didn't want to cause any trouble for him." He paused and sighed deeply. "As sure as I stand here, that little fellow over there is feeling sorry for me."

The merchant returned at this point. He said in an offhand way, "And what is the decision?"

With a suddenness which surprised himself William made up his mind. He walked to the rack where the overcoats were hanging and began to look them over. "I guess, Lockie," he said casually, "we'll just pick out one of these for the boy."

The merchant said under his breath, "Hurrah!" He joined his neighbor at the rack and began to go over the stock.

Tilly did not move, but after a long pause she demanded, "Have you forgotten, William Christian, that I have the money from the newspaper right here in my purse? I won't give any of it up for any such sinful extravagance! How do you intend to pay for this coat?"

William pulled one out from the rest and examined it. This gave him a splendid excuse for not facing her. "This one might do. I like the cape on it. . . . How will I pay? Well, I'm sure Lockie will trust me until Saturday. I'll come right down from the shop and give him the money."

"William Christian!" There was horror and incredulity in his wife's voice. "You wouldn't use your wages! Why—why, you haven't any right. I wouldn't let you. Have you forgotten how we've portioned out every cent of it for the next three months? Have you gone insane?"

"No," answered William. "But I'm afraid we won't be able to follow out that whole portioning plan. There won't be anything to add to your principal, Tilly."

So a very fine overcoat was selected for Ludar. It was made of chinchilla cloth and had a cape which covered the shoulders, and buttons of bone which the boy thought handsome in the extreme. Then, after doing some figuring and discovering there was still enough in the fund for a hat, William surrendered to a mounting mood of princeliness and picked out one of gray Persian lamb which could be pulled down over the ears and so was very comfortable as well as good-looking. When Ludar caught his reflection in a mirror in all this grandeur, he could not believe he was seeing himself. Why, this looked like a boy who had always had a family and good clothes of his own and had been treated as though he were wanted. Perhaps William had some idea of what was passing through his mind, for he came up behind him and whispered, "Well, sonny, we'll show 'em a few things now, won't we?"

4

But the day which thus threatened to end in collision, conflict, bitter words, and sure disaster was brought to a quite different conclusion by something close to a miracle. After returning her son's damaged Fauntleroy suit and driving a hard bargain with the owner of the store in the matter of the cash allowance, Mrs. Tanner Craven had remained to make some purchases. She was wearing one of the Russian overcoats with long velvet capes which had become the rage, and when the Christians met her near the front entrance her arms were filled with small parcels. She stopped and indulged in a curious look at Ludar.

"So this is the boy," she said. Then she nodded to Tilly. "You seem to have done nicely with the money. That's a fine overcoat and a very fine hat. Yes, Matilda, you've shown splendid judgment. Have you spent all the money?"

"A little over," was the answer, in a tone compounded of pride at this favorable notice and gloom over the way the money had flown. "I met the deficit out of the money he brought from the passengers."

Mrs. Craven then looked at William with an eye which became less friendly. It must be acknowledged that he was cutting a poor figure in his homemade round hat and his overcoat which had passed the stage of shabbiness years before and was now well advanced into disreputability. "I never see you in church, Mr. Christian," she said.

This frontal attack was so unexpected that William stammered in replying. "Well, no, Mrs. Craven. You see, I'm not a Baptist. I'm Episcopalian."

"I hadn't heard that you worship at the English church, Mr. Christian."

"He doesn't," declared Tilly. "His father was Episcopalian and he takes advantage of it. Ah, the arguments we've had! What he really likes, Mrs. Craven, is the Salvation Army."

"The Salvation Army!" Mrs. Craven, it was evident, was surprised and even horrified. She looked at William as though he were some curious and new type of heathen. Then she frowned and said, "I feel it my duty, as you *are* connected with our church in a sense, to have a talk with you about this. Our sleigh is outside, and I'm going straight home. I'll drive you up Wilson Street. It will be only a block or so out of the way, and we can talk."

William had a strong desire to refuse. He said to himself, "She's only willing to be seen in public with us because she thinks the boy belongs to a good family." In this he was wrong. Mrs. Craven was not thinking about Ludar. Her interest in church matters was so great that she felt she must probe into his state of mind and, if possible, bring him to a more reasonable view. He was right in another respect. Like most men who are easy and accommodating on the surface, he had the habit of bitter inner judgments. Looking at the slight droop to the corners of her mouth, he said to himself, "This woman's as tough as the whalebone in her stays."

The sleigh was a new and imposing one, with a seat up in front for the driver. There was a driver, moreover, an Irishman with a good-natured face and a stage brogue. The latter got down from his high perch and ensconced the ladies with their faces to the horses, and then spread a buffalo robe over their knees. With a sly wink he did the same for William and the boy on the opposite seat. Then, with the whip cracking in the air and the snow crunching under the runners and the sleigh bells jingling, they were on their way.

"And now," said Mrs. Craven, "why do you like the Salvation Army? You must not misunderstand me. I like the Salvation Army too. I think it does wonderful work."

William, cornered and at a loss for words to describe his feelings, made use of the first reason he could think of. "I like the way they pitch into the singing." Then, collecting his senses, he fell back on one which came closer to explaining him. "If you put no more than a nickel in the hat they thank you—and they mean it."

"I don't see any difference yet. Our members throw themselves into the singing too. I've seen tears in their eyes often when the hymn selected has touched their feelings. Yes indeed. And if someone is not in a position to contribute more than five cents we accept it thankfully." She went on talking with great earnestness about the duty of all good citizens (conceding him to be one) to belong to a church and become a regular worker. It was on the value of regular work that she laid special emphasis, citing the good which was being done by the members of the First Baptist Church. William listened respectfully and nodded his head at intervals.

When they passed Winchester Street she brought her dissertation to a close. "That's all we have time for, Mr. Christian. I sincerely hope you agree with what I've said and that we may expect to see you oftener in church."

William said, "Yes, yes, of course."

The talk then veered to immediate churchly interests and was confined to the ladies for the three blocks of the ride which remained. As the sleigh came to a stop Mrs. Craven said, "Matilda, I'm happy to have had the chance to become closer acquainted. And," with a slight hint of reserve, "with you too, Mr. Christian."

When they alighted Tilly was in a mood of great delight. She had acquitted herself well; she had said just the right things; all the neighbors had seen them arrive in the Craven equipage and had heard the cordial note of the great lady's farewell.

"Well, William," she declared as they made their way to the side door, "I guess you were right. For once. Mrs. Craven thought we had used the money well. Now, then, get the range going. I'll make potato cakes for you for supper."

William was particularly fond of potato cakes—crisp, thin biscuits in which mashed potatoes were used in place of flour. He knew that all was well in the Christian family.

5

Ludar had asked if he might take a look at the back yard before going in and had been told that he might if he did not stay out too long. It had grown quite dark, but he trudged down the slippery slab walk with high expectations, for what he had seen of the yard from the windows of the house had aroused his interest. As he passed the summer kitchen Old Cat was forcibly ejected from the back door and went streaking across the snow, to disappear over the fence into the McGregor demesne.

Ludar discovered that the yard possessed a number of fruit trees which had so intertwined and grown together that even in their present leafless condition they hid much of what lay beyond them. In one corner was a concrete-and-frame building which, he was to learn later, represented an unsuccessful effort of many years back to raise chickens on a commercial scale.

The boy walked slowly down the path shoveled through the snow, looking about him and taking everything in. His progress led him soon to an exciting discovery: that part of the view ahead was cut off by a building on the other side of the fence, one of white brick with a metal weather-cock atop it (he recognized this at once as a stable), but that to the right of this, where only a tangle of bare shrubbery stood, the outline of a large house could be seen. It was a very tall house of white brick with red facings. It had slate-covered gables and many chimneys, round and spiraled like corkscrews, and, most interesting of all, a tower jutting up above the line of the roof. It was faintly reminiscent of the great house where he had once lived, and he began to make his way through the snow for a closer look.

But the space between two pickets of the fence which he selected for observation had an occupant already: a girl, a shade smaller than himself, with a pair of solemn dark eyes under a hat with pompons of brown fur.

"Hello," said the girl when she realized that her presence had been detected.

"Hello," said Ludar. He was sure he had no right here, that he was intruding and should go back to the house at once.

"Don't go," said the girl when he started to put this resolution into action. "Are you the boy who came on the train yesterday?"

This made a reply necessary. Ludar stopped, stared down at the snow, and said in a voice so low that it was hard for her to hear, "Yes, I was on a train. For a long time. Days and days."

"Father said a whole week." She had a nice voice and she was smiling at him, through the narrow space in the fence, with an unmistakable desire to be friendly. "My, it must of been exciting."

The boy sensed the friendliness and decided to remain. However, he did not look at her but showed himself very much interested in his own feet. "I was on a boat too."

"Oh, were you on a boat?" The girl thrust her face as far as it would go through the fence in order to see better. She was a pretty little girl.

Ludar nodded. "Well," he said to himself, "I better go." But instead he remained where he was and even stole some quick glances at her. She was a *very* pretty girl.

"What's your name?" she asked.

"Ludar. Ludar Prentice."

"Mine is Antoinette Milner. Everyone calls me Tony except my father, who says it's a boy's name. I like to be called Tony. I think it's a nice name. Do you?"

"Oh *yes!*" Then shyness overtook him again. "I—I must go in now."

"Why? Dinner won't be ready for nearly an hour yet. Or do you have it earlier?"

"We have supper." Ludar sensed that this was in the nature of an admission of inferiority, although he had no idea why.

The girl, it seemed, was full of questions. "I think your hair must be naturally curly. Is it?"

"I guess it is. It's hard to brush in the mornings."

"It's nice to have naturally curly hair. Mine isn't. It's very straight. And it's dark. Yours is fair, isn't it?"

"Yes." He broke the brevity of his answers by adding: "Uncle Billy says I'm a cotton-top. I don't know what that means."

"I do. If you were a girl it would mean you were a blonde. Sometimes I wish I was a blonde instead of having this straight dark hair." She was studying him with sober and intent eyes. "That's a very nice hat, Ludar. I like Persian lamb. I've a coat of it. Gray, like your hat."

"It's a new hat." As the clothes he had on were the first new things he ever remembered possessing, he felt disposed to enlarge on the fact. "I got it today. This overcoat is new too. And in the house I've got a new suit. It has brass buttons, and braid on the sleeves."

Most girls would have said that the money for his new clothes had been raised through her father's newspaper. Tony Milner did not say it. She did not want to hurt his feelings. All she remarked was, "I guess it's a very nice suit, Ludar."

The desire to be expansive was growing in him. "I'm coming out tomorrow and build a snow man. It's going to be big. A giant, I think, like Blunderbore. I'll make a nose on him as big as my whole head, and teeth long enough to eat little children. When you see it you'll be frightened."

"No, I won't. I won't be frightened. I don't get frightened at things." Her eyes were shining through the fence with intense interest. "Could I come over tomorrow, Ludar, and help you build the giant?"

He perceived difficulties. "You couldn't climb this big fence."

She was honest enough to acknowledge that the climb was beyond her powers. She had tried it once and had fallen and hurt herself. "But," she said, "I could walk around and come through the front gate, couldn't I?"

"Would your mother let you?"

"I haven't a mother. Just a father. My mother died years ago, and I don't remember her." It was on the tip of her tongue to refer to the death of his father, but again her early-developing sense of tact prevented her. "Is your mother alive, Ludar?"

The death of the stranger who had been his father but whom he had never seen had not affected Ludar to any great extent. He had still only a vague idea of what death meant. Tony's question, however, brought into his mind all the emotion which the tragedy might have been expected to create.

"I——" he said, finding it hard to keep back his tears. "I think so."

"You don't know for sure!" The girl spoke in an incredulous tone. "Why, you *must* know if your own mother is alive or not."

Ludar was thinking of the lady at the big house who had taken care of him and had been so kind. The memory seemed to recede with every day, and he could not recall anything she had ever said to him. And yet she must have been his mother, although Aunt Callie had seldom made any reference to her. His thoughts ran particularly to the day when he had been taken away and had not seen the lady before he left, the day when the house had been so quiet. "I lived with an aunt," he said finally. "But she wouldn't answer questions."

Tony nodded her head. "Aunt Mona May doesn't like to answer questions. I guess aunts are like that. She's Father's sister and she keeps house for us. She's strict. It's hard not to have a mother, isn't it?" Her mind hurried off on another tack. "Ludar, are you a lord?"

"I don't know."

"My," said Tony, "there's lots of things you don't know, aren't there? What is a lord?"

It was on the tip of his tongue to answer, "I don't know," but he restrained himself in time. "It's," he began, trying to think of something to say which would not make him sound too ignorant, "it's someone important, I think."

"How much important?" she wanted to know. "As important as having a father who owns a newspaper?"

"No," answered Ludar. "Not as important as that, I guess."

She smiled and nodded at him from between the pickets. "I always thought lords were up in heaven," she remarked.

At this point a voice from the back door of the large house was raised in a summons. "Tony! Where are you, Tony? Come in at once!"

The girl said: "That's Aunt Mona May. I'll have to go now or she'll be very sharp with me. She's *very* strict." She did not turn to leave, however. "Were you sorry to leave your friends and come out here?"

This touched the boy in the tenderest spot of all. He said in a thoroughly humble tone, "I had no friends."

"No friends? Why, I've *lots* of them." She became aware that his feelings had been hurt and she added quickly: "I think it's nice you are living here, Ludar."

The boy realized suddenly that seeing this friendly little girl had made things different for him. It had done more to raise his spirits and his esteem of himself than all the talk about what he might have been in that dark and mysterious life he had left behind him. For the first time he smiled. "Oh *yes!*" he said. Then he added, "I hope you can come over tomorrow and help me make the snow man."

6

William Pitt Milner came back into the dining room, after answering a telephone call, with a red and angry face. The imperfections of the new device always upset him, but on this occasion the nature of the discussion over it had been enough to send him into a rage.

He seated himself at the head of the long dining-room table, which accommodated twenty people when all the leaves were used, and glared down at his sister. "It was Daniel Cape," he said. Then he let his feelings explode. "The doddering old nincompoop!"

"Brother!" Mona May Milner seemed almost as angry as he was. "How dare you use such language before your little girl? How many times must I protest about this temper of yours and the way you let your tongue go?" She turned to the only child of the family, whose seat in the middle of the board made her look very small. "Tony, I've said to you ninety-nine times that you must never repeat anything you hear your father say. Now I tell you for the hundredth time."

Tony had been disappointed in the dinner, which was stewed veal. She was doing no more than nibbling at the food on her plate while intently studying a steel engraving of the capture of Mary Queen of Scots which hung over the fireplace. Without turning she said, "I never know what Father's saying, Aunt Mona May."

"And it will be a good thing if you never do!" Miss Milner helped herself to salad with a gesture which was like a flounce and began to crunch the lettuce, which formed a large part of it, with indignant teeth.

The publisher was paying no attention to his sister's humors. "That mental scarecrow, that old miff cheating the undertakers! He had the audacity to call me up and say he didn't approve of the kind of paper I'm publishing. He thought he could be of use to me and said I should consult him about things. Why, he would be of as much use to me as a seasick gull! Compare the up-to-date, all-around newspaper I'm giving the people of Balfour today with the old-maidish rag he turned out, with its surfeit of local news, its countrified make-up, and its dismal editorials!"

The publisher got to his feet and began to pace about the room. It was the largest and most ornate dining room in the city. The walls were paneled in bright curly maple, and there were beams of the same yellowish wood across the ceiling. The fireplace was huge and ugly. There were gas fixtures along the walls in the form of bronze Roman warriors, and an immense light was suspended over the table in the shape of a gondola.

"Just take, for instance," he began, stamping his feet as he walked, "how we covered in today's issue this story of the suicide and the boy. Two columns of good human-interest stuff. If there had been time to get the cuts back from Toronto I'd have had pictures of the boy and of that family group. Today I've spent a small fortune in cables to England to find out about the family. I'll have word in a day or so about where and when the boy's to be sent back——"

Tony suddenly laid her head on the tablecloth beside her plate and began to weep loudly and unhappily. Her father stopped his pacing immediately and hurried around the table to her.

"What's happened, Mona May? Did she bite her tongue? Has she swallowed a bone? My poor little pet, what's wrong? Are you sick?" As the child continued to wail as hard as ever, he waved an impatient hand at his sister. "Mona May, for God's sake, don't sit there like a stuffed owl! Do something! Phone for Dr. Barton at once!"

"What's wrong, Tony?" asked Aunt Mona May, without stirring from her chair.

Tony raised a grief-stricken face. "I don't want him sent back!" she wailed.

Milner looked at his sister with a puzzled frown. "She doesn't want him sent back! Who? This English boy?"

"I don't see who else she could mean." Miss Milner went on eating her salad and looked neither concerned nor curious.

Milner leaned a hand on the table beside his distressed child. "Antoinette," he said, "tell your father what you mean. Who are you talking about?"

"About Ludar, Daddy," she said between sobs.

"Ludar! So it *is* the English boy. What do you know about him, my little pet?"

"He's a nice boy. I don't want him sent back."

"But, Antoinette," said Milner with a patience which would have surprised anyone who had business dealings with him, "the boy's relatives are in England, and now that his father's dead he must go back and live with them."

"I want to help him make a giant!" cried the child. "I want to see him in his new suit with the brass buttons!"

The publisher looked at his sister uneasily. "Our Antoinette and the young man seem to have become well acquainted." He patted the child's shoulder. "There's nothing to worry about. The boy won't be sent back for weeks. You'll have time to build a giant with him, if your fancy runs that way."

Tony was a very tidy and particular young lady. She took out a handkerchief, dried her eyes, blew her nose, and then sat up straight again in her chair. She looked at her father and gave her head a sympathetic shake. "Poor Ludar never had any friends, Daddy. I'm going to be a friend to him."

Chapter VII

I

At the western end of Holbrook Street, where that important artery of trade and commerce dipped down to the crossing of the river at the Iron Bridge, there was a sprawling and somewhat dilapidated building which had been a tavern in the old days when cattle drovers and horse traders and land speculators swarmed into the town. As only hotels could hold liquor licenses, a sign still swung over the front entrance with a painting of a fabulously large caravansary and the words "The Shropshire Arms," and there was a current fiction that, if one so desired, it was possible to find accommodation there. Just where guests would be put was something of a mystery. The ground floor was given over to the bar and to a small workshop where the present owner of the property carried on his trade, and the second story was rented out for law offices. To maintain the fiction, however, there was a sign in a window of the bar which said, "Rooms two bits. All you can eat, ten cents. Provender for beasts." This, clearly, was a survival from much earlier days.

William Christian entered the building by the side door, which did not connect with the bar and over which hung a second sign:

SHOES PATCHED AND SOLED,
SINCERELY, Z. X. GOLD.

He was making for the stairs which led creakily up to the lairs of the legal ogres when he heard himself called in a high-pitched voice.

"William Christian, William Christian, William Christian!"

The proprietor was standing in the door of his shop, a bent old man with a sad and sallow face under a dingy plug hat which was pulled down tightly to his ears and wearing a black coat buttoned up so tightly that no vestige of shirt or collar could be seen. Z. X. Gold, whose real name was something quite different and quite difficult of pronunciation, was one of the characters of the town. Although he owned the Shropshire Arms and several other buildings in the business section and was reputed to possess stocks and bonds in such quantities that no ordinary safe would hold them all, he continued to ply his trade as a mender of shoes as though his daily bread depended on it. He charged moderate prices for expert work. The leading men of the town liked to drop in on him and listen respectfully to his opinions and the expounding of his philosophy. Edwin George Laird, the smartest lawyer in town, was a special crony of the old man. "Why read Kant and Hobbes and Descartes and the rest of them?" he used to say. "Go and listen to Z. X. Gold instead."

"Come in, William Christian," said Mr. Gold. "I want a word with you."

Unable to think of any reason for this sudden interest in him, William obeyed the summons and stepped into the shop. It had once been much larger, but the increasing business of the bar had been cutting into the space. All that was left now was a mere hole in the wall, with a counter, a rack for shoes waiting attention (the repaired ones were placed in tidy and well-polished pairs along the counter), and a shelf of books which the proprietor kept for his use if the time should ever come when he had no work to do.

"So," said Z. X. Gold, "there is mystery about this boy? There is so often trouble and mystery when families are divided and have to travel. I know it well and so I am interested in this boy who arrives with this sign on his back. Have the police any word about him yet?"

"No," said William. "They don't seem able to learn the real name of the father."

The old man nodded his head several times. "It often happens that way. There is need for secrecy and so new names are taken, and then the parents lose track of the children and the children find themselves lost in a world which can be very unkind and cold. But, William Christian, you will find out all about this boy in time. Oh yes, I am sure of that. It happens that way often too. In my family, William Christian, there was a boy who went all alone to London to find his parents who had gone first while he stayed behind and was hidden by some kind people. At first he couldn't find his father and mother and so he had to look out for himself. He swept crossings and he ran errands and he nearly starved, that poor lost boy. But in time his parents found him, and after that all was well."

"I hope we get to the bottom of it in this case," said William. "But I'm not sure we will."

The old man's habit of repetition took hold of him whenever his feelings were sufficiently engaged. "Don't give up, don't give up, don't give up!" he exclaimed. "It is very sad thing when children do not find their parents."

He had ensconced himself on a high stool behind the counter and had resumed work on a dilapidated tan shoe. The work kept his eyes down most of the time, but at intervals he would raise them suddenly and disconcertingly.

"You will need money," went on the old man, holding the end of a waxed thread in his teeth, which, in spite of his advanced years, had no break in the ranks. "To make more inquiries, perhaps to hire lawyers, to go to court for this boy, to send him back someday soon to claim his rights. You cannot do this yourself, William Christian. You must make up a list and get people to put down their names. Yes, make up this list at once, and at the top of it put 'Z. X. Gold, twenty-five dollars.' Oh no, this is not for profit. It is not even for interest or for paying back. It is for the memory of this other boy who had to travel from the East to the port of London and all alone. Perhaps I was this other boy, but it was a long time ago and sometimes I am not sure. My memory gets bad, William Christian, and then I remember things which did not happen and I forget things which did happen. Most of the time I am sure I was this other boy who became lost." His fingers continued to ply the needle, but they trembled a little. "You must do as I have said to you, William Christian. You must get up the list and at the top you must write, 'Z. X. Gold, twenty-five dollars.' And you must do it at once."

2

There were five doors opening off a dingy anteroom on the floor above. Three were reserved for members of the law firm which rented the front half and were marked respectfully, "Mr. Thompson," "Mr. Laird," and "Mr. Fine." Mr. Fine was dead, and his office was given over to the young men of the second generation of Thompsons and Lairds and Fines, who were being harnessed early to the reading of law against the day when they would join the firm. The door had been removed from the fourth room, and here sat, elbow to elbow and chair to chair, a staff of clerks and copyists, and nearest the window an old legal hack who resembled Sydney Carton in the amount of liquor he could consume and in no other respect, and who was paid not too generously to write briefs. The fifth room had been rented out by the firm, which believed in economy in small matters. This door was closed and conveyed the information to all comers that within could be found Abimelech Newstead, Public Stenographer and Typist, Court Work a Specialty. Thompson, Laird and Fine had an ulterior motive in keeping Mr. Newstead near at hand, for he did all their stenographic work.

William Christian tapped lightly on Mr. Newstead's door, and a reedy voice from within bade him enter. The room into which he stepped was perhaps the dustiest and least orderly room in which business had ever been transacted. It contained no fewer than six dirty flat-topped desks, five of them piled so high with broken-backed lawbooks and torn pamphlets and piles of the carbon copies of evidence and legal rag, tag, and bobtail generally, all of it overlaid thickly with dust of long accumulation, that only the fact of possessing four legs made them recognizable at all. There was something ghostly about the place, as though occupancy had ceased long before and the spirits of those who had created the mounds of material were still hanging about.

The truth of the matter was that Abimelech Newstead had a fear of destroying any kind of paper or document and an even greater dislike for the labor of tidying up; with the result that once a desk became so heaped up that it was no longer possible for him to work there, he left it and had another cheap one moved in.

The sixth in this long succession was now about half covered, and there was still room for him to work on it. His typewriting machine had been shoved to one side, however, and he had to swing around in his chair in order to use it. It was a double triple-bar machine of the very latest design; and, in spite of his slovenliness in every other respect, he produced work on it which was as clear and clean as copperplate.

"Huh, it's you, is it?" said the occupant of this strange place of business, swinging his bald head around to get a glimpse of the visitor over his shoulder. "What ye got now? An idea for a flying machine to reach the stars?"

"No," said William. "But I've been doing a little thinking along that line. It can be done, Bim, and, who knows? I may be the one to do it."

They had an interest in common which led to many excited talks between them, an interest in new things. William's absorption in the invention of machines was matched by the public stenographer's schemes for the making of much, and easy, money. They spent long hours on Sundays and holidays revolutionizing the world and everything in it, even the people and the laws.

"Now, don't say a word," commented the stenographer, turning around in his chair and planting a hand on each knee. He was fat and mild and so untidy in his dress that people wondered why he had never been carried off by mistake in a dirty-clothes bundle. "Let me tell you my idea first. I've got a scheme, a brand-new scheme, and a blazing good one which is going to make so much money for me that I'll go to Paris and live in a palace and have wine for breakfast if I want it. Well, as a matter of fact, it's not a new scheme at all. It's a kind of an old one. I thought of it years ago, but I laid it aside. Last night it popped into my head again and I said to myself, 'Bim Newstead, you great ox, you don't deserve the brains you got when you neglect a wonderful plan like this.'"

He drew a red handkerchief from a rear pocket of his shapeless trousers and mopped his high bare forehead. It was chilly in the room, but talk of ideas always brought out perspiration on him.

"Here's what I'm going to do," he went on. "I'm going to make a list of all the best authors in the world. Then I'm going to take each one in turn and I'll get two books he's written out of the library, and I'll read them. I'll read them right to the end. After I've done the lot I'm going to decide which one of them's the best, and then I'm going to write books exactly like his. William, it's as simple as that, and yet there's a fortune in it. I'll be rolling in money in no time at all."

William did not read books, and so the weaknesses in the plan did not occur to him. He nodded and said, "It's quite a scheme, Bim, quite a scheme."

"You see," said the now thoroughly enthused copyist, "I don't have to experiment or waste a lot of time writing up the wrong alley. What the public likes to read has been decided already. If I find it's Dickens they want, I'll give 'em Dickens. If it's—well, if it's someone else, I'll give 'em someone else." He got to his feet, snapped his braces up over his shoulders, and tucked his shirt back into place all around. "And it's not only books. There's plays too, William. After I've written a few books I'll go to New York, see all the plays, and decide about them. I'll write a few plays. And popular music's just the same. Great Scott, William, the way they sell popular music! Invent a catchy air, write some verses about your true love dying, or someone being cased up in a coffin, or a villain luring a young girl away, and there you are!"

"Do you have that list of authors?" asked William, who found his interest growing.

"Of course I have it. At least I *did* have it. Six, seven years ago. Let's see." He placed a hand on his forehead and concentrated on the business of remembering. Then his face lighted up and he walked to the third desk in the row. Plunging a hand unerringly into the mass of papers, he brought out a sheet triumphantly. Blowing the dust from it with a vigorous puff of the lungs, he held the document out for William's inspection.

"There it is," he said. "There's the proof of what I was saying. The key to wealth is in that list of names, William Christian, the road to fame and fortune. Look here. Scott, Dickens, Balzac—that's a funny name—Thackeray, Mark Twain, all the rest of them. Do you know what I'm going to do? I'm going to close my eyes, take a pencil, and poke it at the list, and the name it points to will be the one I'll start in reading first."

He put the list down beside the typewriting machine and placed a sponge on top of it to anchor it there. Then, in a more subdued tone, he asked, "What's brought you in today, William?"

"Well," said William, getting rather self-conscious because he had come to ask a favor. "I need some help. It's about this boy. They haven't been able to find out who he is yet, and I'm beginning to think they never will. And if they don't find out who he is, then what happens?"

The stenographer gazed up at the ceiling and thought. "Well, what *does* happen?" he asked finally.

"A new home has to be found for him. I've become fond of the little codger and I want to keep him. I'll want to adopt him. And that's where I'll need your help. What do you have to do to adopt a child?"

Newstead poked a thumb over his shoulder in the direction of the law offices. "It's a good thing you didn't go in there, William," he said. "They'd have got a week's salary out of your pocket before you opened your mouth. And adoption's a simple matter when you come right down to it. You don't need lawyers for that. All you do is go before the proper authorities——"

"And who are they?"

"I'm not just sure. I tell you what, William, I'll have a quiet talk with someone in there—perhaps old Fogarty—and I'll ask a leading question or two, and I'll get the whole thing out of him. Then you'll know what to do and it won't cost you a cent."

William nodded and smiled. "That's fine, Bim. I—I don't like going to lawyers. I don't like *them*. And of course I don't have any dollar bills to toss around."

Newstead gave a hollow laugh. "*You* don't like them! How do you think I feel about them? *I* have to work with them all day long."

The stenographer conducted his friend out through the anteroom, saying in a loud and satisfied voice, "Well, Mr. Christian, I guess we've got your problem solved." On the way back, as a result, he was buttonholed by Miss Adelicia Craven, who had been waiting for a talk with Mr. Laird.

"Did you say that was Mr. Christian?" she asked. "Mr. William Christian? I've been wanting to meet him for a long time. I—I've a matter to talk over with him."

"Well," said the stenographer, "he's halfway up the block now. I guess it'll have to be the next time."

Miss Craven's face showed that some deep emotion had taken possession of her. In a low voice so that no one else could hear she said, "Mr. Newstead, I may as well confess to you that I'm responsible for the death of the boy's father."

The stenographer had turned in the direction of his office. At this remark, however, he swung around and regarded her with slack-jawed amazement. "Huh! What did you say?"

"Oh, I don't mean that I actually had anything to do with it. But I was told about the man and I could see he was in a desperate state of mind. I was busy and I delayed doing anything. Then it was too late to help him." She paused and shook her head solemnly. Then, her manner resuming its more normal air of briskness, she asked, "Are you a friend of Mr. Christian?"

"Yes, you might say I am, Miss Craven. We have interests in common."

"Is he a good man to have the care of the boy?"

The answer was emphatic. "The best possible. He's honest and hardworking and he thinks a heap of the boy. Of course he's only a workingman and he doesn't have much money. They live simply, Miss Craven."

Adelicia Craven nodded. "Everything seems to fit in, then. I must talk to Mr. Christian at once. You see, I feel responsible for the boy in a way and I want to do for him whatever I can. When the time comes I'll probably send him to college. I don't know anything about his real father——"

"Huh?" Mr. Newstead looked startled. "Do you mean to say you didn't know this man who killed himself?"

"No, I never happened to lay eyes on him."

"And still you say you're responsible for what happened?" The stenographer shook his head to show that this was completely beyond his comprehension. "I'm glad, Miss Craven, I don't have your kind of a conscience. It would keep me in trouble with myself all the time."

3

It looked as though Mr. Edwin George Laird had his hands in his lap. Actually he was shuffling and reshuffling a deck of cards. Having a knack for magic and sleight-of-hand tricks, he spent a part of each day practicing, and on this occasion he was trying, after countless failures, to accomplish the second-card deal, the very hardest feat of all.

"Well, what am I to do?" asked Adelicia Craven.

"Nothing," answered the lawyer.

He was a tall and spare man who never used two words to do the work of one in conversation but who could suddenly blaze into flights of eloquence when addressing a jury. That he was a patriotic citizen and a solid Conservative was apparent from the pictures on the walls of his rather spacious office: Queen Victoria, the late Prince Consort, the royal children, Disraeli, Sir John A. Macdonald.

"But, Ted," protested the spinster, "I can't just do nothing. It's against my nature to sit still and let things happen that I don't approve."

The lawyer felt a sudden exultant glow. He had done it! He had succeeded in dealing the second card from the top! Now that he had actually accomplished the feat, he was sure that after a few months more of steady practice he would be able to make it a part of his repertoire. Content with what he had done for the day, he sat up in his chair, drew out a drawer of the desk, and dropped the cards inside.

"For two reasons," he said. "First, you can't afford it. You realize, of course, that hard times are ahead of us? And you have too many charities, Lish, particularly the upkeep of that absurd household of yours which is enough to send you into bankruptcy. I'm telling you straight, you can't afford the luxury of going to court to fight the Craven Carriage and Furniture Company. You have no case. Your father's will provided that the older son, as president of the corporation, was to remain in the house, which belongs to the corporation. Since his removal as president he has continued to live there, and his occupancy has been tolerated—I'm speaking in legal terms, you understand—pending determination of what should be done with the property. The will does not cover this point, but I assure you, Lish, that the courts would back the directors in their determination to incorporate the building in the plant now that the need for expansion has arisen. You'll lose if you fight them—and have the costs assessed against you.

"But the second reason," he went on, "is more important. The whole town will be against you solid. This expansion will mean more dividends for several hundred citizens who happen to own a few shares of the stock. The new department will provide employment for twenty-five men, according to what I've heard. There you have it, Lish: you can't stand in the way of progress. The Board of Trade will pass resolutions in favor of the scheme. The Trades and Labor Council will have bitter things to say about you if you go against the sacred principle of more employment. Your popularity—and right now you're the best-liked person in Balfour, bar none—would melt away overnight."

"I wouldn't let my popularity stand in the way of doing my duty," she protested.

"But is it your duty? Aren't you letting your temper get the better of your common sense? Langley should have given up the Homestead long ago. This is a case where it would be an expensive and useless folly to fight them."

Adelicia gave vent to an angry grunt. "I was all set to give Tanner the scrap of his life!" she said. "What if I do lose? What if it costs me a lot of money? I don't intend to let Tanner ride roughshod over Langley without raising a finger."

The lawyer said briskly: "Now you're talking like a fool, Lish, and you don't mean a word of it. You have a cool head, and when you get home and think it over you'll see that I'm right."

Lish said, "Damn it!" in an explosive tone. Then there was a pause, a long pause. At the end she sighed gustily. "You're right, Ted. But I'm mad as hops at you just the same. Right this minute I'd like to spit in your eye and in the eyes of every judge and lawyer in the whole world!" Finally she nodded her head in grudging acceptance. "Well, I can't fight. But what about Father's collection? Is Tanner going to work some sharp scheme about the things too? I've told you how Father left it, haven't I?"

"Yes. A number of times. I have your sworn statement about it in the office safe. I'm not in Tanner's confidence, but I fully expect he'll be up to something. Never be surprised at anything where Tanner's concerned. As smart as paint, Tanner. There's a number of things he could do about the collection. He might say that Langley has not followed out your father's instructions and that a settlement must now be made. Pending the reaching of an agreement, he might apply for an injunction against the disposing of any of the articles or even to prevent their removal from the house. Certainly he could demand that an inventory be made of everything and an appraisal of values before the stuff is moved. If the question got into litigation the court might hold that, since the wishes of your father could no longer be followed out in full, the collection should be sold as a unit—what would meet his wishes in one sense—and the proceeds be distributed among you."

"It's a family matter, just among the three of us!" declared Adelicia. "If I back up Langley and swear that Tanner's selfishness made it impossible for him to make the settlement——"

"But would you do that? You'd be going against your own best interests. Of course if you did——" The lawyer paused and gave the matter further thought. "It wouldn't prevent Tanner from getting his inventory. I approve of having one made anyway. In one respect you're in a strong position; Langley has possession and you acquiesce in it. You're two to one." He leaned back in his chair and squinted down the length of his rather long nose. "Tanner seems very much concerned about the furniture. He was in the club late yesterday afternoon. You know, of course, that he likes his nip as well as the rest of us good churchgoers? Usually he slips down the hall to the steward's quarters and comes back chewing Sen-Sen. Yesterday he ordered a scotch and soda and drank it down in full view of everyone. Then he had two more. He got a bit talkative and finally started spluttering about how badly he'd been treated over the furniture. There were tears in his eyes when he mentioned that drum table. Just then Langley put his head in at the door—he's given up his membership, but he comes in often as a guest—and Tanner started to get to his feet as though he intended to have it out there and then. We shoved him back into his chair, and Langley took the hint and vanished into the cardroom." He held his right hand up in front of him and flexed the fingers several times. "We can take it for granted that Tanner will be up to something, feeling the way he does. His lawyers won't miss this chance to get the matter settled. They're clever, Swayze and Carrington. Very clever."

"You talk as though this is a game of chess. You make a move and then they make one. Then the lot of you sit back and admire each other's cleverness."

The lawyer, knowing he had said everything that was necessary and that his advice would be accepted, said nothing more. He was thinking triumphantly, "There are only three men in America who can do the second-card deal!"

Miss Craven got to her feet. "You've convinced me, damn it!" she said. She swept up her bag from the seat of her chair and turned toward the door. "Somehow, Ted, I don't like you as well today as I usually do."

Chapter VIII

I

It snowed on the Monday after New Year's. There is a quality about a Canadian snowstorm which baffles description, for beneath the complete silence with which the flakes fall thickly out of the sky and settle down slowly there is a hint of voices and of sounds too low of pitch, or perhaps too complex, for the human ear; as though nature is grumbling at the liberties man has taken and hinting of a purpose to plunge the whole globe back again into the glacial period. William, returning from work at noon, did not read anything of this in the steady drop of the snow which had turned his brown cap white and had caught in his mustache and had already turned the trees into ghostly white tents. He clucked disgustedly when he saw that the sidewalk was becoming rapidly impassable, and did not enter the house, going instead to the summer kitchen for a snow shovel. With this he was proceeding to clear the walk when Ludar came out through the front door with a shiny new schoolbag over his shoulder.

"Hello, young fellow," said William. "Aren't you staying to dinner?"

The boy was too full of news of his own to answer. "I'm going to school this afternoon," he said excitedly. "Mr. Willing came and said I could go. He spoke to Aunt Tilly."

William leaned on the shovel and rested from his exertions. "That's fine," he said. "I'm glad to see you like the idea. Boys who like school will make something of themselves."

"Aunt Tilly says I'm to wear the new suit for once. And I'm wearing my new overcoat and hat."

"My aunt Sally Ann! You'll be a regular dude, Ludar."

The boy's health had been showing some signs of improvement already. There was no change yet in his weight, but his cheeks, which had been pale, were sometimes almost rosy. He possessed, at any rate, a great deal more energy.

"Antoinette won't be at the school," he said. "She goes to a private school. What are private schools, Uncle Billy?"

"Institutions," said William shortly, "which should be abolished." He tapped the snow off the end of the shovel. "It's packing. You'll be able to build that snow man now."

Ludar took off his coat and hat for dinner, but he kept the bag over his shoulder, being afraid that the signal to start might catch him unprepared if he dispensed with it. Some late contributions had been turned in at the newspaper, and so the schoolbag was of good leather and had a brass clasp. He felt reassured as a result. Other children would surely not look down on a boy with such a fine bag.

He took so little food that William shook his head and said that if he did not look out he would turn into a "galloper" (by which he meant a consumptive), and Aunt Tilly became very short with him. Nothing, however, could dampen his enthusiasm for the great adventure ahead of him. He was going at last into the world of boys and girls, the playland he had heard so much about but had never seen, the realm of games, of shared toys, of wild excitements and curious fears and extravagant hopes, where everything was new and exciting and not ruled by the funny ideas of grownups.

He was waiting at the side door with his hand on the knob fully ten minutes before Aunt Tilly had finished tying on her bonnet and adjusting her coat and fiddling with her gloves. The front of the school was empty as a result when they arrived, and Ludar's heart sank. He looked up at Aunt Tilly and asked in a tense whisper, "Will I be punished for being late?" William had been delivering lectures on the importance of punctuality and the terrible things which happened later in life, such as becoming hod carriers, to boys who were tardy for classes. Aunt Tilly said impatiently, "Don't be silly!" and, taking him by the hand, marched him in through the front entrance.

Through a half-open door the boy caught a glimpse of rows of small heads and the hum of low voices. Suddenly his courage deserted him. He felt like the fledgling which clings to the side of the nest and twitters not to be shoved out into the world. "Let's go back," he whispered. "I don't think I want to go to school, Aunt Tilly."

"There will be no more of *that* kind of nonsense," she answered grimly.

Shortly thereafter he found himself in a square classroom on the second floor to which he had been escorted by Miss Williams, the principal. He stood just inside the door, acutely aware that every pair of youthful eyes in the room had turned in his direction, while Miss Williams, who was a brisk young woman with a high color and a high voice, talked with Miss Lindley, the teacher of the class. Miss Lindley had no color at all and a voice which seemed to mew. She was a sway-backed lady with a pompadour like a tea cozy and a chin which trembled when she spoke.

"Here's another for you, Vera," the principal said in a half whisper. "I didn't expect so many to start this term, but you never can tell. What have you done with the fighting cousins?"

Her eyes swept the left side of the room, where the boys sat in pairs at small slant-topped desks: desks which had seen much hard usage and were scarred with initials of long-gone occupants and from which the inkwells had vanished, leaving empty gaps. She frowned when she saw that the two new scholars she had referred to were occupying alone the desks at the end of each row, with nothing separating them but the aisle.

"Do you think that wise?" she asked in a worried tone. "I took it for granted you would have them safely separated. When they're close as that they may start fighting again."

Miss Lindley's face puckered up into an expression of distress. "The seats at the end were empty," she said, "and I thought it would look too pointed if I rearranged things just to keep them apart. Perhaps I was wrong. I so often am."

"Well," conceded the principal, "you had a good reason, Vera. Of course we *can't* find other seats for them now. *That* would be too noticeable. Fortunately you have this new boy, and you can put him with one of them and make that much of a barrier, anyway."

Miss Lindley saw more difficulties ahead. "Which seat are we to put the new boy in?" she asked. "Would Mrs. Tanner Craven be upset if we put him with Joseph and left her son to sit alone? Or would she consider it a privilege if her son was allowed a seat all to himself? Oh, why must there always be decisions like this to make?" She seemed almost on the point of tears. "I'm just a teacher, Miss Williams, and not a—not a diplomat."

"It's fortunate we have only one mother to consider," said the principal, who found her own capacity for decision strengthened by the spinelessness of her assistant. "Mrs. Langley Craven won't complain, no matter what we do. Now there's the kind of mother teachers should insist upon for all children in their care. I think, Vera, it would be better to put the new boy in with Joseph Craven and let Norman have the privilege of a seat all to himself. If Norman's mother has anything to say about it I'll tell her we didn't know how the English boy would turn out and that we thought it best to spare her son any contacts at first."

While this whispered conversation was carried on Ludar remained just inside the room, standing first on one foot and then on the other, and keeping his eyes on the floor. Now that he found himself in company of his own age he was realizing that it was a terrifying experience. There was no indication of friendliness in any of the eyes which watched him. Rather they seemed hard, curious, and even cruel. They were looking him over and deciding whether they would like him or not. He sensed this and turned so that his schoolbag would show, hoping its newness would impress them.

"Children," said Miss Lindley after the principal had left, giving Ludar an encouraging pat on the head, "we have a new member for our little circle. This is Ludar Prentice, a little boy from England, and I want you all to be very, very nice to him. Repeat after me, please—Welcome, Ludar!"

"Welcome, Ludar!" shrilled the class.

"I think," said the teacher, "we will put this new little boy with Joseph. Come this way, Ludar."

The newcomer found himself seated beside a boy of about his own size. He did not dare look directly at him, but out of the corner of his eye he saw that his desk mate was dressed in a neat gray suit and was wearing a wide bow tie of blue silk.

Sitting perfectly still and looking neither to right nor left, he watched the teacher, who was scribbling with chalk on the blackboard which covered the whole wall at the front of the room. It was a thin back she had turned to the class, swathed in brown and topped off with a tight starched collar.

His desk mate began to speak in a whisper. "Say, my name's Joseph Craven. Everybody calls me Joe."

Ludar turned then and looked at him. Joseph Craven, he saw, was a pleasant-faced boy with friendly gray eyes and crinkly yellow hair cropped close to his head. There were freckles on his nose and under the rosy tone of his cheeks. There was about him a brightness, an easy gaiety, a brisk willingness to accept everything which cropped up in the life of a boy of seven or so and to take all things in his stride. It was apparent at one glance that he carried a big stock of smiles and would have to dig deep under the counter to produce a frown.

"You better call me Joe too."

Their eyes met, and each knew instantly that they were going to get along well together.

"Yes, Joe," whispered Ludar.

"Never knew anyone named Ludar before. That an English name?"

"I don't know. My aunt said my mother read it in a book."

"Huh!" said Joe. "Mothers are always doing things like that to boys. Me, I'm named after my grandfather. I got half his name, and my cousin Norman got the rest. That's him over there. That frozen stump with the red hair."

Ludar had already taken one quick glance at the boy who sat across the aisle from him. He was thin and wore glasses, through which he had been watching them intently. Taking another look now, the English boy saw that everything about Norman Craven was spick and span, that his shoes were new and not scuffed, that he had a fine pencil in the ledge in front of him.

Joe became confidential. "I don't like him. We had a fight a few days ago. It was a good thing we did, because I had my hair cut off after it. My mother made me have it long, right down to my shoulders. Wasn't it lucky I got mucilage in it during the fight?"

"Yes," whispered Ludar.

"It wasn't much of a fight. It was at the minister's house. We broke some things, and they stopped us." He gave Ludar a nudge with his elbow. "Look at that pencil! He always has things like that. His mother gives them to him. She spoils him. His father doesn't give him expensive things. He's stingy. His mother gives him money all the time. My father says it's wrong for boys our age to have money like that." The eager eyes smiled. "Still, Ludar, it's kind of nice to have nickels in your pocket. That cousin of mine has dimes and quarters. That's the way his mother spoils him." There was a moment's pause, and then he asked, "Weren't you the boy with the sign on his back?"

Ludar nodded. "I wore it over a month."

Joe gave his head an admiring shake. "And you crossed the ocean all alone by yourself?"

"Yes. I was all alone on the ship. And on the trains too."

Joe's interest was thoroughly chained now. "Was it a very big boat, Ludar?"

"I guess it was the biggest boat in the world, Joe."

Joe was willing to accept this estimate of its size. He nodded his head eagerly. "I'm going to be a sailor when I grow up. Once I thought I would be a soldier like the Iron Duke, but now I think I'll be an admiral like Nelson instead."

The ambition to be a sailor was something new, but Ludar found himself thinking the idea a mighty good one. Perhaps it was better than his own plan, which was to be a poet. If Joe went to sea he would want to go too.

The redheaded boy leaned out across the aisle and whispered, "He'll get you in trouble with his talking, new boy."

Joe sat forward and glowered at his cousin. "Go away back and sit down, Norman Craven."

"You're a regular gabbler," declared the redheaded cousin. "I heard the teacher say so."

All trace of good humor had left Joe's face. "I guess I'll have to sock you again," he declared.

"Joseph Craven!" came the voice of the teacher from the blackboard. "I didn't put our new little scholar beside you so you could talk to him. You *know* it's against the rules. There will be no further talking in class on pain of punishment."

2

The boys who lived in the northern reaches of the ward and so had to pass through the danger zone on the way home were gathered in a group after school was dismissed, waiting for Peter Work to appear. There were signs of worry on every face. What kind of a mood would Pete be in today?

"He got good marks in arithmetic," said one who was in the same room with the boss of the school. "At least good marks for Pete."

"Miss Thurlow let him alone in history," contributed another. "She didn't ask him a single question. Gee, that was lucky! The lesson was about Samuel D. Champlain, and Pete wouldn't know anything about Samuel D. Champlain."

"Yeh," said a third boy, "but don't forget what that Hoyt said to him."

Mr. Hoyt was the music teacher who came to the North Ward School one day a week. He had never become popular with the boys because his subject was, in their estimation, a sissy one. An uneasy attitude was noticeable at once.

"Well, Harry," asked one of the smaller boys finally, "what *did* Mr. Hoyt say to him?"

"He said he had a voice like a crow."

This was serious. Pete would be very angry, and everyone in the gang would have to walk on eggs if they did not want to be jumped on by the bad-tempered leader. That there was good reason for apprehension was apparent as soon as Pete put in an appearance. He was frowning and he had already taken the whip end from his belt and was swishing it about as he walked.

"I suppose you're all giggling your silly heads off over what that Hoyt said to me today," he declared. "Well, I may have a voice like a crow, but I'll tell you one thing. It's the kind of voice that's good at giving commands. What's more, it's giving one right now. Fall into line at once, you galoots! The tallest ones up front this time. I hear the Dogans are as mad as hornets at us, and we got to be ready for business today. The kind of business I like."

But his followers, usually so docile, did not jump into their places at once. One of them cried in expostulation: "Pete! There's the new boys. Three of them. Don't they get their breaking in?"

Pete had raised the whip in the air in preparation for a wide flourish. It remained suspended over his head.

"By grab, I forgot them," he said in a tone of disgust with himself.

"Gee whiz, Pete, we're all ready for 'em," cried a second boy, brandishing a shingle.

Others produced their shingles and started to clamor for action. "What will we do with 'em, Pete? Will we give them Old Blister? Will we make 'em bend over and each of us give them a good loud smack on the round robin?"

"Wait a minute," said Pete, lowering the whip. "There ain't three. There's only two."

"No, Pete, three," declared the boy who had spoken first. "Another came this afternoon. And who do you think it is, Pete? It's the English boy. Gee whiz, we certainly will be kind to *him!*"

Pete did not enter into the zestful talk aroused by this bit of information. He had fallen into serious thought, apparently, for he was frowning and his eyes were fixed on the ground. After a few moments he wound up the whip and attached it to his belt.

"So it's the Duke, is it?" he said finally. "This will take some thinking over."

"Why, Pete? All we got to do is grab him as soon as he gets down and give him a good warming."

"Now you kids better not be so fast," protested the leader. "There's never been a duke in the North Ward school before, has there? Well, you just can't start in warming a duke. By grab, there may be some kind of a law about it. They might put us in jail for ten years."

Another boy, who was apparently of a studious turn, had an idea. "Do you think they'd put us all in the Tower of London for socking him?" he asked.

Pete nodded his head sagely. "That's what I meant, Garnet. It's too dangerous. Lambasting a duke might be as bad as lambasting the Queen."

"Here they come now."

Ludar and his desk mate had come out at the end of the line which had wound quietly down from Miss Lindley's room. Joe was asking questions about Ludar's trip, and they were so engrossed that they did not notice the interest being taken in them until they found themselves close to the circle of boys surrounding the great Pete Work.

"They're going to break us in," whispered Joe, stopping short.

Pete made a lordly gesture. "Come over here, you two kids." When the order had been obeyed he looked at them hard and then nodded his head. "I've got to decide what to do with you beginners," he announced. "You're new boys, and we always see that new boys get a good lamming with shingles. I'll think of something else later." Suddenly he became belligerent again. "Someone told me there was three. Where's the third one, eh?"

Unfortunately for Norman Craven, he was detected at this moment in the act of starting up Caledonia Street by himself.

"Here, you!" shouted the leader, outraged at this disregard of established authority. "Come back here. What d'ye mean, anyway, walking off like this?" When the third of the new boys went right on and seemed blissfully unaware that he was being addressed, the leader became really angry. "I mean you, redhead! You with the nice little white collar, and the nice little clean suit, and the nice little yellow shoes with nice little yellow buttons! Come back here if you don't want this whip of mine playing tunes around your ears."

There was no mistaking this. Norman Craven turned back.

"You're new and didn't know any better, so I'm not going to be hard on you this time," said Peter Work. "What's your name, new boy?"

"Norman Craven." Although uncomfortable at being thus singled out, the son and heir of the Craven family spoke as though sure that the mention of his name would make everything right.

If the school boss was startled at finding he had picked on the son of the richest man in town, he did not let it show. In fact, he went to a far extreme to prove his lack of interest. "Craven?" he said, winking at those near him. "Are you sure? There's no family in this town named Craven. Did you just move to town?"

"No, we didn't just move here!" The face of the heir to the largest fortune in town had turned as red as his hair. "The Craven family is the best family and the richest family in town. My father's the most important man in town. And you know it!"

"So! The Craven family is the best in town, is it?" The boss stepped over to the new boy and glared down at him. "Did you ever hear of the Work family? The Works of Whisky Hollow? Well, I think they're the best people in this town, see? If you think you're better than I am, just tell me about it and then see what happens to you."

"You dare hit me," cried Norman Craven, "and my father will attend to you! You'll see!"

"Yes," said Peter Work, "your old man could get me licked all right. But do you think that would bother me? I'm tough and I don't mind getting a licking. Not by a jugful! I would hit you again, and then I'd get licked again. I'd hit you still harder, and I'd get the tar whaled out of me. I could keep that up until Christmas. I could keep it up for a year. But could you, rich boy? You better think it over before you go squawking to your old man." The leader gave his whip a preliminary flourish. "At this school you're just one of the boys, see? If you've got any gumption at all, you'll do just like the rest of them."

"Well——" began Norman. He hesitated and gulped. "Well, perhaps I will."

Pete seemed disappointed that the third newcomer had surrendered so easily. "Well," he said, "I guess I can't give you a licking then. I had my mind made up to give you a real out-and-outer, Master Norman Craven." His eye fell on Ludar, who had moved up into the line with Joe, his schoolbag hanging on his shoulder at a proud angle. "Here we got three beginners and, by grab, I haven't licked one of them. I guess I must be getting soft. But," addressing the whole group with suddenly aroused savagery, "don't any of you muttonheads go counting on it!"

Chapter IX

I

The snow man took several days to build. Ludar labored on it from the moment he returned from school until the light began to fade and the sound of whistles and factory bells announced that it was six o'clock. Each day Tony Milner came over to help him. Her part consisted of bringing snow in a small wheelbarrow, but she worked at it valiantly and was very unhappy whenever her aunt's voice from the other side of the fence summoned her home.

When the tall white figure of Blunderbore had been finished, Ludar stood back and looked it over, and acknowledged that it was good. Tony was more outspoken. "It's wunnerful," she said. "Simply wunnerful."

Pete Work seemed to share her opinion. He and some other boys happened to come by and caught a glimpse of it. Pete came to the gate and stared into the yard.

"By grab!" he said. "That looks real. It's got teeth and everything. I've never seen a snow man like that before. Did you do it, Duke?"

Ludar answered, "Tony and me."

"That girl?" Pete's voice was scornful. "She wouldn't be any help."

"She was," said Ludar loyally. "She did a lot."

"Well, anyway," said the school boss, "I guess you're kind of smart, Duke. I guess you got to be smart to be a duke in the first place, haven't you?"

Ludar had no time to savor the sweetness of praise from the great Peter Work because Catherine McGregor chose this moment to come over the fence, and the boys promptly moved on. She inspected the work of art with a nod of approval. "So you've finished it, Stiddybuttons," she said, indulging in her habit of calling him queer names she had made up. "I've been watching you at it and I bet Father a nickel you would keep on until it was done. He's lost the bet, but now, of course, he'll say gambling is sinful and won't hand over the money." She studied the figure, her head cocked to one side. "You were right not to use coal for the eyes. I suppose, Ludar, you'll make figures for a living when you grow up?"

"Perhaps. Or perhaps I'll be a poet."

"A poet? Isn't that a new idea? I wouldn't have thought you knew what a poet was."

"Bobby Burns was a poet. Uncle Billy has a book and he read me a lot about him." Ludar suddenly developed an extreme touch of diffidence. "I wrote one. It's—it's about you. I wrote it in my mind."

"About me?" Catherine was all interest at once. "I think I should hear it then. Can you say it for me?"

He nodded his head. Then, standing up very straight and in a voice which excitement made high and quavery, he recited:

"The lady next door will make a nice bride,
And the name of her husband, I hope, will be Clyde."

Catherine hugged him ecstatically. "Oh, I hope you are right! Oh, how I hope you are!"

Tony had felt out of things while all this was going on. When Catherine climbed back over the fence, Tony announced her intention of going home. "I'm having a party on my birthday," she said. "It's just three weeks away. All the boys and girls of my dancing class are to be invited, and I want Aunt Mona May to ask you too. She says it depends." She sighed. "It won't be any fun if you're not there, Ludie."

Aunt Mona May had a habit of breaking up all their talks. She appeared now at the back door and called sharply: "Tony! It's time to come in."

"I think Aunt Mona May is mean," declared the small girl. Nevertheless, she fitted her hands into her mittens, which hung on a knitted cord around her neck, and took the wheelbarrow by the handles. As she proceeded to the gate she said over her shoulder: "I don't like parties much. The boys are so rough. But I guess I got to give one."

"Good-by, Tony."

All the sense of pride which had stirred in him because of his success with the snow man had vanished. He went into the house with a despondent air and took off his overcoat and mittens slowly. He did not believe he would get an invitation to the party.

A thaw came, and one morning when Ludar climbed out of bed and ran to the kitchen window to have a look at his creation the snow man presented a new aspect. He appeared caved in and seedy and despondent. He had, it seemed, shrunk seriously during the night. The next morning there was nothing left of him but a rapidly dwindling column of not very clean snow. Ludar was unhappy, feeling that he had not only lost a friend but that he had been robbed of his right to shine in the light of a great achievement.

Time rolled on. He learned to spell "cat" and "rat," and his self-esteem began to revive because of the standoffishness most of the boys displayed toward him. They seemed convinced that he was not as other boys, and they treated him with a half-reluctant but unmistakable respect. He needed this bolstering of his pride because no invitation to Tony's party had been received. Tony herself was still hopeful. "She's thinking about it," she explained once. "If my real mother was alive, Ludar, she wouldn't have to think about it so long."

Then the important men who decided things in the city stopped paying the two dollars a week for his board. Ludar's first intimation of this serious disaster was when he heard William say to his wife, "There won't be any more to add to the principal after this, Tilly."

She received the news with a stony face, not comprehending at first the full extent of the disaster. Then she flushed angrily.

"Do you mean they're cutting off the two dollars a week?"

William nodded. "Sergeant Feeny came by the shop this afternoon and told me the news."

After a moment of silence Tilly lifted her arms in the air. "The rascals!" she cried. "The thieves! Do they think *we* can go on paying for his keep?"

Later that evening, when a grimly silent Aunt Tilly was washing the dishes, William took the boy aside and whispered some advice in his ear.

"Your aunt has had a blow," he said. "I'm not just sure what she's going to do about it. You had better be very nice to her, Ludar boy. Be as obedient and helpful as you can. If you do, I think everything will come out all right."

Ludar was as obedient and helpful as any small boy could be. He jumped at the word of command and he even thought up ways of making himself useful around the house. He carried in kindling wood and he took out ashes, using a small box for the purpose and making many trips with it. One morning, before it was time to start for school, Tilly found him trying to wield a broom on the sitting-room floor.

"And now what are you up to?" she demanded sternly.

"I'm—I'm trying to help you, Aunt Tilly."

"Well"—with a frown and a shake of the head—"you won't help me by stirring up dust with that broom and knocking things off the table with the handle of it. You just give that to me."

Because of this strained situation at home William began to watch with greater anxiety for news from England. Each evening when he returned from work Tilly would ask, "Have they found out yet who he is?" The answer was always in the negative. Uncle Alfred dropped in a good deal, and his manner got frostier all the time. Once he said, fixing Ludar with his eye, "Take my word for it, 'Tilda, it's a hoax." The boy did not know what a hoax was, but the word had an ominous sound which disturbed him.

And then a very sad thing happened which served to clear up the situation. He was carrying in kindling from the old chicken house late one afternoon when his eye lighted on a suspicious bundle of gray fur on the ground under the crab apple tree. He stopped at once.

He had never seen death in any form, except for flies and beetles, which did not count, but he knew at once that there was no life in the gray shape by the trunk of the old tree. And he recognized that it was Old Cat. She had finally lived out the last of her nine lives.

Ludar had become fanatically attached to the ancient tabby. She had taken to sleeping on his bed, and it had been his greatest pleasure to waken in the morning and find her there. He would say, "You crazy old thing!" and would stretch out his arm to pat her head. She had acquired the habit also of sitting beside his chair at meals, with her tail curled around her paws; and whenever Aunt Tilly got up to take something off the stove or to go to the pantry he would hastily smuggle a piece of meat down to her. Uncle William would pretend not to see.

Ludar stood beside the body of the dead pet in silence for several moments, then his tears could no longer be held back. He sank to his knees on the ground and, covering his face with his hands, gave way to a loud and dismal outburst of grief.

Aunt Tilly came out from the summer kitchen in a great hurry, asking in a worried tone: "What is it, Ludar? Are you hurt?"

When he did not answer but went on weeping so hard that his back shook with the violence of his sobs, she came over to him. She fell into a long silence while she stared down at the body of the cat, then she reached down a hand and drew the boy to his feet. "Come in, Ludar," she said. "You mustn't cry any more or you'll make yourself sick. When Uncle Billy comes home he'll bury Old Cat for us."

Once inside the house, however, she behaved in a curious way. She first pulled down all the blinds on the windows and then, sinking into a chair and throwing her apron up over her face, gave way to a grief as unrestrained as his had been. She rocked the chair as she sobbed, and she told him in broken tones that Old Cat had been hers, that she had acquired her as a kitten the day before she married William Christian. "I loved her," she said. "Just as much as he did, though I didn't let myself show it. I—I always hold my feelings in, Ludar. I can't help it. But I feel things as much as others do."

It seemed to grieve her particularly that the cat had died alone. "The poor old lady!" she sobbed. "I knew she was sick and I should have watched her. I shouldn't have let her go out into the yard like that."

This was too much for Ludar, who started to cry with more vigor than before. The pair of them wept together for several moments in complete abandonment to their grief. Finally Tilly removed the apron from her head, dried her eyes with it, and said to the boy: "Now, Ludar, we mustn't go on this way. All the crying in the world won't bring Old Cat back." Her face had become red and mottled. "I wanted to call her Matilda, but William said one in the family was enough. That made me angry, and so I wouldn't consent to any other name. At first she was just

Cat, and then, after she had so many families, we began to call her Old Cat."

The grief she had given in to had, as it now developed, served a valuable purpose. It had broken down her reserve. As soon as her eyes were thoroughly dried she smiled at Ludar with so much affection that he was taken by surprise completely. "I can talk now," she declared. "There's something I want to say to you.

"I'm what they call self-contained," she stated, tucking her handkerchief away. "I don't say what's in my mind and I don't show my feelings, not the way your uncle Billy does. I suppose I seem cold and unfeeling to people. But I'm not, Ludar." She even managed to smile. "I'm fond of you. Just as fond as your uncle Billy, I guess. I—I liked you right from the start, Ludar, but I didn't know how to show you. You thought I was hard and stern." She nodded as though what she wanted him to know had been said and that she would now go back to more normal ways. "I'm glad I've been able to talk to you once like this. I don't suppose I ever would if Old Cat hadn't died and got me started this way."

When William came in there was an air of great peace and understanding in the kitchen. The fire in the range was blazing away with the promise of a fine hot supper, and the kettle on top of the stove was steaming and humming. Ludar sat in front of the fire and was most contentedly munching on a cooky with raisins. The boy was looking more at home than at any time since he had arrived.

"Well," said William, very much pleased. "This is a cozy scene. A very happy little family, I call this. Is that a fresh cooky you're eating, young man? Am I to understand that a new batch has been made?" The plainest of eaters, he seemed to have one weakness only, a liking for crisp cookies. "I hope we're going to have them for supper."

Receiving no answer, he looked curiously at his wife. "Hello! Something wrong here, I see. What is it, Tilly? Has there—has there been a death in the family?"

"Yes, William," she said. She gulped and added, "Old Cat is dead."

All the pleasure went out of William's face. "Oh!" he said. He sat down in a chair and looked as though he, too, might very easily succumb to grief as they had done earlier. "I've been afraid it was going to happen," he added after several moments. He got to his feet again. "Well, where is she?"

"Outside. Under the crab apple tree. Ludar found her when he was bringing in kindling wood. We didn't disturb the body."

William sighed. "I saw her try to jump into a chair the other night.

When she didn't get there I knew the end was coming soon. Well, I'll do the best I can for her. After supper I'll make a wooden box and I'll put that old shawl in she liked to sleep on. Then I'll bury her under the tree. She must have liked that spot, since she went there to die. I'll make Old Cat as comfortable as possible."

Tilly began hustling about from stove to table and back, putting the supper on. "Come, now, you two," she said sternly. "Sit down and eat your food. There's no call for all this carrying on."

2

The task of clearing the supper table had not been completed when the sound of quick, light steps on the walk at the side of the house announced the arrival of Catherine McGregor. She tapped on the door and then came in without waiting.

"Mr. Willing of the *Star* telephoned to find out if you were home," she announced. "I told him you were because I could see lights in the windows. He said to tell you he was coming up to see you this evening."

William said, "There's been some news!"

"Mr. Willing didn't say. But he must have something to tell you." Catherine planted her elbows on the table and smiled across it at Ludar. "Well, Young Easy-Kneesy-Nosy, have you been writing any more poetry about me?" Her face suddenly lighted up. "Father is simply furious with me! He wouldn't speak to me at noon. Just finished up his dinner without a word and never gave me as much as a look. But before he returned to the store he got an old school slate of mine and wrote on it. Then he came stamping into the dining room and propped the slate up against the tea cozy. He stamped out without saying a word."

"Goodness gracious, Catherine!" said Tilly. "What have you done to upset him like that?"

The girl was only too pleased to give a full explanation. "Well, it was this way. I went last night to the opera house to see the Marks Brothers in *Struck Oil*. Father seems to have had the idea I was going with Minnie Sims. In fact, I—I said something which planted that thought in his mind. Clyde met me in the lobby with the tickets. We had a wonderful time! We were in the third row of the orchestra, and Ernie Marks kept us laughing all the time. But when I got home! There was Father, who'd heard about it from someone on the telephone—and he glared at me and said, 'Ye're an undutiful daughter!' He told me I would come to a bad end. I asked him if he would turn me out into the storm the way the father does in *Way Down East,* and he said he wouldn't allow any daughter of his—I'm the only child he has, as far as everyone knows—to make fun of him. Poor Father! I'm such a trial to him."

After a short pause she began to laugh again. "I was going to bring the slate over so you could see what he had written. Instead I copied it out. And by the way, who is the Reverend Elmore Nash?"

Tilly obliged with all the information available about the Reverend Elmore Nash. "He's a Presbyterian minister from up Goderich way. Knox Church is thinking of giving him a call. In fact, I hear it's just about settled. He's thirty-two years old and kind of solemn-looking." She glanced significantly at Catherine as she added, "He's a widower, but he hasn't any children."

"That explains it then." Catherine produced a sheet of paper and held it up. "This is the message he left me at noon."

> *You are an ingrate and someday you will break my heart.*
> *I am bringing the Reverend Elmore Nash for supper.*
> *Have cold lamb, ciscoes, muffins, dandy-jack with bananas.*

"He must think the Reverend Elmore Nash is a real catch," she said, "or he wouldn't have ciscoes as well as cold lamb. And imagine bananas at this time of year!"

"Did you get them?" asked Tilly.

"Of course. If he wants bananas for the Reverend Elmore Nash, he's going to have them. But they cost a fancy figure, I can tell you."

"Your father is a great one for his supper," declared William.

Everyone knew that Lockie McGregor put great store by his late supper. He had a cup of tea and a sweet biscuit in his office at five o'clock and then remained after the establishment was closed and the salespeople had departed, in order to check the cash and the books, study the sales slips, and make a thorough inspection of the store. He never arrived home before nine and he wanted a good meal ready for him, which he would attack at once with a furious appetite.

Catherine nodded confidently as she walked to the door. "I know how to handle him. He'll bring his hat salesmen and his Reverend Elmore Nashes up to the house, and I'll be polite but distant with all of them. And then one day he'll come to me and he'll say, 'Catherine, my child, I've reached a decision about ye. That young fellow, Clyde Carson, now, he seems a steady lad. I've been thinking about it and I want ye to take him and stop running around with all these other scatterbrains.' He'll be convinced the decision is all his own, and that will make everything right. You watch if it doesn't come out that way."

"From what I hear," said Tilly, shaking her head, "this young minister is a very fine man."

3

It was fortunate that Ludar had been tucked in bed and was already sound asleep by the time Mr. Willing arrived. The editor came in and seated himself with an air of the deepest gloom.

"Mr. and Mrs. Christian," he said, "you see before you a failure. Perhaps you have thought of me as a man of ability and promise. On the other hand, perhaps you haven't. But—if you have, you've been very, very wrong. The way I feel about myself right now, I should resign my position and take up something more fitted to my capacity. Such as being the brain power behind a broom or a third assistant gravedigger."

"Then there's been bad news?" asked William.

"Worse. There's *no* news. The shipping people haven't been able to trace the parties who bought the boy's passage. I was so sure of the stuff I dragged out of the youngster about the clock tower and Old Mr. Cyril that I insisted I was the one to handle the case. Well, I wasn't. I've nothing to show for all my labor and all the money I've spent. I've been baffled by Old Baffle, and I'm completely at sea over Dear Fishface. The golden moment has passed and, because I've failed, this boy may never learn who his relatives are—or who he is himself, for that matter." He shook his head despondently. "Why did I have to be so sure of myself? Why didn't I let them put a good lawyer on the case?"

"Then you think it's likely to remain a mystery?" asked William.

"An impenetrable mystery," affirmed the editor gloomily. "As deep and unfathomable as the Man in the Iron Mask. There's only one chance left. We've a new printer at the office, an Englishman named Bert Lamb. He lived in Shropshire as a boy and he has some recollection of a big place with a clock tower. He's writing to an uncle of his, a lawyer, to make some inquiries. It's a thin clue, but it's the best this poor excuse for a sleuth has to offer at the moment."

"It will take some time to get an answer, I'm afraid."

"Months, at the rate we've been going. This uncle's name is not Lamb. It's Ratticlipper. John Ratticlipper. Do you like the name, Mr. Christian?"

"Not much," answered William.

"Neither do I." The editor got to his feet. "Does it seem to you to have just a hint of dishonesty about it? Does it set up a whisper of doubt in your mind and make you apprehensive about the change in your pockets? I may be wrong about it, of course. He may be the honest and hard-working father of a large family of plain girls, a church warden, and the trusted treasurer of mission funds." He reached for his hat. "Well, that's all I have to report. I do so in full abasement of spirit."

At the door his manner changed. He became completely serious. "Mr. Christian, what will you do now about the boy?" he asked.

"We'll keep him," was the answer. "That is—well, I think we'll decide to. But of course it will have to be talked over first."

"I hope you do decide to give him a permanent home. He's a nice little fellow. I wouldn't want to see him sent to the orphanage."

William looked shocked. "But surely, Mr. Willing, there would be others ready to take him if we didn't. There's been so much interest in him."

The editor shook his head. "There was. But it was at the start, when it looked certain that he belonged to a good family and might even come in for a fortune. They were ready to scramble for him then. But now it begins to look as though he's just another immigrant boy and that, Friends, Romans, and Countrymen, is a different matter. The frost is on the punkin. If you don't keep him, Mr. Christian, he will end up at the foot of that long line of lugubrious youngsters who go to church in files every Sunday, rain or shine."

When the editor had left, William returned to the sitting room. The gravity he had maintained during the conversation had left him, and a smile was puckering up the corners of his eyes.

"Tilly," he said briskly, "I'm being selfish, I know, but I can't help being glad over this news we've just had. Aren't you?"

"Glad? Have you gone crazy, William Christian? Of course I'm not glad. Here we've been hoping the boy would turn out to be someone of importance. Why should I be glad because it's certain now that he isn't?"

William frowned, as though he could not understand her failure to see the matter in the same light that he did. "But, Tilly, it's the best news we could have. I mean, it's good from our standpoint, if not from the boy's—and at the moment I'm selfish enough to look at it our way. I've got a confession to make. I've never tried to get him to remember what his real name is. I've watched him whenever anyone has started to ask him questions, and I've been sure it was pounded into him so hard that he mustn't tell that he never will, no matter who asks him. Perhaps he really has forgotten it by this time. I hope so." He looked at his wife as though appealing for her understanding. "Don't you see this means we'll have a son of our own?"

Tilly said flatly, "We can't afford it."

"Can't afford it?" William's face fell. He stared at her in sudden bewilderment. "Tilly, are you serious? Why, it never occurred to me that you could have any real objections. Why do you say we can't afford it when—well, look at the Clancys on the next street. They can afford eight children, and Harry Clancy only works when he feels like it."

Tilly had taken out her knitting needles and a ball of yarn from a workbasket. Now she stuck the needles back into the yarn as though she proposed to burst the bubbles he had been blowing in the same way. "If you see no difference between us and the Clancys!" she said bitterly. "Let me tell you this, William Christian. I have to supply all the common sense in this family, and the decision about the boy must be made by me. Do you understand that? Do you think I'm going to rush into this blindly? If you do, you're badly mistaken."

Since his victory in the matter of the overcoat William had been less prone to step down when their wills clashed. He looked intently at his wife for several moments and then asked in a quiet tone, "Do you know what Mr. Willing said would happen to the boy if we didn't keep him?"

"No, I do not. It wouldn't affect my decision if I did. It's all very well for little whippersnappers of editors to say, 'Keep the boy.' Why doesn't he take him himself?"

"He has three small children now, and his wife's hands are full."

"And I suppose mine are idle!" Tilly got to her feet and held out the workbasket for his inspection. "*That,*" she declared, "will always be twice as full if we *do* keep him. And let me point out that you don't have the work to do. I do." She started for the kitchen door. "I'm fond of Ludar. Just as fond as you are, Billy Christian. But because of that I'm not going to rush in and sign away all my leisure time. This is going to take some serious thought."

On a shelf in the pantry, behind the sugar bowl and the cruet stand, was a pile of small black notebooks, nearly a dozen of them in all. Each evening, without fail, Tilly took the current book and carefully noted down in its pages the expenditures of the day. Sometimes she added comments such as, "Six cents a pound for granulated sugar! What *are* we coming to!" When the last line had been filled she would put the book on top of the pile and start to use a new one.

Perhaps she drew some conclusions from this accumulation of daily entries; perhaps the mere fact of writing down what she had spent sharpened her in the matter of household economy. No one had ever seen her open one of the old ones, however, and it was more likely that she was indulging a habit and nothing else.

Her own explanation was given on one occasion to William when he questioned the value of filling up books which were then left to gather dust. "I consider it absolutely necessary," she said, "to keep a record of what we spend. Then you know where the money has gone. If you are questioned" (let anyone dare question her!) "the answer is there. You can never be unjustly accused. You have the proof of everything set down in black and white for the whole world to see if necessary. Why, I couldn't get to sleep if I knew that I hadn't posted up my little black book for the day."

In addition to the current account book, she was keeping a second one now, which hung on the wall with the carving knife, the colander, and the corkscrew. When she reached the pantry she took this one down and carried it back to the kitchen. Seating herself at the table, she opened it, with a sense of satisfaction, at the first page. At the top, in clumsy large capitals, she had written BOOK OF LUDAR, and under this had kept a complete record of every cent which had been laid out for him since the day he had arrived. An impartial observer might have pointed out that the record was misleading in that it not only charged a full third of all household expenses to him but included as well the cost of his new clothing, as though the money spent on his outfit had come out of the Christian purse.

She dampened the end of the pencil by putting it in her mouth and then proceeded to reckon what the keep of the boy had amounted to for the day. "Sometime," she said to herself, "the point will come up as to what he has cost us, and then I won't have to depend on guesswork. I'll have the whole thing to show them."

Rather painfully, in a childishly large hand and following a system of spelling which was all her own, she made the following entry:

The boy has the best of everything. I give him plain but wholesum food and he has gained two pounds and six ountses. I am doing my duty by him.

She nodded her head with sly triumph. "Just because I gave in about the coat, Billy Christian needn't think he can tell me what to do," she said to herself. "Of course we'll keep the boy. This would be a dull place if he left us now. But I won't tell that husband of mine yet. I'll just let him stew about it awhile."

4

Perhaps it was because of the lack of news from England. Perhaps the social caution of Aunt Mona May would have prevailed in any event. The fact remains that Ludar did not receive an invitation to Tony's party.

Aunt Tilly watched through the door of the summer kitchen when sounds of activity began to reach them from the house on Grand Avenue. Several times she returned to report progress to Ludar, who was pretending to study his *First Reader*. She had heard a sleigh arrive and a whole parcel of children alight. The wagon from Gilkie Brothers had come and had delivered the ice cream at the side door. Gertie, the house maid, had shaken a tablecloth at the back door.

"If you go out to the back yard you'll be able to see a lot of what goes on," she said to him on the last trip with news.

Ludar rejected the suggestion scornfully. "I wasn't asked to go," he said. "I won't stare at them and have them all laughing about me. I should say not!"

He repeated to himself several times that he would rather be dead than watch them, but after a long period of dismal indecision he put on his coat and hat. Usually when he donned them he felt a sense of pride in the softness of the fine cloth of the coat and the splendor of the gray Persian lamb fur of the cap, but this provided no consolation now. He strolled casually into the back yard, as though unaware of anything special happening in the neighborhood. Taking advantage of a moment when the Milner yard was empty, he whipped behind the thick trunk of the oldest of the astrachan apple trees. Here he remained for half an hour, stamping his feet to keep warm and sometimes rubbing a mittened hand across his face to make sure that the end of his nose did not fall into its uncomfortable habit of freezing.

Tony had said she had lots of friends, and Ludar decided that all of them had come to the party. He saw well-dressed children running in and out of the house and heard shrill voices raised in delight and in dispute. He heard one small boy, wearing a paper grenadier's cap, say to another with a clown's hat, "Gee, Horace, they've got two kinds of ice cream!" Once he saw Tony. She was wearing a frilly white dress, and she came out from the back door in a great hurry and ran down to the fence. She came to the palings and looked through, hoping, no doubt, to catch a sight of him. He remained very quiet and tried to shrink into smaller space behind the trunk of the tree. After a moment she turned and ran back to the house. When she had disappeared he rubbed a knuckle into one eye and said to himself, "Well, Tony wanted me asked, anyhow."

Finally he walked quickly back to the house and hurried through the summer-kitchen door. The feasting had begun by this time, and his retreat was unobserved. Aunt Tilly was insensitive about matters like this, and she failed to notice the tensity of his face as he took off his overcoat. "Well, Ludar," she said cheerfully, "I guess that was almost as good as going to the party."

He answered "No!" in a sudden passion and ran out of the kitchen.

Aunt Tilly was putting chocolate icing on some cupcakes. "Now what's got into the boy?" she asked herself without pausing in her work. "When he gets his teeth into one of these he'll forget all about the party."

Ludar had gone to his bedroom and hidden himself behind the curtain which took the place of a wardrobe. He did not appear when William returned from work, and the summons to supper brought no response. William looked questioningly at his wife. "Has something happened to upset him?" he asked in a whisper.

"I suppose it's about the party," she answered. She was busy making the tea. "He seems to be bothered because he wasn't asked."

William got up from the table and walked into the bedroom. Without making any effort to entice the boy from his hiding place he sat down in the one chair and began to talk.

"Now about this party," he said. "You shouldn't feel badly about it, because as time goes on you'll go to parties so much more important that this one will look like nothing but a silly little bun fight. Think of it this way, Ludar. Someday you may be an admiral like Nelson"—he had kept track of the veerings of the boy's ambition—"but more likely you'll be a writer, or perhaps an artist. Perhaps you'll have a newspaper of your own like Mr. Milner. Perhaps you'll have several newspapers. I think you're going to amount to something in life and you'll be invited to lots of parties and people will be glad to have you. But most of these boys who are at Tony's will be pretty small potatoes then. They'll be running little businesses they've inherited from their fathers, or they'll be counter jumpers. All you've got to do, my boy, is believe this, and then suddenly you won't care about this party any more."

A few moments passed, and then Ludar emerged from his hiding place. He gave William a shamefaced smile.

"Yes, Uncle Billy," he said. "I don't care about the old party. When I grow up I'll have a lot of my own. They'll be big, and I'll have three kinds of ice cream." He added with an almost fierce determination, "But, Uncle Billy, I'll aways invite Tony."

Chapter X

I

Spring came. The great piles of snow, which had been growing higher and higher on the lawns each time the walks had to be cleaned, melted away quickly. In the mornings, when Ludar charged out of the house to have a look at the world, he would find that the snow had receded many inches below the level of the windows during the night. One evening when the snow had completely vanished Uncle Alfred arrived, Bible under arm, and as gloomy of mien as ever, and said it was expected that the ice would go out on the river in a few hours. "Lots will be there with lanterns all night to watch," he declared scornfully. "What is there to see in the breaking of the ice on a little river? Why don't they watch for the great thing which will happen soon, when the heavens will open and the mountains will fall and the great seas and all the little rivers will dry up—and the Lord will come in His might to pass judgment on men?"

The river *did* go out that night. Nothing else was talked about at school next morning. Pete Work, who had been up before dawn and had gone with his father to watch, reported that all of West Balfour was flooded and rowboats would be needed to get the people out. All of the boys announced their intention of going to the safe side of the river as soon as school was over that afternoon.

As Ludar and Joe Craven walked downstairs at the foot of the line the latter began to talk. He was a brisk little fellow, quick in all his movements, and with an almost staccato way of speaking. "Look here," he said. "You going down to the river, Ludar?"

Ludar shook his head. "Aunt Tilly said at noon I mustn't. Uncle Billy said so too. He said I didn't know my way around and it wouldn't be safe."

"I can't go too. Look here, why don't you come home with me? There's going to be lots to watch. I think we're moving."

"Moving? You mean from your home?"

Joe's mood became more serious. "Yep," he said. "We got to move. That uncle of mine is turning the Homestead into a factory office. This morning my aunt Lish arrived with some men, and they started to pack things in boxes and barrels. Aunt Lish winked at me and said I mustn't tell anyone. But I guess the moving is nearly done, and it won't hurt to talk now."

Ludar wanted to accept the invitation, but he was under strict orders from Aunt Tilly to return home always as soon as school was out. He tried to explain this to his companion, but Joe brushed the objection aside.

"Oh, look here," he said. "Aunt Lish is coming by to pick me up. She'll be driving Boney and the Iron Duke. There'll be lots of room. And I guess she'll drive you home afterward. She likes boys—even my cousin Norman." When Ludar still hesitated he went on in coaxing tones: "Aw, come on, Ludar. We'll have lots of fun. You and me are friends, aren't we?"

That settled it. The news from England, or rather the lack of it, had reduced Ludar to a humble state of mind. Tony's party had added to his unhappiness. But now he had been invited to go to the new home of his desk mate and he had heard the latter say they were friends. This was the kind of salve needed for the wounds he had suffered.

"I'll go with you, Joe," he said. "But I guess I'll get into trouble about it."

"Pshaw!" said Joe. "I get into trouble all the time and I get out of it. I don't care."

As soon as they emerged from the front entrance they saw that there was a surrey waiting before the school. A lady, sitting in the front seat, noticed them at once and beckoned.

"That's Aunt Lish," said Joe. "And she's got Boney and the Iron Duke, just like I hoped. Gee whiz, this is going to be fun! Come on, Ludar."

Ludar, seeing that the surrey was already filled with people, held back. "There isn't any room, Joe. Not for me, anyway."

"Aw, there's always room for boys," declared Joe. "Aunt Lish will pack us in. Up under the top or down over the springs."

He started to run toward the carriage, and Ludar followed reluctantly. The lady on the front seat, who was wearing a funny-looking saucer bonnet of gray felt which sat flatly on her head and had ribbons tied under her chin, said gaily to her nephew: "Hello, Joe. And have you been busy all day? I'll inform you that the rest of us have. We've really done a job of moving in a hurry. I'm going to take you now to your new home." Her eyes lighted on Ludar. "Who's this young man? A friend of yours, Joe?"

"Yep, Aunt Lish. I asked him to come home with me and see the fun. He said he would."

"That's a good idea, Joe. You'll keep out of our way if you've got someone to play with." She nodded down at Ludar. "And what's your name?"

"Ludar Prentice, ma'am."

"Eh! What's that you say?" Miss Craven looked thoroughly startled. "Boy, did you say your name was Ludar Prentice?"

"Yes, ma'am."

"Well," she said after a pause during which she studied him carefully, "I guess the hand of Providence is in this. Are you and my nephew in the same class?"

"Yes, ma'am. We share a desk."

"Do you indeed? And are you getting to be good friends?"

Joe took it on himself to answer this one. "You bet, Aunt Lish."

"I'm glad to hear it. I hope you two boys will always be the very best of friends." She looked back at the cargo the surrey was carrying and decided it would have to be lightened. She had brought all the members of her household along, and they filled the two rear seats to overflowing.

"We'll have to make a change," she announced. "I want to take the boys up here with me, so I guess some of you will have to walk the rest of the way. You, Shorty, and you, Cousin Hubert."

Shorty Hanley got down at once, saying cheerfully, "Right you are, mum," but Cousin Hubert was not as complaisant about it. He did not budge from his seat but scowled angrily over the top of his cane.

"It wouldn't do me any good, Cousin Adelicia," he said. "I'm giving you warning: A walk would be bad for me. My work has been going slowly"—he was supposed to be slaving away at some stupendous musical composition—"and this will set me back."

"Cousin Hubert," said Miss Craven, "I've noticed this about the work you're doing. You never get ahead. Perhaps it would be a good thing if you were set back. Motion of some kind might be useful."

The reluctant musician climbed slowly out over the rear wheel. He was small and dark and the possessor of a face which could display resentment easily and immeditately, which it was doing now by wearing a darkly indignant frown. Somewhat of a dandy about his clothes, he was on this occasion all togged out with a brown bowler hat, spats, and a cane.

"Very well," he said. "I'll walk. It's three and a half blocks from here, and my feet will suffer. You may laugh when I say that a blister on the foot will affect the workings of the muse. Would it have helped Bach to make him tote a rifle? Would the genius of Schubert have blossomed in a salt mine?"

He raised his cane in resignation and followed in the steps of Shorty

Hanley, who was already well up the block and whistling cheerfully. Miss Craven, muttering to herself, had the two boys take their places beside her on the front seat. "Rifles and salt mines!" she was saying in an indignant whisper. "And did you ever see such a getup for helping move furniture? Spats and a cane. I've a notion to break that cane over his lazy back."

2

They drove slowly up Caledonia Street. The owner of the carriage, forgetting quickly the passage with Cousin Hubert, began to address her conversation to Ludar. "I'm glad to see you and little Joe together like this," she said, waving the whip at the driver of a passing grocery delivery wagon who happened to be an old schoolmate. "It relieves my mind a little. It doesn't make me feel right, mind you, because nothing will ever do that. You boys must wonder what I'm talking about, but I'm afraid I can't explain it all yet. I'll tell you when you get older. *That* would relieve my mind a great deal. You see, Ludar, I was careless once—wickedly careless—and something terrible happened because of it. There was a duty laid on me—and well I knew it—and I should have attended to it at once. I didn't. I was tired and I had just been through a scene and I—well, I decided that next morning would do just as well. But next morning was too late, and I'll regret to my dying day not stirring myself to act as soon as I heard he had been in the house. It's a terrible thing to have on the conscience, and I'm going to see if there are ways to—to make up for it."

They had progressed to Hanover Street. On the corner was a house with a high brown stone fence around it and two huge furniture vans in front, marked in large black-and-red letters, *Creek and Stapley, Livery and Moving*. Several men with black arms and dusty faces were coming and going between the front door of the house and the vans, and resting between trips to decide which pieces should be carried in next and how best they could be handled.

Miss Craven cried briskly, "Everyone out!" When the surrey had been emptied she tied the lines around an iron hitching post and injected herself into the unloading like a cyclone. In the course of a few seconds she had taken over command of the whole operation and had the movers running. "Things have got to be done in a hurry," she kept saying. "I want every stick of furniture inside that house in half an hour at the *very* outside."

Thereafter she seemed to be constantly in several places at once, directing the movers, carrying in chairs and bed rails and boxes of books herself, keeping after the auxiliary crew she had brought and sometimes shoving them along with impatient hands. Her voice was constantly raised in admonition. "Heave-ho!" she would shout. "Shake a leg, Shorty!" and "Don't go to sleep, Hubert. Just imagine you're in a salt mine and the supervisor is holding a whip over you!" Finally, in a sort of desperation as the minutes ticked off, she played her trump card. "A round silver dollar goes to each of you if we get everything into the house in twenty minutes."

The two boys had been instructed to go into the yard and keep out of the way. As they turned in by the side gate a third member of their class came up the block, whistling quietly and swinging his schoolbag in one hand. His name was Arthur Borden, and he was well built and very dark of hair and eye. He smiled seldom but he gave the suggestion of being steady and forthright.

"Hello," he said, stopping.

"Hello, Arthur. We're moving in here."

"That's good," said the newcomer. He turned quickly as a red bicycle went by. Wetting two fingers of one hand he pounded them into the palm of the other. "Seventeen. Pretty good for one day, but I've done better. Do you count red bikes, Joe?"

"Nope. I count yellows and whites. Two punches for each because you see so few of them. I've had fourteen so far today."

Arthur was studying the house. "Some of the fellows were saying you'd probably move to the East Ward. Pretty big house, this. It's got a fine yard, the best in all Balfour, I guess. I played here once with Eugene Lester. He was older than me and he bossed me so much I never came back."

"Let's go in and look things over," suggested Joe.

When the trio had made their way in through the gate Joe looked about him with the sharp curiosity of one who sees for the first time the new home where, presumably, he will live for many years. He seemed both relieved and satisfied.

"Say, this is all right," he said. "I guess it's as big as the Homestead."

The house was of brown stone and it covered a lot of ground. There was a large square porch in front with the name Lester in a scroll over the door and colored glass in each side. The glass panels had views of a very large place marked "Hawarden." At one side of the house there was a single-story wing which looked like an architectural afterthought, and there was a square tower at the back.

"Look at the yard!" cried Ludar in a sudden ecstasy.

It was an unusually large yard and it had many different varieties of fruit trees and a massive frame barn in one corner. It was the kind of yard which has an air of mystery about it and which rewards search with all manner of discoveries. In five minutes the trio found a dial stone in a clump of berry bushes, a cast-iron deer with broken horns lying on the ground in a corner, a carriage wheel, a rusty hatchet.

"What a place for hunting, Joe!" said Ludar. "You might find rabbits and foxes, and even bears, in this yard. You could play cowboys and Indians here." He stopped and his eyes lighted up with excitement. "Joe, you could have an army!"

His two companions became equally excited at once. "Gee whiz, Ludar!" cried Joe. "That's a swell idea. You're just full of swell ideas."

"My father was a major in the Fife Rifles once," said Arthur. "He's got two helmets at the house. Perhaps he'd let us use them."

They squatted down on the edge of the path for a council of war. "My father was an officer too," said Joe. "He has his sword still and a bugle. One key's broken, but I guess it could be fixed easy enough."

Ludar had the habit of listening to every scrap of adult talk he could hear and of remembering. His imagination, which needed little to feed on, was now busily at work. "You could call it the Army of the North Ward and use the Union Jack as its flag. If you got fifteen privates and three officers you could hold this yard against any attack, even if the Dogans were in full force." These were expressions he had heard used, and he had stored them up in his memory. "The army would have to drill every day."

Joe was watching him with fascinated eyes. "Say, you know a lot! I tell you what, we three will be the officers. I'm one because this is my yard and I can use my dad's sword and bugle. Arthur's one because his father was an officer, too, and he has the helmets. Ludar's one because it's his idea and he's probably a duke or something, anyway. Everyone else will have to be privates."

The trio then visited the barn. Through the partly open door they could see the body of an ancient carriage inside. Before they went in, however, a small girl in a very short accordion-pleated skirt came skipping down the path from the house.

"Hello, Arthur," she said. Then she stopped and stared hard at Ludar. "Who's this, Joey?"

"A boy from school, Meg," answered Joe casually.

"What's his name?"

Joe looked at her sternly. "See here, Meg, we're busy. We've got lots of planning to do. Why do you bother us with so many questions?"

"I just asked what his name is, smarty!" cried the small girl in shrill indignation.

"Well, his name is Ludar. Now you know and you can go and peddle your papers!"

"Did you ever get left!" said the girl, showing no inclination to leave. She stared at Ludar and said: "My name's Meg. Short for Margaret."

Ludar, uncomfortable under her sharp scrutiny, said, "Hello."

"You're kind of skinny. Arthur's kind of fat. I don't know what Joe is, but he's only a brother, anyway. I might like you, Ludar, but I haven't made up my mind yet. I guess I'll have to think it over."

"You like me, don't you, Meg?" asked Arthur, grinning at her.

"Sometimes I do and sometimes I don't." She lifted one foot off the ground and swung herself around. "See that, Ludar? I can dance. I'm a *very* good dancer."

Her interest in them evaporated. Moving on tiptoe and occasionally launching herself into attempts at pirouettes, she went skipping across the yard.

"Meg's kind of crazy," said her brother. Then, feeling he had not been fair, he added, "But she's not a bad kid."

They investigated the barn thoroughly and realized that it was perfect. "It's better than the stables at the Homestead," declared Joe. "It was too clean there. This is just full of stuff. And it's fine and dirty, isn't it?"

They made one discovery which pleased them immensely. There was a small room filled with harness and without a window of any kind. In one corner there had been a drawer sunk into the bottom of the wall. This apparently had been lost, for in its place now there was a deep and dark cavity.

"Gee whiz," said Joe. "We got to find some use for this."

"Use for this!" cried Ludar, his eyes glowing. "Why, this is where we'll keep the secret papers."

"What secret papers?"

"The papers of the army. All the plans for battles. The papers for our spies to carry."

Joe nodded. "Of course. I didn't think of that." He looked at Ludar admiringly. "Say, you think of a lot of things."

As they emerged into the lightness of the yard from an exhaustive survey of haymow and metal-framed stalls they heard Miss Craven's voice raised somewhere in the house and demanding to know, "Where's Langley? I haven't laid eyes on him since I arrived." A lighter and more melodious voice chimed in, "Yes, where *is* Langley?" Miss Craven appeared at a side door and called: "Langley! Have you fallen down a well?" After a brief wait for a response which did not come she turned and vanished inside.

A few minutes later the long face of Cousin Hubert appeared in the back door. He was very angry. Raising his arms in the air, he addressed the world at large. "I'm expected to help move heavy furniture! I, with my hands of a musician!" After glaring at the boys he turned back into the house and slammed the door after him.

"Who's that?" asked Ludar.

"Oh, that's Cousin Hubert. He plays a piano."

They had found their way to a corner of the yard which was much higher than the rest of the ground, when Langley Craven emerged from a small door leading into the cellar. He was wearing a faded blue coat and shapeless gray trousers but in spite of this managed to look well dressed. Seeing the boys, he walked over in their direction.

"Well, Joseph, I see you are surveying your new domain," he said. "What do you and your young friends think of it?"

"It's fine, Dad!" exclaimed Joe. Then, realizing that a father and a new school acquaintance would need some kind of introduction, he said: "This is my father. He'll tell us how to get up an army. He knows everything."

"How do you do, boys? I am happy to have my son praise me, but you must pay no attention to him. I know just enough to realize that I know—very little indeed." Langley Craven turned to his son. "How are you getting on at school?"

Joe frowned indifferently. "Oh, all right, I guess. I don't think I'm going to like school very much, Dad. You have to sit still so long. I'd rather stay in this yard and learn to be a soldier or a sailor. You could teach me, Dad."

His father's manner changed to one of severity. "Now see here," he said. "That's the wrong way to look at things, Joseph. You have fifteen years of schooling ahead of you, and I want you to make a real success of it. I want you to go out into the world and earn a name for yourself, my son." The severity of his gaze vanished and he began to smile. "So! You young paladins want to organize an army, do you? An excellent idea. It will teach you how to organize yourselves for the battles of life later." Ludar was watching Joe's father and was surprised to note how quickly and frequently his expression changed. He looked pensive now, as though there was something to distress him back of the advice he was giving. This mood did not last more than a moment or two, however. He glanced about him and smiled when he perceived the high land in the corner. "The exact spot to fight the Battle of Waterloo! See, boys, this is very much like the land where that great battle took place. Here's the ridge where the redcoats stood. Here's where Wellington was stationed. This broken barrel is Hougoumont. And here, by Jove, is a break in the ground which will be the Ohain Road."

"My dad *does* know everything," declared Joe proudly.

Miss Craven's search for her brother brought her out to the yard again. Determination and purpose showed in her stride as she tramped over to them. "Langley, where have you been? Pauline and I have been looking all over for you."

Langley Craven made no attempt to conceal the fact that he was annoyed. "Now see here, Lish," he said, "what's the meaning of all this running around and shouting at me? Am I a servant? A delinquent office boy to be ordered about?"

"But, Langley, there's *so* much to be done if we're to get the furniture in before"—she lowered her voice—"before the troublemaker arrives."

"My dear Lish, relax for once. Big-Ears hasn't the least notion that we're moving today. How could he? I didn't know it myself until you came before dawn and hauled me out of bed with the news that the movers were in the house."

"Langley, the vans have been on the streets all day——"

Her brother smiled and patted her shoulder. "Lish, all the stuff which counts is already inside. I had it put in the first van and then I checked each item as it was unloaded. Everything on my list has gone in. Having thus cut away any ground which might have existed under the feet of the enemy, I proceeded to the cellars to see that the wine was safely stowed away. By Jove, Lish, the cellars are enormous! While you were all halloing for me I was wandering around with a candle in my hand, looking like a Trappist monk, no doubt. They spread out in all directions. In the mood they produced in me, I swear that if I had stumbled over a cask of amontillado and had seen a set of handcuffs attached to the wall I'd have gone right over to them and chained myself up resignedly."

Her sister said, "Yes, Langley," in a somewhat tart tone. Then she relented to the extent of remarking, "It's a relief that all the good stuff is safely stored inside."

A voice from the direction of the house, the same pleasant one which had joined in when the master of the household was missing, said, "Ah, I see the truant has been found," and Mrs. Craven came out into the yard to join them.

"Now see here, Pauline," said her husband, "I object to all this chatter about truancy——"

"Do you, Langley dear? Well, never mind about it now. We've other things to think about."

Ludar was looking at her in wonder, thinking her the loveliest lady he had ever seen, lovelier even than Catherine McGregor, who had been enjoying all of his allegiance.

"There is the matter of dinner," went on Mrs. Craven. "I confess that I haven't given it a thought."

"I've seen to it," declared Lish. "I was sure you babes in the wood would forget, so I brought up a hamper in the surrey. There's cold ham and a tongue and plenty of bread and butter. Also there's an orange layer cake and some gooseberry tarts. As soon as we have things shaped up a little inside we'll all sit down to a picnic supper."

"Lish is a goddess in the clouds, raining down manna on less provident people," said her brother affectionately.

Mrs. Craven became aware of Joe's friends at this stage. "I think this is Arthur Borden," she said. "And who is this little boy? I don't think I've ever seen him before."

"I'm Ludar Prentice, ma'am," said Ludar.

"I should have guessed. Joe has been talking about you a great deal."

Meg came running down the path, crying out in her piping, excited voice, "Hello, Daddy, and where have you been?"

Mr. Craven frowned. "Even my youngest child seems to feel the need of chiding me." Then his face cleared and he smiled down at her. "May I have the next dance, Miss Short-for-Margaret?"

"Oh *yes!*" cried the child.

He took her hands, and they went dancing down the path while he sang:

"My name is Shifty Sadie,
And I am a perfect lady.
But if I come down on your feet in the dance,
You'll have to go home in the ambu-lance!"

When he came to the last line he hopped high into the air and came down on the ground with a loud clatter of heels.

His wife laughed. "I really believe Langley would have been a great ballet dancer if he had gone in for it. He amazes me the way he does everything so well. Chess and whist and the violin. You should have seen the sketch he drew of me on his table napkin at dinner the other night. I hadn't the least notion what he was doing."

"Yes," thought the sister. "He's wonderful at everything, poor Langley, except checking expenses and watching invoices."

Ludar was watching and listening with a feeling that he had been translated into a different life. He was almost sure this was the place which Uncle Billy had told him about one night when the pains in his stomach had come back and he had not been able to sleep, where there were kings and queens and princesses and fairy godmothers, and everything was upside down. Perhaps little Meg was a princess. Certainly Mrs. Craven was beautiful enough to be a real queen.

He would have realized that these strange people could be serious, however, if he had heard the talk which Langley Craven had with his sister after the dance with Meg had been brought to a somewhat breathless finish. The head of the family took his sister by the arm and led her back into the corner of the yard which so closely resembled the battlefield of Waterloo.

"Now, Lish," he said, "we must get down to cases. Two days ago you came to me and said you thought we could get the Lester place. I was disposed to laugh at you, knowing how tightfisted all the Lesters are and the kind of renting proposition Frank would make. Then this morning you turn up at the crack of dawn and say that the Lester place is ours and that we're moving at once. You brush aside all my demands for explanations by saying you're too busy to talk." His tone became most businesslike. "Now see here. To what have you obligated me? Here we are in the Lester place, and it's larger than I thought. Am I likely to be shoved through a bankruptcy court in trying to pay the rent? In God's Name, Lish, what is the arrangement?"

Lish smiled reassuringly. "It's going to cost you very little, Langley, because I'm involved in the deal too. I've got myself in a fix, and you must help me out. You'll have so much space here, and I'm getting so crowded I don't know which way to turn. I would like it if you would take Old George off my hands. What he needs for the rest of his life is a comfortable stall and a big yard like this to be turned out into when it's warm. I'm willing to pay a little something each month if you'll take the poor old fellow off my hands. Then there's Miss Fitch's dog, Yancey. Most everyone in the house hates him, and she's finally agreed to let him go if a good home can be found for him. I'm willing to pay more for Yancey than for Old George. And then—well, this *is* a situation, and only you can get me out of it, Langley."

She looked about her anxiously to make sure that they could not be overheard. "It's Cousin Hubert," she said. "Oh, it's not his continual whanging away at the piano. There are so many of us in the house that Hubert has involved himself in no less than six different feuds. Things have reached such a stage that I've got to saw him off on someone. Langley, I'll pay you *handsomely* if you'll let me saw him off on you for a few years. Here you have this big south wing with its two rooms on the ground floor. The larger one could be his studio, and he could sleep in the small one. He could have his meals in his own rooms if necessary." She paused and then went on, eying him shrewdly: "I'm asking so much of you that the cost of it would kind of run up. I'm inclined to think that what I would expect to pay you would just about take care of the rent."

Langley gave the matter some thought. "Well," he said, "I would like to have Old George, and the two children would be simply enchanted. I've a way with dogs and I might be able to train that miserable Yancey. But Hubert! Do you suppose Pauline would object very much? I hate to propose it to her, even though she has such a happy way with people like that. Of course she might turn Cousin Hubert into the semblance of a human being if she cared to undertake it."

"I'll speak to Pauline about it," declared Lish. She knew that Pauline, on whose shoulders rested the responsibility of making the family purse provide for the household expenses, would jump at a chance like this. "If she says yes, then I'll come to you, old boy, with an itemized proposition. If the deal goes through I'll be the most relieved person in the world. My worst troubles will be taken off my shoulders." She took her brother by the arm and turned toward the house. "We haven't been able to decide where the Barbudo-Sanchez should hang. You had better come and settle it."

"Great Scott, I hadn't given a thought to *that!*" exclaimed Langley. "Let's get it settled now."

3

But the hanging of the Barbudo-Sanchez was not to be attended to at once. A round freckled face appeared in the front gate, and the voice of Peter Work asked, "Does Mr. Langley Craven live here?"

Mr. Craven walked toward the gate. "Yes. What is it?"

"If you got your telephone in, you're to call a number, Mr. Craven. If not, will you please go to Nimm's grocery store? There's someone on the line there asking for you. It's a man's voice. The store's two blocks up and one over."

"I know where it is." Mr. Craven fumbled with the change in his pocket as he turned out to the street. He produced a dime and handed it to the messenger. Pete grinned delightedly and said, "Gee whiz, thanks, Mr. Craven."

"Hello, Pete," said Joe, peering around the corner of the house. "Come on in."

"Hello, Joe." The school boss came into the yard. "The Duke's here, too, and Arthur, eh? By grab, this is a big place. I could get up a real tough North Ward gang with a barn like that."

Joe was on the point of saying that they were going to have an army, but an inner sense of caution caused him to keep the information to himself. Anything that Peter Work had a hand in became a dictatorship at once. It would be wiser not to let him know about the army.

Pete continued to inspect the yard. "Say, if we lived here my old lady would make me cut down all the weeds. That would be a job for you!" He brought out the coin and looked at it proudly. "Say, come on over to Nimm's with me and I'll stand treat."

The boys would have liked to go, but a sense of duty this time prompted Joe to say: "I can't, Pete. We've just moved in and there's lots of work for me to do."

At this point a vehicle of some sort, which they could not see because of the height of the brown stone wall around the property, stopped outside. They heard a rough and urgent voice say: "Here we are. Hop in, Dave. You get that round table he talked so much about. Remember now, it's got brown leather on top. Don't go bringing out any kitchen table instead. I'll get the blasted picture."

Two men came through the gate and vanished into the house by way of the open front door. "Say, what's going on?" asked Pete. "I know that big one. He's Bud Hames, and my old man says he's no good."

He ran toward the front door of the house, and the other boys followed as fast as they could go. They heard a sound of loud voices raised in argument inside. Then the front door, which had closed in the meantime, banged open again and a man emerged in a great hurry, his face red above the drum table in his arms. Behind him came another man holding the prized Spanish painting above his head, while bringing up the rear was Cousin Hubert, looking both puzzled and indignant.

"What is this, anyway?" demanded Cousin Hubert in a loud voice. "Who gave you orders to take these things away?"

The tap of Adelicia Craven's heels sounded in the hall. "Stop them, Hubert!" she called. "They're robbers. It's all a plot. Langley was called away on purpose, and they didn't think there would be any men left here."

All Pete Work needed to hear was the word robbers. Before Cousin Hubert could open his mouth again, the boy flung himself at the second man, who was carrying the painting, and wrapped himself around one of his legs. The man, taken by surprise, looked down and gave his leg a shake.

"Great gobs of muck!" he said. "Where did you come from?"

Ludar looked at the other leg and hesitated. It was certain that the man was strong and determined and that he was not going to let small boys stand in his way. He had withdrawn one hand from the picture and was trying to dislodge Pete with it, shoving and punching angrily. The sight of Pete's red face as he clung determinedly to the intruder's leg and absorbed the punishment convinced Ludar that, although he had no stomach for this kind of conflict, he must take part. He plunged forward and wrapped himself around the other leg.

Joe had chosen the same moment to join the fray, and Ludar heard him saying in a shrill voice, "Hang on, fellows, and I'll pound him in the back!"

Arthur had wrapped himself around the arm which held the prized picture. The thief, in fact, resembled Gulliver in the power of the Lilliputians.

"I know you, Bud Hames!" panted Pete. "You'll go to the coop for this."

Bud Hames, finding himself unable to take a step, called out in an aggrieved tone to his comrade, who in the meantime had placed the drum table in an express cart with a single horse attached: "Dave, they've got me nailed. Come here and take this damned chromo while I get rid of them!"

His companion glanced back and then looked hastily up the street. A look of alarm took possession of his face. He leaned over and began to pull in the iron weight attached to the reins which he had cast over on arriving.

"He's coming back on the dead run!" he called. "Kick them off and get that picture out here or we're dished!"

Bud Hames decided it would be better to get away without the picture, and lose part of the pay, than to stay and be caught with it in his possession. He dropped the canvas, used his arms to dislodge Arthur, and then flailed with both fists at the boys clinging so stubbornly to his legs. The blow aimed at Ludar glanced off the side of his head, but the force of it was sufficient to send him sprawling backward. His head struck against the edge of the strip walk, and he lay there half dazed.

In the meantime the school leader was demonstrating the stuff of which he was made. The alarmed and angry Bud Hames struck him in the face with both hands and, when this had no effect, tore at the boy's arms until they gave up their hold. Another blow in the face sent Pete back on the ground beside Ludar.

The house had spilled out its occupants by this time. They were calling for help and waving their aprons frantically. Neighbors were pouring out into the street and demanding to know what was wrong.

Langley Craven came through the gate with a red and angry face. "Now I understand," he said, panting from his exertions. "I don't know who called me. He gabbled away about something or other and then laughed and said, 'Good-by, my friend,' and rang off. I knew something was wrong then and ran all the way back."

His sister nodded to him grimly. "Not quite soon enough," she said. "The drum table's gone. But nothing else as far as I can tell."

Langley Craven took his sister by the arm and led her back into the yard.

"Tanner had his carriage down in the next block," he said. "He paid no attention when the cart passed him but had his driver start off around the other corner."

"There's nothing we can do about it," said Lish. "He's made his play and he's got the table. And that ends it."

"This," declared her brother vehemently, "is just the beginning. Do you think, Lish, I'm going to lie down and do nothing?"

"Listen, Langley. He had his application in for a writ to compel an appraisal. We removed the stuff from the Homestead and brought it all here. Tanner has torn a leaf out of our book and taken the table from you to his own house. Oh, I know you can go to law to get it back, but he can use the same grounds to claim you must give up what you've taken. The best you can hope for is that if you do nothing he'll let matters rest too."

Langley shook a fist in the air. "Do you realize," he demanded, "that everyone in town will know about this and have a good laugh? At my expense! They'll think Tanner got the better of me. Gad, how they'll roll this morsel over their tongues!"

"Certainly they will," agreed Lish. "They've always been interested in us. We were the Cravens, and they gossiped about us all the time. You ought to be used to it by this time. But what makes you think they'll say Tanner got the better of you?" She laughed scornfully. "You whisked everything out of the Homestead under his very nose. All he's got back is a measly table that I for one never gave a hoot for. People will laugh at Tanner when they get the whole story—and I'll see they get it. They'll say he used a whale gun to kill a sprat.

"I don't think he got his lawyer's advice on this," she went on. "Ted Laird says they're smart, and I guess they are, though I don't like either of them. They'd have insisted he must wait and take legal steps. I'll bet, Langley, that Tanner got so damn mad when he heard you had stolen a march on him by moving out that he hired this pair without saying a word to Swayze and Carrington. And now he's dished himself."

They returned to the front of the house, Langley gradually coming around to the belief that she was right after all. He stopped and looked at Peter Work, whose face had been badly battered by the Hames fists. One eye was black, and the boy's nose was swollen and bleeding.

"This lad has been badly hurt," Langley said anxiously. "Who is he?"

The school boss spoke up for himself. "I'm Pete Work," he said. "That Bud Hames socked me in the eye."

"Aren't you the boy who brought the message I was wanted?" asked Langley.

Pete nodded and touched an exploring hand to his injured eye. "Old Alvin Nimm sent me. We live just back of his store."

Suspicion subsided. "I think we had better get a doctor right away. That eye needs attention."

Miss Craven looked at Pete with equal anxiety. "Does it hurt you much?" she asked.

"Some," answered Pete. "I don't mind that, mum. But my old man will think I've been fighting again and he'll give me a whaling when I get home."

"I'll attend to that. I'll drive you home, Peter, and tell your father the whole story. You were a regular hero." She looked at the covey of excited females twittering around the front door. "Someone run to the nearest butcher shop and get a piece of raw beefsteak before they close. This eye needs attention at once. You, Cousin Hubert. And get it back here fast."

Ludar had managed to get to his feet. As a souvenir of his share in the conflict he had a slight cut at the corner of his eye. Pete Work grinned at him approvingly. "Say, all you fellows pitched in like good ones," he said. These words of praise were sweet, and the boy thought, "I'm glad I tried to help." He knew, however, that his share had been small and that all the credit belonged to Pete.

Miss Craven put one arm around Ludar's shoulders and the other around Arthur's. "You're all brave little boys," she said. She straightened up suddenly. "Where is Joe?"

Mrs. Craven, who had been engaged in a far part of the house and had missed all the excitement, arrived at the front door at this moment. "Can't you find Joe?" she asked.

Her sister-in-law emitted another startled cry. "Where's the picture? That man dropped it on the ground. I saw him do it. But where's it gone to?"

Joe supplied the answer. He came running up the path from the back yard, shouting, "Have they gone?"

"Yes, they've gone," answered his father. Langley Craven looked his son over with an anxious eye. "Are you all right, Joe?"

Joe nodded easily. "Sure, Dad, I'm all right. I didn't get into much of the fight. I was just going to jump on his back and then I saw he'd dropped the picture, so I grabbed it instead and ran."

Langley patted his head approvingly. "Did you take it back into the house?"

"Nope," answered Joe. "I thought it had to be hid, so I took it to a place where those men wouldn't have found it in ten years."

"But where, Joe, where?"

Joe frowned. "Gee whiz, Dad, won't it be all right if I just bring it to you? If I tell you where I hid it you'll know the place where we're going to keep all our secret papers."

Chapter XI

I

During the long summer holidays, which began near the end of June and continued through Labor Day, Ludar tasted for the first time the full sweetness of a normal boy's existence. He was always wakened by the closing of the door early in the morning, which announced William's departure for work. He would spring from bed, give a hurried peek out of the window to see what the weather was like—it was a warm and beautiful summer, and so he was seldom disappointed—indulge in a quick but adequate wash, eat a small breakfast, and then go out to the back yard, a picture book in one hand and some toys in the other. Here without fail the new cat awaited him.

It was a poor stray cat with a damaged tail and a coat of oddly assorted colors which Ludar had christened, for no reason he could explain, Skeek. As though aware of his shortcomings, Skeek was always ready with an ecstatic welcome. He would rub his head around Ludar's ankles and purr loudly. It was easy to guess what he was trying to say. It was something like this: "I know, Little Master, that I'm a poor specimen of a cat and that I'm not much to look at. But just give me a chance to show you that I'm more affectionate and obedient than any of these sleek, handsome cats around this neighborhood. I like this place and I want to stay. Oh, I want to stay so much!" Ludar knew exactly how the cat felt. Once he said to the new pet: "You and I, Skeek, we got to be kind of careful or we'll be sent away."

Tony paid him a visit one morning. "Aunt Mona May had an argument with Daddy at breakfast," she announced. "I heard them. That old aunt of mine pretended to get a headache just to make him feel bad, and went back to bed. I don't think it made him feel very bad. He went to the office whistling. And now she has her door shut, and I can play as long as I want."

It happened that on the previous evening William had presented Ludar with some ingenious toys he had created out of matchboxes and straws from old brooms. They were made in the form of churches and stores and houses. Ludar had brought his new acquisitions out and had already made considerable progress in the building of roads and fences with the rich loamy earth.

"What are you making?" asked Tony, studying the results of his labors with round and intensely interested eyes.

"A village, that's what." Ludar had retained his voice inflections but had begun to talk in the argot of the North Ward school. It made a strange combination, the Canadian colloquialisms spoken in soft English tones. "Uncle Billy made them. He can make anything."

Tony squatted down tailor-fashion and studied the plan of the village. She indicated one of the houses with her forefinger. "Who lives there?"

She had set him the kind of task he liked best. He gave some thought to the problem and then began to describe the household. "The mother in that house is beautiful," he began, "and she has long dark hair——"

"Oh, Ludar, don't you mean golden hair? I thought all boys liked girls with golden hair."

"This mother," firmly, "looks like you'll look when you're all grown up. The father's an admiral, and he only gets home once a year. He brings lots and lots of presents. There are so many children—oh, ten or twelve, I guess—he has to bring the presents in a wheelbarrow pushed by a sailor with a long pigtail."

Tony said longingly, "It would be lovely to live in a family where there's a mother and ten or twelve children." Then honesty compelled her to add an amendment: "But when there's presents, it's nice to be the only child."

Ludar began then to talk about the church, being very much interested in churches. Aunt Callie had taken him once to a wonderful one. It had been as high as the sky and very dark and mysterious, like the house he had once lived in, only much larger. All the light got in through painted windows.

"The minister in this church," he said, "doesn't have a long black beard. He has a nice face instead. He doesn't scare boys and girls when he preaches. He smiles and he makes people pitch into the singing like the Salvation Army. This church has all the fine funerals."

Tony asked, "Who lives in this little house down near the station, Ludar?"

Ludar gave a scornful sniff. "Oh, that's just one of the boys who got asked to all the parties and then grew up to be a counter jumper."

"Which house do you live in? Are you—are you the admiral?"

Ludar hesitated. "I don't know, Tony. I guess it all depends. I may *try* to be an admiral."

She seemed curiously reluctant to leave when the time came for her to return home. She pulled at the strings of her straw bonnet and looked in the general direction of the weather vane on the Milner stables. "Making a village is fun, Ludar," she said. "I wish—I wish——"

She did not have a chance to tell him what she wished because the servant girl came to the fence and said: "You had better get on home here, Antoinette. She's stirring and in a few minutes she'll be down and rarin' around."

The next morning a boy who was about the same age and size as Ludar came through the gate into the yard. He was freckled and had a turned-up nose, and his hair was cropped so close that he seemed bald. Ludar was under the tree, hoping that Tony would be able to get over and in the meantime working on plans for a more ambitious village, one with a cemetery, a race track, and a grain elevator.

"I'm Mickey Dewire from up in the next block," said the freckled-faced boy.

Ludar got to his feet and said, "Hello."

The newcomer seemed at a loss. "I got a message for you," he said. Then he grinned. "I'm a Dogan and you're a dirty Protestant. What's that you're making?"

"A village."

"Gee, I wish I could stay and help you—if you wanted me to help you." The visitor was smiling all over his face now in the most friendly way. "D'ye suppose I could lick ye? You're not very strong, so I guess I could. But I dunno. I'm not very strong myself. I'm the poorest fighter in the gang."

This frankness won Ludar's good will at once. "I'm not a good fighter," he confessed. "But I grabbed the leg of a man who was stealing furniture."

"I know. It was Bud Hames. He lives up near us. He's a drunken bum." Then the visitor got around to the business which had brought him. "My aunt Gertie Shinny is servant girl to the Milner house. She told me four things I was to tell you. The little girl wanted you to know they left town yesterday. That's one. They went to Grimsby Beach." He was counting on his fingers. "That's two. They won't be back till Labor Day. That's three." He hesitated and frowned. "Let's see now. Oh yes. Four was that the little girl was very sorry." He added on his own: "My aunt Gertie Shinny said this little girl was crying because she didn't want to leave without seeing you. I guess she's kind of stuck on you, Ludar."

Ludar began to demolish the fence of small stones he was erecting around the church. His feeling of loneliness lifted a little, however, when he remembered what Tony had said the day before, or rather what she had started to say. He was sure she had meant to tell him that she wished she did not have to go, or perhaps even that she wished he could go with them.

Mickey Dewire volunteered another piece of information. "We're moving away next week. For good. My father's a printer, and we move around a lot. We're going to Hudson."

"Will you like it there?"

"Naw!" There was strong conviction in the boy's voice. "I never been in Hudson, but it can't be a nice place if the Dewires are moving there."

Ludar looked up at him, feeling very sorry suddenly that he was going away. "Say, Mickey," he said, "why don't you stay and help me with this?"

Mickey sat down so abruptly that he almost rolled over on his back. "Say, gee whiz, this will be fun!" he said. "If any of my gang see me playing with a dirty Protestant they'll give me double knuckles. But I don't care. Say, Ludar, what ye want me to start at?"

"Why don't you find a lot of small white stones for the cemetery?"

The afternoons were always spent with Joe and they became, from that time on, the golden moments of the day. Joe was the sun around which all Ludar's thoughts and hopes and plans revolved. Joe, always cheerful, always ready for any kind of sport, always full of fine, robust boy talk, was the ideal friend. Ludar would make his way to the new Craven home immediately after his noon dinner and would generally find Joe waiting for him at the gate in the high brown wall. "You're late," the latter would say. "Doggone, Ludar, I've been waiting for hours. We got another boy for the army today."

The army, however, was proving a difficult project. Arthur had been taken to Grimsby for the summer, so they were short an officer. The helmets were too large for any of the heads of the young soldiers, even Ludar's, which was somewhat larger than the rest, and the belts were not usable around such thin waists. A kettledrum which Langley Craven had owned as a boy had lost all resonance and was useful only for appearances. The recruits, moreover, had a habit of not turning up. They would have to stay home and weed gardens or go on errands for their mothers, or they would be taken away on visits out of town. Occasionally Meg could be induced to fill in, but she was never very satisfactory. She would giggle at the orders sung out by Joe and Ludar and say, "Silly!" Or when the order was "Company, right!" she would generally wheel to the left. Even on the few days when she was on her best behavior she seemed to consider that her enlistment was for a limited period only. She would leave the ranks after a brief time, say "Ta-ta!" and be off on interests of her own.

Once Mrs. Craven helped out, wearing a peaked soldier's cap made out of a newspaper and proving as satisfactory at drill as Meg had been the opposite. Watching her march and wheel and turn about, handling her broomstick with snap and dispatch and a complete sobriety of expression, Ludar thought that on this day he would rather have been a mere private so that he could stand beside her in the ranks. She even took part in a charge against an Indian ambuscade (in the elderberry bushes) and shouted "Hurrah!" and waved her hat when the enemy fled.

Once also Cousin Hubert, who had been "sawed off" in accordance with Miss Craven's desire and had his piano now in the big wing, was captured by a press gang and inducted into the service. Surprising enough, he proved a very good soldier, obeying commands promptly and accurately with a loud slapping of his thighs and a vigorous way of handling the hoe he carried as a gun. He obliged at the finish by bringing out Mr. Craven's regimental sword, which hung on the wall in a small nook off the dining room, and placing it on the ground. Then, with arms folded on his chest and his trousers pulled up over his knees, suggestive of a kilt, he went through the steps of a Scottish sword dance. At the finish he said, "Didn't know I could do that, did you?" and returned to his wing, where he proceeded to make a lot of noise with Chopin.

The best of all times in that golden, exciting summer were the afternoons when the two friends went swimming in the company and care of several older boys from the neighborhood of the Homestead. The world is full of rivers of all lengths, widths, and kinds, but there never was one to equal the Great. It began, as far as the youth of the town were concerned, about a mile from the last house on Fife Avenue, where it rolled ominously over a wide dam. Here the water was cold and deep and only the more daring spirits actually went in. From the dam, however, the river described a wide arc through flat and unoccupied country and was approachable only by way of obscure paths winding along patches of scrubby trees and high rank grass. It was not an easy trip for seven-year-old legs, but once the beaches had been attained, white with clamshells and flat stones, the rewards were great. Here the river was wide and indolent and never deep, so that it was perfect for small boys whose best attempts at swimming were a few wild flurries of the arms. After meandering slowly around this pleasant bend, the Great River became narrow and much deeper, and it always claimed a victim or two each summer in its most formidable phase when it rolled under the Iron Bridge near the center of town. After that it widened again and proceeded on its leisurely way, no longer sand-shingled but banked with willows and twisted alders, the solid green shot through with the yellow blooms of wild asparagus and the bright pink of mallows; not as dangerous now, but filled with deep holes of which boys spoke with awe at school, and abounding with unexpected twists and turns and sudden currents. Almost immediately then it became foreign territory, and the boys of the East Ward and Eagle Nest claimed it as their own. It was known that the Great finally became quite beautiful and mysterious and romantic, but this section was sacred to canoeists and of no interest to small boys who were not allowed in canoes. None of the group who went swimming together had ever seen the sign on the banks which announced:

Men may come and men may go,
But the Great River goes on forever.
Like the bargains at the store of
The Remarkable Man.

2

The end of the summer holidays came, and more than one hundred boys of the North Ward went back to school, their faces brown from vacation suns, their hair smacked down tight on their heads by maternal brushings; and at least as many small girls, their faces soberly aware of the importance of the studies ahead of them, their dresses neat and new. Arthur was back from Grimsby and in a somewhat pensive mood which stemmed from the fact that he had become interested for the first time in a girl. He was disposed to talk about it at great length. Pete Work was missing. He had graduated to the Central School (a classic example of leniency in marking his examination papers), but it was rumored that his parents would put him to work instead.

On the evening of the school's opening William heard a tap on the window of his workshop. He walked across the room and saw that Catherine McGregor was leaning over the fence with a cane in her hand. He opened the sash.

"Uncle Billy!" Her eyes were shining with excitement. "There's news! Straight from England. News about Ludar, Uncle Billy. I think they've found out about his family. Isn't it *simply* wonderful!"

William was silent for a moment. His face did not reflect any of her enthusiasm over what had happened. "Wait a minute, young lady," he said. "I haven't heard anything about this. How do you happen to know so much?"

"There's to be a meeting. In your house, at eight o'clock. Father's been asked to go. That's how I came to hear about it. And the news isn't all wonderful, because it seems there's a catch somewhere. That's why the meeting's being held."

"When did you hear this, Catherine?"

"A few minutes ago. Father telephoned from the store to have supper put on early. Margaret's trying to hurry up the kidney stew now. Father said to tell you."

William was in a thoughtful mood when he gave the news to Tilly as she made toast at the kitchen stove. She became highly excited at once, however, and saw visions of wealth for them all. "No more tinkering with kitchen stoves!" she cried jubilantly. "I'll just say, 'Toast, Mary Jane, and see that mine is good and brown.'"

"Tilly," said William, "we mustn't stand in the boy's way. If he has the chance now to go back to England and live in that fine big house with the outdoor clock, we must let him go. And we must do it willingly."

Tilly turned to stare at him. "Stand in his way! I should say not. Billy Christian, have you lost your senses?"

"I realize something that you don't seem to. We're going to lose him."

"Oh no, we're not! We have a share in him now. If these snobs in England think they're going to take him back and give us the go-by, they'll soon find out their mistake."

"What could we do?" William shook his head soberly. "We've got to face it, Tilly. If the family has been found, there won't be a son in this house any more. They might offer us something for his keep, but I wouldn't want to take it."

His wife closed the lid of the stove and began to butter the toast with an agitated hand. "I would! It would be no more than our right. I'd have a statement of expenses all ready for them. And what's more, Billy Christian, I'll knock this meek-as-Moses spirit out of you with a poker handle if necessary!"

"Tilly, listen to me." William spoke firmly. "This isn't a matter of dollars and cents. It's a question of doing the best we can for the boy. If it turns out they want him back, we mustn't put any obstacles at all. We mustn't mention expenses unless they do. The boy hasn't been a boarder with us, he's been a member of the family."

Tilly paid no attention. "I'm going to get out the Ludar book right away and total it up," she said. She looked at her husband scornfully. "Look out for Number One! This is where we do it."

Allen Willing arrived a few minutes before eight, accompanied by Albert Travers, the manager of the Balfour branch of the United Provinces Bank, who was also a deacon in the Baptist Church and a man of substance. Lockie McGregor came in immediately after, and the meeting started. The banker produced a paper bag of cigars and handed them around. "Two-for's," he explained. Willing took one, but William and Lockie McGregor, not being smokers, declined.

The editor produced a letter and smoothed it out on his knee. "It's from that lawyer in England, Mr. Christian," he said. "It's to his nephew, and it arrived late this afternoon. It's long, but I guess I had better read it all out."

John Ratticlipper, lawyer, whose address was Little Brilling in Surrey, had located the boy's family, it developed from the reading, but had been most careful not to mention the name or to give a clue as to where they lived. They belonged to the lesser nobility and were liberally endowed with the goods of this world. The head of the family was still alive and had a reputation for two things, stinginess and a furious temper. The mother had been dead for some years. The older of the two sons disappeared some years before, and it was generally supposed that he was dead, although his name was never mentioned by the other members of the family. There had been some scandal when he left. There was a second son named Cyril and two daughters, both married, named Poppy and Nancy.

At this point Willing glanced up from his reading. "Everything seems to fit in with what we know," he commented.

"A bit too well, perhaps," said the banker. "Isn't it possible he's repeating back what he has been told?"

"I don't think so," declared Willing. "He has too much other information which fits the case. The family seat, according to Mr. Ratticlipper, is an old manor house with a moat, an ivy-covered gatehouse with a clock facing the park, and quite extensive offices and stables. It hasn't been used much lately because, with all the children married off, the father prefers to live in town, meaning London. Most of the help have been discharged as a result." He glanced around the circle of faces. "Well, that's the evidence we have to go on. It seems conclusive to me."

The banker held out a hand and gave his fingers a snap. "Let me have the letter," he said. "I want to see the handwriting. It is, I've found in my work, an infallible indication of character." He studied the sheets closely for several minutes, then said, "Humph!" and handed them back.

"Well?" asked the editor.

"Cupidity," declared Travers. "Slyness. Resourcefulness. Lack of scruples, up to a point. Ability and shrewdness. A highly developed degree of suspicion."

"Then you regard the man as dishonest?"

"Not at all. Not at all. These qualities don't add up to actual dishonesty. In fact, they are the qualities which might prove most useful in the kind of arrangements which may have to be made with the boy's family."

"He's a lawyer," declared William. "Couldn't we leave it at that?"

Lockie McGregor had listened to this brief discussion with considerable impatience, and now he injected himself into it. "There's more to the letter," he said. "We wouldn't have been asked to come here if there wasn't. I'm sure this lawyer, whose character doesn't seem to faze Mr. Travers but is regarded with very little favor by my friend Mr. Christian, has had something to say about terms."

"He has indeed," said Willing. "Most of the letter, in fact, is devoted to what our worthy Mr. Ratticlipper regards as his due. He asks a payment of ten pounds down to cover his expenses to date—stamps, paper, duly enumerated, and his time, which he regards as quite valuable—and another fiver for the expenses he anticipates in making contact with the head of the family. He says flatly that all negotiations with the family must be entrusted to him and that he's to receive ten per cent of any financial arrangements which may be reached."

"Excessive!" snapped the banker. "In view of the fact that he's a rank outsider, acting solely on a suggestion from some whippersnapper of a printer, it's absurd!"

"I grant you the terms are excessive," said Willing. "Unfortunately, however, he's the only one who possesses the information about the family."

"Wait," said Lockie McGregor. "I've a conviction, gentlemen, that it may not be necessary to have any dealings at all with this grasping Mr. Something-or-other of Little Brilling. I've always had a suspicion that the boy could tell us his real name if he cared to do so. Now, William, before we try to make up our minds about anything, let's have the laddie in and ask him a question or two."

Ludar was doing his homework in the kitchen. He came in reluctantly, knowing he would be asked questions he could not answer.

"My boy," said William, "we have some word about your family. Unfortunately the man who has written us has seen fit not to tell us who they are. It's very important for us to know the name of the family. I've asked you many times if you remember it and I've believed you when you said you didn't. But, Ludar, don't you think, if you tried very hard—extra-special hard—you might get hold of it?"

Ludar gave his head a shake. "No, Uncle Billy. I've tried hard to remember. Sometimes I think I do and then it—it gets right away from me. I don't seem to remember much about anything that happened when I was a small boy."

"Do you recall, Ludar," asked Willing, "when they first began to call you Prentice? Wasn't it just before you got on the boat?"

"Oh no, Mr. Willing. It was a long time before." His forehead tightened in a determined effort to recapture the past. "We went to an office. There was an old man sitting in a chair with a bag over it, and he said my name was Prentice and not what—what it was before. He said I mustn't ever think of the name again or say it. I didn't, Uncle Billy. I never said the name once." He made another effort to regain a hold on the past. "The old man gave me a shilling. I think it was taken away from me."

The editor asked, "Was it as much as a year before you got on the boat that you talked to this old man?"

"I don't know, sir. It was a very long time. My aunt"—he paused and groped for the name—"my aunt Callie wrote a lot of letters."

"Ludar," said the editor, "you want to go back to England, of course, and live with your own family?"

And then something happened which surprised them all. The boy's eyes filled with tears, and he replied with a passionate negative. "No! No, Mr. Willing. I want to stay here. I'm afraid to go back. I don't want Aunt Callie to get me."

"But you wouldn't be going back to her. You'd go to your own family. To your grandfather, who owns that house you told us about—Old Mr. Cyril. You remember about the big clock and the stables and—and Old Baffle and the rest of them?"

The boy rubbed a hand over his forehead. "No, Mr. Willing."

"Ludar, you would have a fine life, a very exciting life, if you went back. You'd have servants to wait on you and you'd always have lots of pocket money. You would probably have a pony of your own and a whole pack of dogs. Wouldn't you like a pony?"

The boy nodded. "Y-yes, Mr. Willing. It would be nice to own a pony."

"Then why do you say you don't want to go back?"

There was a long pause. "Please, sir," he began, keeping his head lowered, "I'm frightened about it. I don't want to live with strangers. They might not like me any more than Aunt Callie. I want to stay here. I like Uncle Billy and Aunt Tilly. And I have friends of my own now. At school."

"Ludar," said Lockie McGregor, "if ye wanted to go back, would it be easier to remember what your real name is?"

The boy turned to him and answered in a tone of greater passion than he had used before: "I don't remember it! I don't! I try to, but I can't do it. I'm telling the truth, Mr. McGregor. It's gone out of my mind, and I don't think it's ever coming back."

William, finding it hard to suppress the note of satisfaction in his voice, said at this stage: "He's telling the truth, gentlemen. I'm certain of it."

"So am I," said Willing. "Most reluctantly I bow to the inevitable. He isn't able to help us, and so we're left at the mercy of this very mercenary Mr. Ratticlipper."

"Off to bed, sonny," said William. "We won't bother you any more about it."

When the boy had left the room the editor said, "I suppose the next step is to decide what's to be written to the lawyer." He turned to William. "It's felt, Mr. Christian, that every effort must be made to push the boy's case. You can't bear the cost of investigation and, perhaps, of litigation later. It's no one's responsibility, and yet we feel it would be a blot on the fair name of the city if we sat back and did nothing. Each of us here is in a position to circulate a list among the people who should be the most interested, and we want to start at once and raise the necessary money. The citizens of Balfour are generous and they'll be glad to help again in such a good cause. If the boy comes into money it would be only right and fair for the contributions to be handed back, and we'll have that part of it understood. It might even be arranged that there would be a little bonus in that event. If it's hard to raise enough we might hit on some fair basis of expectation to grease the wheels of philanthropy."

William was listening with a feeling of aversion for the plan. He nodded his head at each point made by the editor, but to himself he was saying: "I don't like this at all. There must be some other way. This will make trouble for us in the end." At the conclusion of Willing's explanation he remained silent so long the visitors realized there was some unwillingness on his part; which, needless to state, came as a surprise, for they considered the plan both generous and helpful.

"I think," said William finally, "that I'd better speak to my wife."

Tilly was sitting in the kitchen with a copy of the pink-covered *Family Herald* in her hands. Ordinarily when she settled down to peruse this most popular of penny weeklies she became so concerned with the inevitable marriage of the poor heroine to a belted earl that nothing could get it away from her. At the moment, however, she was holding it upside down, and it was not difficult to deduce that she had been listening.

"I don't like this," whispered William, seating himself beside her.

Her eyes opened wide in surprise. "And why don't you like it? Don't you want the boy to have his just rights?"

"Of course I do." William frowned uncomfortably. "But this makes it a public affair. It's like charity. Everyone who puts up a dollar will feel he has a claim on him for the rest of his life. I think it's a troublemaker, Tilly. Surely there's some other way of doing it."

Her voice had taken on the familiar ominous note. "Will you tell me how else it can be done, please?"

"Well——" As usual when he was embarrassed his face had become flushed. "Perhaps some one person would advance whatever is needed. It would be so much easier to deal with one person. And there wouldn't need to be publicity that way."

"What person have you in mind?"

"Well—perhaps Miss Adelicia Craven. You know how interested she is in the boy."

"Miss Adelicia Craven," declared Tilly, "is sensible enough to let others share in the cost when they show their willingness to do so. Miss Adelicia Craven is not a fool." Her manner and expression became more hostile. "Billy Christian, I don't know where you get such notions. Here's a chance to get the boy his rights, and yet you hold back, you great galoot. Do you think *we* can do it? Well, we can't!"

William sighed. "I suppose it's because I've got into the habit of thinking of him as our own son."

"And what difference, pray, would that make?"

"All the difference in the world," he declared, feeling himself on safe ground here. "If he was our son I wouldn't step aside for a committee of citizens. I wouldn't let him be made into a public charge this way."

"Well, as it is," said Tilly, "you just get back in there and say we agree to whatever they want to do."

William returned slowly to the sitting room. "Well, gentlemen," he said, "if you're going to get up this list I have a name which must go at the top. 'Z. X. Gold, twenty-five dollars.' Mr. Gold spoke to me some months ago and said he wanted to help. In his case it will be a—a donation."

"I've another which will be a donation also," said Willing briskly. "This one's for fifty dollars and will simply be listed by the initials F. G."

The mind of each man in the room started to work at once on the identity of this mysterious and astonishingly generous donor. Travers was the only one to emerge with anything concrete. "The only F. G. I can think of is old Frank Gottschalk," he said. "It can't be him. He's so mean he hates to throw away the scales of last Friday's fish."

William had a sudden flash of understanding. Fairy Godmother! That was it, of course. No one but Miss Craven would be so generous, and she had curious ways of doing things. He looked at the newspaper editor and then nodded and winked. The latter winked back.

3

The fall merged into winter, Christmas came and went, the snows melted, the river went out, and with the arrival of spring buggies and bicycles appeared again on the streets and crocuses flaunted brave yellow and purple flags in all the well-tended gardens on Grand Avenue. In the meantime a succession of letters had passed back and forth between Balfour and Little Brilling, hopeful on the one side, demanding on the other. The demands were always for more money, expenses for this, compensation for that. No satisfactory information had been received in exchange for the drafts sent eastward. "I have talked with a member of the family," the lawyer would report in his neat handwriting, but he would not say what the outcome of the conversation had been, or he would state that, in the opinion of one of the daughters, the head of the family said this or thought thus and so.

One night William came home in a state of very low spirits and said to Tilly, "Well, it's started."

Tilly was warming up potatoes in a pan on the top of the stove. "What's started?" she asked.

"The questions. The wanting to know what's happening in England and when the boy will be going back to claim his estate. Two people stopped me today, and they hemmed and hawed and then came out with it: what had been done with the money they chipped in? Harry Findlay at the shop was one of them. He chipped in a whole dollar and he talks as if he owned a half interest in the estate—if there is an estate. I wouldn't be surprised if that fool wife of his went out and bought a whole new dining-room set on the strength of their prospects."

"I can't say I blame them." Tilly looked up from her task and fanned herself with her apron. "I think about it all the time and wonder when the good news will come. There's one thing I'm going to insist on. I'm not going to fry another mess of potatoes as long as I live. When we get a servant girl, Billy Christian, I'm going to sit down in a rocking chair and watch her at it—and I'm going to enjoy every minute of it and every move she makes."

"There won't be any news for weeks," said William. "I'll be glad, too, Tilly, if things turn out so you can take it easy."

"I've never been able to take things easy, and that's the truth." She came to the table and sat down beside him. "I've finished that book about Lord Fauntleroy. Do you know, it's the first book I've ever read right through in my life. I always like the stories in the *Family Herald* better. But I read this one word for word. It made me think of Ludar, and after I finished I was sure he would be summoned back to England soon to step into a big fortune." She sighed ecstatically. "Things must be wonderful in England. Oh, the homes I've read about in the *Family Herald!* I think, Billy, you and I will go over with Ludar and settle down there. We won't want a big estate like the one he'll have. No castle, just a nice home with a few acres of land and a married couple to do all the work for us."

William was taken with this picture and began to help her in the elaboration of it. Then abruptly he regained his former sober mood. "There were twenty-seven people contributed a dollar to the fund," he said. "Seven put in two dollars each. What I'm afraid is that every last darn one of them is thinking and talking like us." He shook his head soberly. "No fortune in the world could provide servant girls for thirty-four families."

"It's not like you to look on the dark side of things."

After a moment's silence William became more cheerful. "We may not depend on this fortune in England to get you a servant girl, Tilly," he said. "I've got an idea in this old head of mine. It's kind of vague yet, but it keeps churning around in my mind, and after a while I'll have it all worked out. It will be a gold mine this time." His eyes began to glow. "A roller boat. What do you think of that idea?"

"A roller boat?" Tilly clearly was very much puzzled. "What *is* a roller boat?"

Having allowed himself to talk about his new plan, William was getting excited. "It's to be made so it will roll over the surface of the water. It will roll right ahead in any kind of weather and beat all the steamships hollow. My boats will never sink or get wrecked."

"Will the passengers roll with it?"

William looked scornful of such ignorance. "No! There will be an inner section which won't move. The outer part will revolve around it. I think about it all the time and I'm beginning to see how it can be worked out."

Tilly was still skeptical. "Do you mean we'll go to England in this kind of boat you're going to make?"

"No. We'll go to England on the profits I'll get by having other people travel in the boats I'll make. Tilly, there's a fortune in this one."

She heaved a deep sigh. "You've thought of so many things with fortunes in them, but I haven't seen a single dime come rolling uphill for us."

"You'll see whole silver dollars rolling uphill this time." He had been speaking with smiling confidence, but now he reverted to his earlier mood. "I wish they'd stop pestering me. It's the ones who put in the least that say the most. I passed Mr. Gold the other day, and he said, 'Don't get impatient, Mr. Christian. It takes a long time to get anything done when the selfish interests of people are concerned.' It would be fine if they were all like him, but I'm afraid these people who contributed a dollar are going to get meaner about it all the time."

4

Things were developing more rapidly in another direction.

William worked late at the shop one evening in early summer. He was finishing a pair of swords, destined for use in the army, and had gone to great pains to get the blades long and smooth and had even fashioned gauntlet hilts for them. He was proud of his handiwork as he finally emerged from the rear door of the shop, the weapons tucked under his arm.

As he came out on the street he glanced across at the small factory building where Langley Craven had established his furniture repair business. As it was a relatively new structure, it was much cleaner-looking than most factories, and the sign over the entrance, "The Balfour Specialty Company," was fresh and even artistic.

At the same moment that William reached the street Abel Jack came out of the Craven shop. He was a tall and rangy man with nostrils flaring at the end of a conspicuous snout, an untamed pair of eyes, and undisciplined hair wisping out from under the brim of a broken straw hat. An opinion was pretty generally held in town that Ab Jack should have been on the stage.

"You're pretty busy over there nowadays, Ab," called William.

The other man flapped an arm in the air with a long forefinger at the end of it as an invitation. William crossed the street.

"I'll tell you how busy we are," declared Abel Jack in dramatic tones. "We've taken on two more hands. We're getting a larger delivery truck. When only the Gentleman was here—bless him, he's a fine man, but, great Caesar's nightgown, he's not cut out for this kind of thing—we didn't think it dignified to get out after business. But now we have Miss Craven with us and we go to houses and grab stuff right out of the parlors. Shirttails crackle around here these days. Even the Gentleman's, which crackle more than any because they've got so much starch in them."

"I'm glad to hear it, Ab."

"Billy," said Abel Jack, waving both arms in the air, "I'll explain the difference this way. Our saws used to move to waltz music. *We've-all-the-time-in-the-world-to-finish-tra-la-la.* But now"—he demonstrated with a vigorous movement of his arms—"the saws move to jig time. *Get-busy, keep-busy, get-busy, keep-busy!* I tell you, we get orders filled nowadays. In one day, finished and out the next, money in the till. That's our new motto, Billy. Money in the till."

He lowered an arm for the purpose of placing his forefinger on the lapel of William's coat. "But now, Billy, I'm going to tell you something to chill your blood. You have a fine, gentle soul, and it will be hard for you to believe such cold-blooded villainy can exist in this fine, gentle world. Billy, it's hard to get materials, velvets, velours, reps. Do you know why? Influence! Hints that our credit is no good. Offers to pay higher prices. Sometimes people in this town are afraid to let us have their stuff to fix. The same influence. Whispers that jobs might be lost. I'll tell you what it is, it's diabulical!

"But let me tell you what I did." He paused, raised his head slightly, and gazed down at his listener as though he had just walked on a stage. "I went to the residence of Tanner Craven. I went right to the front door, mind you. I rapped. The butler came. Family must have been expecting company, because the Bow-and-Scrape had on white tie and tails. I said to the Bow-and-Scrape, 'Any furniture to be repaired?' He glared. 'Come to the rear door at a proper hour and I'll h'inform you that there's nothing for you.' 'What!' I said. 'No drum tables to be polished up so they won't be recognized?' He was going to close the door, but I got in a final word. I said, 'Inform your master, minion, that there's one man in this town who isn't afraid to let Mr. Tanner Craven know that he's no better than a little house behind a big one.' Billy, that's as far as I got. I had lots of choice insults to launch, but the thin-skinned fellow slammed the door in my face."

William was worried. "See here, Ab," he said, "you'll get yourself in serious trouble with this sort of thing."

The "Hah!" with which this suggestion was received was a masterpiece of scornful intonation. "I'm Abel Jack and I'm as good as the next one.

Why should I be afraid of Tanner Craven? Will he get my landlord to throw me out of my house? There are plenty of others to rent. Will he run me out of town? The world is full of towns." His eyes focused for the first time on the swords William was carrying. He reached out and took one, thrust with it at an imaginary foe, his nostrils flaring wide. "Gad, how I could have ripped into Shakespeare if I'd had the chance!"

Chapter XII

I

On the morning of Dominion Day, which was ushering in July with a burst of midsummer heat, an unprecedented thing happened. Margaret, the McGregor "help," came to the fence and asked William, who was inspecting the crop of apples on the astrachan trees, if he would be kind enough to step over. "Now what in Sam Hill!" he said to himself as he made his way to the front door of the McGregor house. Not once during the many years the two families had lived side by side had he placed a foot in his neighbor's house.

He was escorted down a long dark hall, off which opened equally dark double parlors, to a tiny cubicle beside the dining room which the Remarkable Man called his study. It could not have been larger than ten feet by eight, but it contained a small desk, one chair, and a marble bust of Sir Walter Scott on a pedestal. There were open shelves on one wall, and here he kept, in addition to some files and personal books, a set of the life and works of Robert Burns; an early edition, bound in rich red leather by the great Zaehnsdorf himself, as he was always ready to point out.

Mr. McGregor was in his shirt sleeves, and his eyes were fairly boiling with triumph.

"Come in, William," he said. "Take the chair. I'll stand, man. I don't feel like sitting down. I've news for you. Such news, William! My little Catherine is going to marry the Reverend Elmore Nash."

"Oh!" said William, too startled to make any other comment. He was very much disappointed because he had been firmly convinced that Catherine would not give in.

"My handling of the whole matter was masterly," went on the merchant. "Ay, there's no other word for it. I didn't urge her. I had nothing to say, in fact. I never mentioned the minister to her, but I saw to it that every now and then he was invited up for a meal. One Sunday he came after the morning service and had dinner with us. Roast goose and apple-sauce and a roly-poly pudding. I had arranged for a telephone call to take me downtown as soon as the meal was over and I left them together. When I came back Catherine said to me, 'He's very interesting, Father. I think he has a fine mind.' After that I knew it wasn't necessary to say things against that young gandershanks from the East Ward. Well, no, there was one occasion. When I heard that W. and M. Appleby had given him a raise I made a point of telling her I knew it was fifty cents a week. She was boiling mad about it and said the socialists had something on their side after all. I didn't argue the point with her. No, man, I was too canny for that."

He seated himself on the window sill. "I played my cards with the greatest care. I even told her stories about the hard life of a minister's wife and said it was my considered opinion she should marry a business-man. I stopped inviting him up for meals. But he came to see her, and I could tell she was beginning to like him. Once I said to her, 'Catherine does the Carson store support the family?' knowing full well that it didn't. She said, 'No, Clyde has to help.' I asked, 'How long would it be before he could make a home for a wife?' and she flushed up and said, 'About ten years, I'm afraid.' I asked, 'Would he accept aid from a father-in-law?' and she burst into tears and said, 'No, not if you were the father in law.' All this time books were coming to her from the manse—the novels of Mrs. Humphry Ward, for instance—and Catherine was reading them eagerly and discussing them with him. Well, the minister proposed to her this Monday. She came to me and said she had promised to think it over and give him his answer in a week. This morning she said, 'I'm going to have a talk with Clyde today,' and I knew from the look on her face that she had decided to accept and was going to break the news to the other one."

He slid down from the sill and gave his chest a proud and resounding thump. "Ay, neighbor, it was a canny piece of work! I've got for my pretty little chick the very best husband she could find if she searched the wide world over." He nodded his head several times at his auditor. "What a man I am! Eh, what a man I am!"

It was doubly surprising, therefore, when Ludar, who was on the point of leaving for the fairgrounds after dinner, called excitedly: "Here's Aunt Catherine. She's in a buggy, and Clyde's with her."

William accompanied Ludar to the front to see what it was all about. Clyde, keeping the reins grasped tightly in one hand (he had had little experience at driving), grinned self-consciously. Catherine greeted them with a smile which was both tremulous and happy. Always before she had been just a pretty girl, a friendly and active young creature with a dimple which showed in one cheek when she smiled; but now there was a subtle change which made her much more attractive and feminine. She had taken off her hat and was holding it in her lap, and under the handkerchief which she had bound around her head in its place, her eyes looked much larger and of a deep emotional blue.

"Has Father gone to the store yet, Uncle Billy?"

"He left about half an hour ago."

"Then we've come at just the right time." She stood up, steadied herself with a hand on the whip socket, and then leaped lightly to the ground over the wheel. "I won't be more than a few minutes, Clyde."

"Just as long as you don't change your mind," said Clyde.

"I won't. It's made up for all time." She turned and addressed William. "We're eloping. Clyde went to Mr. Gerson's house and got him to make out the license, and we're driving to Hudson to be married."

William was delighted with this news. "I was afraid you weren't going to, Cathy."

"So was I. I went to say farewell forever to Clyde, but as soon as I saw him I changed my mind. I knew I couldn't live without him, great goose that I am." She turned and began to run to the front door. "I must hurry. I've a bag to pack, and then we must get away."

When the door closed after her William directed a look at the young man. "Well, Mr. Carson——" he began.

"The name is Lochinvar. Y. Lochinvar, for this one day."

"I hope you realize how lucky you are?"

"I do, Mr. Christian. I'm the luckiest fellow in the world."

"That's no exaggeration. Cathy is not only the prettiest but the finest girl in town."

The prospective groom gave his head a shake. "If I can only be worthy of her! And it's going to be necessary to start in such a small way. My employers will let me have a room above the store to furnish, with access to the one toilet in the place. Does that seem all wrong to you, Mr. Christian—Catherine McGregor marrying a man who can't do better than that for her? But you see I have to keep things going for my family and I'll have that responsibility as long as they live. We couldn't be married except for this arrangement with the Applebys."

"You're young and it doesn't matter," declared William. "Cathy has plenty of sense in that pretty head of hers. One room will do you for a while."

"Her father isn't going to take this very well." Clyde had relapsed into a gloomy state of mind. "I don't care what he says to me, but I'm afraid he'll pitch into Catherine. And of course he's going to be disappointed as well because he fully expected she was going to say yes to that long-jawed, long-winded——"

"He was sure of it," affirmed William, finding it hard not to smile.

"The Remarkable Man will be remarkably upset when he gets home and doesn't find his daughter here." Clyde gave an anxious glance at the second-story windows of the house. "I wish Catherine would hurry. We must get away before he returns."

There was no need to worry. Catherine packed a small bag in record time. It was such a small one, in fact, that she was able to carry it out in one hand, with a light blue shawl folded over it. She climbed back into her seat in the buggy and then, quite suddenly, she began to cry.

"I'm sorry, Clyde," she sobbed, applying a lace handkerchief to her eyes. "I realize an elopement should be gay and exciting. But—but when I saw my dear old room and my poor little Louella up on the dresser with the bow I tied around her neck a dozen years ago, and all the lovely pictures I framed and hung on the walls—why, I knew I was saying good-by to everything. Then I—I came face to face with that long portrait of Father, and it had such an accusing look! Let's get away quickly, Clyde. I'll break down completely if we don't. Good-by, Uncle Billy! Good-by, Ludar! I may not look it, but I'm really quite deliriously happy as well as miserable."

Clyde whacked the reins on the horse's back and said, "Giddap!" Then he turned and waved his free hand. "I'm deliriously happy and I'm not miserable at all. And I *do* look it."

He did indeed. The smile he wore on his face would have matched that with which Young Lochinvar went galloping across Canobie Lee.

2

William went downtown for a stroll in the afternoon. Practically everyone had gone to the fairgrounds to watch the sham fight and the Caledonian games. The streets were so nearly deserted that his footsteps echoed on the pavement. He fingered the silver in his pocket and wondered if he could afford to take home a pint of ice cream for Tilly and the boy. He had reached the decision reluctantly that it would be too much of an extravagance, when he saw signs ahead of him on Holbrook of a most unusual activity.

Two men were engaged in taking down the sign above the entrance of the McGregor store and the life-size figure of the founder! One of them was Henny Cobbleworth, the janitor of the store, and the other was Jimmy Hipple, the carter. Cobbleworth was standing on a ladder, with Hipple holding it, and sawing away at the platform on which the figure stood. As William drew near he saw the golden effigy of The Remarkable Man dip forward, quiver, and then go crashing to the ground. It broke clear of the sign in its fall and rolled over on its back in the ditch. There it stayed, staring up at the sky, with one arm across its chest.

Cobbleworth came down the ladder and grinned at the group of spectators which had gathered inexplicably out of the emptiness of the streets. "You've seen it, gents," he said. "The finish of one of the landmarks of Balfour. I've Mr. McGregor's orders to take it to the city dump and burn it."

"But, Henny, what has happened?" asked William. "I talked to Mr. McGregor this morning, and he—why, he was feeling chipper. I'm sure he had nothing of this kind in his head then."

"No, you bet he didn't," said Henny. "He loved this sign. Not a day passed that he didn't look up at the statue of himself and smile and wink at anybody who happened to be around. He said to me a dozen times that it was the best sign in the whole world. But that was before the accident happened. Haven't you heard about it?"

William shook his head quickly, his heart thumping with apprehension. Something had happened to Catherine!

The janitor gave his head a doleful shake. "His daughter was driving to Hudson this afternoon, and the horse ran away. Horse and buggy and all went right over a fence, and poor Miss McGregor was thrown out against a tree."

William grasped his arm. "Was she—was she badly hurt?"

"They telephoned Mr. McGregor and said she was in such shape they couldn't give him any hope. He came here on his way to Hudson, where she's in a hospital. I've worked for him twenty-seven years, but I'd never seen him in such a state. He said to me, 'Henny, you remember I had you put that figure up the day my wife was buried. Now you're to take it down on the day I've lost my daughter. It was put up in pride,' he said to me, 'but it will come down in humility.' "

Jimmy Hipple brought his cart to the curb, and the two men lifted the gilded statue and put it in the vehicle. The sign was tossed in on top of it.

Before getting in himself the janitor paused and addressed the spectators. "You needn't grin, any of you," he declared. "He *is* a Remarkable Man. You should have seen him when he came here to give me my orders. He was heartbroken, but he didn't let a sign of it show. 'My pride has caused my daughter's death,' he said to me, 'and it's my instructions that you stay until there's nothing left of this thing but ashes.' And he said to me too, 'I'm going to burn every photograph of that statue I can get my hands on.'"

"Henny," asked William, "has there been any word since? Perhaps this first report was exaggerated. They nearly always are. I—I can't believe——"

The janitor climbed into the seat beside Jimmy Hipple. "There's been no other word," he said. "God knows, Billy, I wish there was. I was fond of the boss's daughter. Everyone in the store was, from top to bottom."

Later reports were received soon thereafter, however. As William made his way slowly up Wilson Street he saw that a group of neighbors had gathered on the sidewalk in front of the McGregor house. Margaret emerged from the front door as he drew within hearing distance.

"He just called me from Hudson," she said. It was clear that she was overwhelmed with grief and anxiety, and yet she could not keep a note of self-importance from showing in her voice. "He said, 'Margaret, I don't dare hope—but my daughter *is* alive!'"

"I knew it!" exclaimed William. "I couldn't believe she was dead. She'll come through this, you mark my words."

Margaret did not share his optimism. "Not to hear her father talk. He said to me: 'We must be prepared for the worst. Tell our friends they mustn't get their hopes up.'"

William was not convinced. "She'll get better," he asserted. "I'm sure of it, Margaret. Why would the good Lord allow Catherine to be killed? No, no, it would be too great a waste. She's going to have a terrible time, but—she'll pull through."

The group began to thin out as the neighbors returned to their homes. William waited until the last had gone and then asked the servant, "What happened to the young man?"

Margaret shook her head. "I don't know. Mr. McGregor didn't mention him and I didn't dare ask. Perhaps that means he wasn't hurt."

"They couldn't have been married if the accident happened on the way to Hudson."

"No, Mr. Christian. I'm sure they weren't married. Isn't it dreadful that they could go away together so happy and excited and that this—this should be the end of it!"

It developed during the next few days that William had been right about the outcome. The first morning Mr. McGregor telephoned that his daughter was holding her own and that the doctors were beginning to hold out some slight hope. It was reported in the *Star* that she regained consciousness during the afternoon. The next day the advices were still better. The victim, the doctors now believed, had an even chance. When William returned from work that evening he found Tilly full of smiles. "They say she's going to live after all," she announced. "Margaret had another call from Mr. McGregor. She said he was cheerful. But poor little Catherine! She may never be able to walk again."

"Is her back injured?"

Tilly nodded. "I'm afraid so. But Margaret doesn't get things very clear over the telephone. He said something to her about a wheel chair."

William, who had been so optimistic at the start, was now inclined to look on the dark side of things. "How dreadful it will be if she's helpless all the rest of her life! When I think of the way she used to come climbing over the fence, so quick and light and full of fun!"

Tilly had been gloomy at the start, but now she was quite cheerful. "She's alive, isn't she? That's the main thing. Land's sakes, Billy Christian, we should all be very thankful. I hope you'll say your prayers for once and let the Lord know how you feel."

3

It was three weeks later that Catherine came home. She arrived in an ambulance, and two men began to carry her to the house on a stretcher. It was in the middle of the summer holidays, and because the cherry tree was packed with ripe fruit a number of Ludar's friends were with him in the back yard. They, of course, came out in a body to watch.

Catherine, very pale and thin but seemingly in the best of spirits, called, "Ludar!" He ran over eagerly, and she asked the stretcher carriers to stop for a moment.

"Well, young Rustybumpkin," she said. "I wasn't at all sure I would ever see you again."

"I know." Ludar spoke in an awed voice. "I was—*terribly frightened!*"

"Were you?" She patted his head affectionately. "I'm glad you were, Ludar. You saw me start out with Clyde that day, didn't you? We were so happy. I thought we were driving away into fairyland."

"But you're going to get well, Aunt Catherine. Uncle Billy says so."

"And does that make it certain?" She smiled suddenly and gave the top of his head another pat. "Of course I'm going to get well. I'm going to be as strong as ever, and one of these days I'll surprise you by coming scrambling over the fence the way I used to. Weren't you glad that Clyde wasn't hurt?"

"Yes." He had heard the accident discussed many times, so he asked, "How did the horse happen to run away, Aunt Catherine?"

"It wasn't Clyde's fault." She spoke with a hint of sharpness, as though she also had heard the point discussed. "Well, I must go in now. You'll come to see me, won't you? I'm going to need lots and lots of company."

"Oh yes. I'll come in all the time."

Clyde arrived that evening, looking nervous and unsure of himself. Mr. McGregor met him at the door, bowed brusquely, and said, "Come in." They walked into the front parlor, and the merchant seated himself on a horsehair couch, motioning to the visitor to take a chair.

The latter had never been in the front parlor before, and its grand and austere air made him still more conscious of the difficulty of his position. There was a marble fireplace and a crystal chandelier suspended on a brass chain from the ceiling. On one wall was an expensively framed map of Scotland and on a marble-topped table of Eastlake design a shadow box filled with dried heather. The folding doors between this formidable room and the back parlor were closed.

"Will it be possible for me to see Catherine?" asked the visitor.

Catherine's father shook his head. "She asked me to tell you, when you came, that it would be better not to."

Clyde's face became white and strained. "It's natural for her to feel that way," he said, "after what happened. But," regaining some of his natural independence and determination, "I must see her, Mr. McGregor. There are things we must talk about. Surely you agree that we can't just—just drop things without a word being said."

"You've made a pretty mess, young man, of several lives. Don't you think it would be better not to do anything more?"

"No! I don't intend to leave things like this. Do you know that Catherine and I were going to Hudson to be married? I have the license right here in my pocket."

"I know what you were planning to do."

"I'm entitled to a chance to talk to her, even if—if I did involve her in this accident and ruin her life. Mr. McGregor, I *must* see her! I'll keep on coming back until I do. I'll never stop trying, so you might as well give in now."

McGregor asked sharply, "Do you know she can never marry now?"

"I've been told so. But I don't believe it."

"Have you talked to the doctor?"

"I've talked to him several times. He's right, I suppose, in what he says."

The merchant's nostrils flared with a sudden surge of anger. "Young man, you compel me to tell you that my daughter—in spite of everything—still feels a great liking for you. As she can never marry you, it would be painful for her in the extreme if you saw her. Can't you get it through that head of yours that she must be spared?"

The young man's eyes had a new light in them. "She still likes me! Mr. McGregor, if that's true, it doesn't matter about—what the doctor says. I want to marry Catherine."

"Knowing that she can never have children?"

"That doesn't seem important at all. The important thing is that I love her."

The older man's attitude began to show a change at this stage. He seemed puzzled and he frowned at his companion. It was clear, however, that the antagonism in him was lessening. "I've always thought of you as—well, we won't go into that. Can it be I was wrong?" He paused. "It's all very well to talk, young man. You mean it now, but what of the future? Will you be content with this kind of a marriage ten years from now, or twenty? Every man wants a family. And, to speak frankly, the flesh is strong and the devil knows how to make the most of it."

There was a moment of silence. "All I can say, Mr. McGregor, is that I love your daughter and I don't think I'll change as the years go on."

"There's another side to this." The merchant's manner underwent a second change, becoming businesslike and incisive. "You can't support Catherine as things stand. Have you thought of that?"

"Of course. There may be ways to earn more. I've thought of finding some night work to do."

"But it's not only money. She can't be left alone. For a long time she's going to need a nurse. Perhaps as long as she lives. Will you be able to afford one?"

Clyde shook his head. "No. That's an impossibility."

"She must stay here. No other way is possible. The poor child must have the chance for such comfort as can be supplied. It comes down to this: If you married her you would have to live here. How do you like that idea?"

The boy said nothing.

"You don't like the idea. Well, I can't blame you. We've been on pretty bad terms, and this terrible business hasn't helped matters any. I don't think I'll ever be able to like you now, and it wouldn't be easy having you in the house. It would be just as hard for you. Man, man, I'm difficult to live with! At breakfast I'm unbearable. I've always been as I am now, and there's no chance that I'll change. I don't want to change; I'm satisfied with myself as I am." He paused and changed his position on the sofa with a manner of discomfort. "If you want to work in the store and learn something of the business I'll pay you more than you're getting. If you marry Catherine you'll have to move in and make the best of things. It's the only way. Unless you would prefer to get up now and walk out of the house. And never come back."

Clyde hesitated no more than a moment. "Yes," he said. A long pause ensued, and then he added: "I always hoped that Catherine and I would have our own home. That's impossible now. I think, Mr. McGregor, that under the circumstances you are being generous and—and kind. I would be a poor specimen of a man if I didn't agree that this is the best way. Provided Catherine also agrees."

Lockie McGregor got to his feet. "I'll speak to her. This may change the complexion of things. I don't know."

It was some time before he returned. He looked grave, but there was no longer any trace of antagonism in his manner. "She wants to see you. I don't know what she's going to say. But it's in her hands now, poor child, and whatever she wants will be what I want." Then he actually achieved a brief and wintery smile. "She took some time to arrange her hair and to get a nice bed jacket. You're to go up now."

While waiting to hear the outcome of the talk between the two young people the unhappy father went to his study. Shoving a hand behind the volumes of poetry, he produced a bottle of whisky. Holding it up in front of him, he said to himself: "One drink a year—on New Year's Day—is enough for any man. Except for emergencies, of course. This is an emergency."

He had finished his drink and replaced the bottle before the voice of Clyde Carson reached him from the head of the stairs. He walked to the foot and glanced up. "Well, what's it to be?"

"I told you I brought the license with me. Catherine and I see no reason for any delay at all."

McGregor was startled. "Do you mean now? This evening? Man, is there need for such a great hurry as this?"

"Isn't now as good as tomorrow or next week?"

The merchant frowned uncertainly. "I suppose it can be done. I could send for a minister to come in. Have you talked about who you want?"

"Anyone at all," answered the prospective groom. Then, as a sudden and furious afterthought, "Anyone but the Reverend Elmore Nash."

Chapter XIII

I

The eight years which remained of the period known as the Gay Nineties passed slowly and pleasantly. A multitude of things happen in the course of any eight years. Several hundred million people are born and almost as many die, and it goes without saying that Balfour had its full share of births and deaths. Magistrate Jenkinson was found in front of his fireplace one afternoon, his cane gripped in his rigid right hand. Pete Work's father went on one final carouse and died in the stupor which followed it, and on Pete's shoulders descended the responsibility of the household. Daniel Cape finally succumbed to his arthritic disorders, leaving a long memorandum for William Pitt Milner on how he should conduct the newspaper in the years stretching ahead. The latter returned it to the executors of the estate, marking it *Unopened*.

The dynamic and irascible owner of the *Star* was seldom in his office during the latter stages of these years. He was known to have become a power in the Liberal party and he spent much of his time in Ottawa and Toronto. Everyone in Balfour was convinced that sooner or later he would be appointed to the Senate or to the lieutenant-governorship of the province. He liked to have his daughter with him on his travels, and so Tony's education was continuously interrupted.

The fortunes of Langley Craven had risen considerably during these important years. His little factory had been enlarged twice, and there was talk now of building an entirely new plant somewhere in the suburbs. This was partly due to the energy and vigilance of his sister, but the greatest impetus, strangely enough, had been supplied by Langley himself. He it was who sensed the great changes which were coming about and he secured personally the contract for the making of gramophone cases and cabinets, which became the backbone of the business. Langley was now talking of organizing a new company for the manufacture of horseless carriages, and investors were listening to him with attention.

While the star of Langley rose, that of Tanner Craven remained, at the best, stationary. The spectacular advances of his earlier years were not being repeated. It was whispered, in fact, that he had suffered some reverses out of town. At any rate, he discharged his butler and sold some lots off the back of his property. Mrs. Craven continued to call at the paymaster's office every Saturday morning for the envelope she had been receiving each week since her marriage and which contained the money allotted her for the upkeep of the household. She was still careful to speak in friendly tones with Daniel Boucher, the cashier, and to inquire after the health of his ailing wife. It was rumored, however, that the amount in the envelope had been reduced from twenty-five dollars to twenty and that she had not been pleased at all.

Eight years are almost like eight centuries in the lives of the adolescent, and so it becomes necessary to skim briefly over the things which befell the group of young people with whom this story is chiefly concerned. A few incidents require to be told because they influenced later events, but the rest must be taken for granted. It may be assumed that the usual triumphs were enjoyed and the usual tribulations suffered and that all the group grew both in stature and in character, as everyone does in that period of bitterest travail, the teens.

The incidents which require to be told will be set down as briefly as possible.

2

There was, first of all, the arrival of the important letter from the lawyer in England. Ludar was twelve then and had passed his entrance examinations in readiness to start at the Collegiate when the fall term opened. There had been many letters from the lawyer during the intervening years, and his demands had finally exhausted all the funds without advancing matters appreciably. This one came in the afternoon mail, and when William returned from work he found Tilly in what might have been called her Lady Macbeth mood. Her face was grim, and before he had closed the door behind him she shoved the letter into his hand, saying in a sepulchral voice: "There! Read that! Here's a nice state of affairs. You and your Lawyer Ratticlipper."

When William finished reading the letter he knew that the effort to achieve a fortune for Ludar had come to a dead end. It contained no information about the boy's family or the course of the negotiations. It declared that if he, William Christian, could raise the sum of one thousand dollars a settlement would be made at once. If this sum were not forthcoming he, the writer, weary of receiving no more than enough to pay his expenses, would be compelled to relinquish the case. It was brief and very much to the point.

It was a silent meal. William said as he sat down, "I'm certain the man's a crook. He mustn't get another red cent." He began to eat his headcheese and bread and butter and said nothing more.

Ludar's reactions were mixed. He still had some of the fear which had filled him years before when the first letter came. He did not want circumstances to take him away from his friends and put him with strangers. Secretly, also, he was convinced that it would be better not to lift the curtain which shrouded the past. Aunt Callie had once said to him that he "had no reason to be proud of anything," and his memory had refused to relinquish the words. He feared the past.

With this doubt went the natural desire of a young mind, which was being nourished on quite prodigious reading, to solve the mystery. He wanted to know who he was, even though the knowledge might destroy his dreams. If there happened to be property in which he should share, he did not want to be excluded. It would be nice to own a pony and a bicycle and go to an English boarding school (a desire which had grown into an intense longing with his eager reading of stories about them) and to be able to supply all the money Uncle Billy needed for his experiments.

The hardest-hit of the three was Tilly. Her dreams of a prosperous future had been material ones. She had known most definitely what she wanted and she had conceived of the new life in terms of daily comforts about which there had been nothing vague. Even the servant girl she was going to have had become a real person in her mind. Eliza, she called her, and Eliza was going to be hard to give up. Once during the meal she let her temper get the better of her and she flared out at the mysterious Mr. Ratticlipper. "The man's a thief! I knew it from the start." If this were true it was unfortunate she had not said anything about her conviction. "Does he think we're made of gold over here?" she demanded. "How does he think we can get a thousand dollars?"

William looked up at that. "We're not going to try," he said.

Tilly was startled by the determined note in his voice. "But—but if we don't, we'll—we'll never hear anything more from him," she protested. "All that money we've sent will be wasted."

William shoved away his plate. He was looking thoroughly unhappy. "I'm beginning to think he never found out about the family at all, that he's been making the whole thing up. Just stringing us along."

"Oh *no!* How can you say such a thing?" Tilly looked belligerently at her husband. "If what you think is the truth, then the people who've put in their money were cheated."

"That," said William, keeping his head lowered glumly, "is what I'm worried about right now."

Tilly rose to clear the table. There was a suggestion of martyrdom about her. "Must I get up and do the dishes after every meal as long as I live?" she demanded bitterly. After a few moments devoted to a fast and furious attention to her task she walked back to the table with a damp towel in one hand. She said to her husband, waving the towel excitedly in the air: "I'm not going to give in as easy as that. I think the old cheat knows the family are ready to make a settlement and he's just trying to get as much out of it for himself as he can. The skinflint! I think we ought to keep up the fight, that's what I think."

"Fight?" William indulged in a snort. "What can we fight with? Do you think the people of the town will raise a thousand dollars for a wild scheme like this?"

"I don't care." Tilly returned to her work. She splashed the dishes in soapy water with an impatient hand. "I tell you, Billy Christian, I don't care. We shouldn't give in so easy."

Ludar got up from the table and went to his bedroom. On the top of the bureau was a thick pile of foolscap sheets covered with writing in a large and sprawly hand. He had begun a novel a month or so before and, as Uncle Billy put it, had been "pitching into it like sixty" ever since. He glanced at the top page, which carried the title:

MAURICE OF NASSAU
A Novel
by
LUDAR———

He had left the space for his real name, and now he might never know what it was. The sight of the manuscript, nevertheless, seemed to compensate him for whatever disappointment he had felt. If he could get his novel finished (he refused to consider the possibility that it might not be published) he would make enough money to buy the pony and bicycle, to supply Uncle Billy with the sinews of war, to provide Aunt Tilly with some of the comforts for which she longed. One thing he saw quite clearly, that it would be better to acquire money through his own efforts than by inheritance. On the whole, therefore, he was not particularly disturbed by the contents of the letter.

A bank messenger arrived at the shop one morning soon thereafter and left a message for Mr. William Christian. His presence was requested that noon in the office of Mr. Albert Travers. This would mean going without dinner, but William did not care. He asked one of his fellow workers, who lived in his neighborhood, to explain matters to Tilly and betook himself downtown.

The Bank of the United Provinces occupied a space on Holbrook Street which had once been a butcher shop, and Mr. Travers himself had his office where the icebox had stood; a fact which might have accounted for the frostiness of his attitude in dealing with requests for loans. He was eating his lunch out of a metal food box when William was shown in, and keeping a keen eye through the glass panels on everything which went forward in the bank.

Speaking with some difficulty as he consumed a sandwich of hard-boiled egg, the banker explained his summons. "Your wife, Mr. Christian. A fine woman. I admire her. She brings the money in to deposit at the very same minute every Monday morning——"

So this was the bank Tilly used for the custody of her principal. William had never been told. He had stopped asking questions because of his wife's tendency to become "hoity-toity" when he did.

"—and up to this morning she had never withdrawn as much as a dollar. I don't mind telling you, Mr. Christian, that I watched the growth of the account with pleasure. It seemed to me a model account." He stopped, turned his head toward the rear, and called in an impatient voice: "Evangeline! Is it possible my tea isn't ready yet?"

His secretary came in with frantic haste, carrying a cup on a tray. She was middle-aged and the very picture of brisk efficiency.

Mr. Travers took the cup and began to sip his tea. "But, Mr. Christian, I'm taking it on myself to discuss with you personally this decision to send the thousand dollars this rascally lawyer demands——"

It was a good thing that the banker's frowning attention had been drawn to something happening outside. Otherwise he would have realized from the look of amazement which had spread over William's face that this had been news to him. The latter recovered himself by the time Travers resumed the discussion.

"The draft has been drawn up, but I felt I should give you a word of warning before sending it. In my opinion, sir, you are doing a very foolish thing."

William was at a loss as to how this situation should be handled. He had been surprised, shocked, and outraged. What right had Tilly to make such a decision without saying a word to him? But inasmuch as she had done it, what could he say to this banker? After a moment's silence he remarked: "I guess we had better talk it over again. You see, Mr. Travers, my wife has a better head for business than I have, and she—she handles everything about money." He wanted to ask how much would be left in the account if the thousand dollars were withdrawn but knew that to do so would be to reveal that he was kept in the dark about their finances.

Travers leaned back in his chair and frowned judicially. "As your banker, I must advise against doing this. You know, of course, that not one who had put up money before was willing to risk as much as a dollar again? They were unanimous in believing the rascal was gouging us for whatever he could get. Mr. Christian, have you ever figured how long it took you to save this thousand dollars?"

"No," said William. "But it was a long time."

"It represents years of patient saving. I said this to your wife when she was in, but she seemed in a curious state of mind and kept saying she didn't care. It made me wonder, Christian, if this decision was yours and she herself was against it."

"No," said William hastily. "That's not the way of it."

"Let me suggest this—that you authorize me to hold this draft up and that in the meantime you have a serious talk with Mrs. Christian. I most earnestly hope you'll let me know as early as possible that you've changed your minds."

Tilly became instantly belligerent and slightly hysterical when William spoke to her about it that evening. "Well!" she said, giving her head a toss. "That Mr. Travers had a lot to do! What right had he to go behind my back like this? I've a notion to transfer my principal to another bank."

"What I don't understand," said William, "is how you could do this without talking it over first."

She answered fiercely, "I saved the money, didn't I? If it wasn't for me there wouldn't be a cent in the bank. And you know it."

"I'm not a spendthrift," answered William mildly. "And, after all, I—I did earn it. I should know what's being done with it."

Tilly, who was setting the supper table, began to thump each dish down as she brought it from the pantry. "You've never interfered before. You know you have no more sense about money matters than a billy goat. Why do you have to bother me now when I'm almost frantic about the—the whole thing?"

"Because I think this is a mistake. A great mistake. So does Mr. Travers."

"Mr. Travers!" declared Tilly furiously. "I'll have something to say to your Mr. Travers!" She took up the bread knife and began to saw off slices as though she were operating on the banker. "Am I never to have a proper breadboard? Don't you want comforts in this home? Must I go on slaving the rest of my life?"

William had walked to a window and was looking out, his invariable custom when involved in arguments with her. "I want you to have all the comforts in the world. But this isn't how to get them, throwing our money away." He suddenly brought his courage up to the point of asking, "How much would this leave in the account?"

Tilly pretended not to hear. She went to the pantry and rattled dishes about, returning in a few moments with a platter of cookies. William repeated his question.

"Why do you ask that?" she demanded.

"Because I want to know what kind of a position we'd be left in if this money was lost."

"I'm not quite sure." She busied herself at the table. "See here, Billy Christian, why are you asking all these questions? Don't you trust me?"

"Of course I do. But surely I should know——"

"I tell you, I'm not certain of the exact amount." Her voice had risen. She walked over and faced him, her hands on her hips. "Billy, we're going to send that money. Do you hear me? I haven't been able to sleep nights, thinking of the great chance we're throwing away. I want the boy to have his rights. I want to be independent and comfortable. I—I can't bring myself to let the chance go. The thought of it drives me crazy."

"I'm against it," said William.

His opposition had no effect save to harden her resolution. "I know I'm right. I don't care what you or anyone says about it. I want to send the money."

So William paid another visit to the banker and informed him that the money was to be sent after all. Albert Travers looked at him with an icy eye and said: "Now I'm sure, Christian, that you *are* responsible for this. Mark my words, you'll live to regret it."

He was right: they lived to regret it. For two months they existed in a state of anticipation, waiting for further word from the lawyer. William was of too sanguine a nature to remain completely pessimistic and sometimes he would join with Tilly in the pleasant pastime of spending the money in advance. Ludar was giving every spare moment to the novel, but he had plenty of ideas also for the use of his patrimony. But when the second month faded into the past and no answer had been received to any of the letters William had written, the atmosphere at 138 Wilson Street changed. It became a place of gloom, of sighs and averted looks, of meals eaten without a word being spoken and evenings of complete silence. Tilly again was taking it harder than the others. The sharpness of her disappointment had driven her within herself and, as her mental resources were limited to begin with, she had nothing to ameliorate the shock. She became morose. The house began to show a lack of housewifely attention, and the meals were sometimes scanty and unappetizing.

One evening toward the close of summer, when leaves rustled on the sidewalks as people passed, she got out her array of little black books and sat down to work with a sheet of paper and a pencil. After an hour's steady application she called out in a sharp voice:

"Billy!"

William emerged from his tiny workshop. He had a plane in one hand, and there were shavings on his sleeves.

"I've been figuring things out," said Tilly. "It took us over six years to save that thousand dollars!"

William blew the dust from his hands and wrists. "I knew it took a long time," he said. "Well, it's gone. There's no sense taking it so hard, Tilly."

She looked resentful of this easy acceptance. "All those years I scrimped and saved and went without things," she said. Then she repeated a performance which has already been noted. She ran with furious haste from window to window, shutting them down tight. When she returned to her rocking chair in the kitchen she was almost breathless.

"I was afraid my feelings would get the better of me," she said, "and I didn't want the neighbors to hear."

The next step followed, the throwing of her apron up over her head, after which she went into a positive paroxysm of weeping. William watched her with a helpless expression for several moments. Finally, knowing there was nothing he could do or say to comfort her, he turned and went back to his work.

Ludar had been out of the house during the scene just described. That night he was wakened from sleep by a steady murmur of voices. He sat up in bed at once. A light was showing under his door, and the voices came from the living room.

"I've told you this before," he heard William say. "You're a holder-againster, Tilly. You get something in your mind and you won't let go and then you get bitter. You hold things against me."

"I don't!" Tilly sounded indignant. "What do I hold against you?"

"You hold it against me that I married you. I don't know why, because you were willing enough at the time. I suppose it's because I'm a failure. And now you're holding against me that we lost this money it took six years to save. I was against sending it, but you seem to forget that. You were the one who was dead set on it."

"You forget," said Tilly, "that if you hadn't brought this boy home the money would never have been lost."

Ludar did not permit himself to listen to anything more. He drew the covers up over his head and lay without moving. His world had crashed about him. His presence here, in what he had come to think was his home, was resented. Aunt Tilly called him *this boy,* as though she considered him in a class with Skeek, who had been a stray also. He had sensed scorn and dislike in her voice. Perhaps Uncle Billy felt the same way but was too kind to let it show. Well, there was only one thing for him to do. He must go away.

After thinking things over in a state of unhappy desperation, he decided that the solution would be to get a job somewhere. It must not be in Balfour, as that would create talk. No, he must disappear, go somewhere else and never come back. Hudson was a large industrial city and not far away. Twenty-five miles, perhaps. He would go there. He was a good walker and would be able to cover that distance on foot. . . .

He waited half an hour and then sat up in bed again. The voices had stopped, and there was no light showing under the bottom of the door. He dressed with the greatest care and walked to the side door, carrying his boots in his hands.

It was a good thing that the lamps were blazing all along the streets, for there was not a star to be seen. He stopped a block below the house to consider his course. There was a Hudson Road, he knew. He had never set foot on it because it was a continuation eastward of Holbrook Street and ran through territory unknown to him. However, it must be the way to the city he had set as his destination, so he turned his feet in that direction.

It was more than twenty-four hours later that he returned. He came back on blistered and painful feet. To return was an acknowledgment of defeat; no other course, however, had been open to him. After dragging himself wearily from factory to factory and being looked over critically and told there was nothing for him, he had realized he must swallow his pride and return home. The march back over the seemingly endless stretch of gravel roads, between high dark rows of trees and past an occasional country church with a cemetery beside it, had been infinitely longer than on the first coverage, even though there was a hint of moon to light the way and keep his feet from stumbling. The two nights on the road had been similar and unique in several respects: no rig passed him, no other pedestrian was afoot on the road, no dog barked as he went by. There was something ghostly and unreal about it.

He crept up Wilson Street in such extreme pain that he had to stop every few steps and lean against fences or even sit down on the pavement. It was as dark and silent here as on the country roads. When he came within sight of Number 138, however, he stopped, finding it hard to believe his eyes. Light blazed from the windows.

"Are they waiting up for me?" he asked himself with a sudden revival of hope and relief.

For the last few hours he had been trying to decide what he would do when he reached the house, assuming that it would be in darkness and locked for the night. Would he knock loudly enough to waken them and say, "Please, may I come back?" or should he go out to the insecure old hammock under the crab apple tree and wait there until morning? His decision, as far as he was able to arrive at any conclusion in his numbed state of mind, had been in favor of the latter course.

But now things were different. They were awake and waiting. Perhaps they really wanted him back. He began to pause a shorter time between steps and to lift his feet faster.

When he came close to the house he heard William say: "Here he is. I know his step." There was a rush of feet, the front door flew open, and William appeared in the stream of cheerful, welcome light which poured out.

Ludar decided to say as little as possible about what he had been through and not mention that he had been the whole day without food or that each foot was a mass of blisters. He must cause as little trouble as possible. But now the twelve feet of walk to the front door looked as long as the endless black road from city to city. He was sure he would never get to the end of it. He tried to take one step, and his legs crumpled under him.

Nothing was said, no questions asked. William picked him up and carried him into the house. Tilly had a tub of hot water ready in quick order, and he was propped up in a chair while his feet were given a thorough steaming. Then a plate of scrambled eggs and toast materialized, and for the first time in his life he ate with a truly prodigious appetite.

He volunteered one piece of information only, selecting it because he thought it would interest them. "I saw a ghost in a country cemetery," he said. "It was white and it moved. I let out an awful holler."

"I'll bet you ran like sixty," said William.

"No. That was the awful part of it. I couldn't run. I wanted to, but my legs were too stiff. I just hobbled and I thought every second the ghost would have me." After a moment he added: "Of course there are no ghosts, but it looked mighty real."

"You're due to sleep the clock around," declared William, "and I'm going to take you straight to bed now. I'm going to ask just one question. Where have you been?"

"Hudson," answered the boy, whose head was now ready to topple over on his shoulder.

"Hudson? Why, that's twenty-five miles. How did you get there?"

"I walked, Uncle Billy."

"You walked! The whole distance?"

Ludar nodded, too weary to find words.

"And you walked back?"

Another nod.

William forgot his resolution not to bother the boy with questions. "Why did you go to Hudson, Ludar?"

"I wanted to get a job."

"Do you mean you went around the factories when you got there? That you've been at least sixty miles since last night? Come, come! Isn't this one of those stories you invent in that funny young head of yours? Aren't you just making it up?"

Ludar realized he must satisfy his guardian's curiosity before he could hope for the bliss of uninterrupted sleep. He pulled his faculties together and named five or six shops he had visited in his quest for employment. William knew them all and nodded, convinced.

Tilly, who had been hovering about and making efforts to be of help, had now gone to the kitchen. The boy leaned toward William and whispered, "Does—does Aunt Tilly want me to go away?"

"I should say *not!* She's been in tears ever since we found you were gone. She's been frantic."

"But"—Ludar hesitated before saying anything more—"doesn't she hold it against me because it took six years to save the money you sent to the lawyer?"

William was silent for several moments. "So that was it," he said. "You heard us talking. I suspected it." He reached out and patted the boy's head. "Ludar, people have a habit of saying things they don't mean. Especially when they get into arguments. They like to hurt one another. But you just get this into your head, young fellow. Your aunt Tilly loves you." He reached down and gathered the boy up in his arms. "And now you're off to bed. I don't expect to see you at all tomorrow."

He was right. The boy slept for twenty-four hours without moving a muscle.

3

A year later.

The three friends were walking home from the Collegiate a few days after the opening of the fall term. They had gone into long pants. Joe and Arthur were very neat in gray tweeds, with belted coats and caps of the same material. They carried swagger sticks, and Arthur had a wrist watch of which he was very proud. Ludar's trousers did not fit him well, partly because of their cheapness, perhaps more because he continued to grow so fast.

Their books under their arms, they were walking slowly, with serious faces. Life takes on a new dignity when the high school phase is reached. They were talking about books and discussing R. L. S. and Dr. Doyle and Anthony Hope with considerable gravity.

"Norman's gone to St. Paul's," announced Joe, who was perhaps a shade less serious than the others and apt to jibe at too much deep discussion. "High school's not good enough for la-de da Norman. And where do you suppose he spent most of the summer? This will make the eyes fairly pop out of your heads, fellows. He was at a special camp. With a physical and boxing instructor. An ex-boxer from England."

"Now what in Sam Hill——" began Arthur, stopping and looking anxiously at Joe.

"It means," said the latter easily, "that his old man wants him to fight me again. He's been getting ready."

"Say, this may be serious." Arthur, clearly, was disturbed. "If he has a real boxer to show him the ropes he may get himself too good for you. This boxer will teach him all kinds of tricks. Boxing's a science, you know. You're stronger than Norman, but just look what Gentleman Jim Corbett did to John L. Sullivan."

Joe did not seem to care. "If it comes to a fight I'll rough it up with him, and he won't have a chance to use any fancy tricks."

Arthur shook his head. "You've got the wrong attitude, Joe. You're overconfident. I want to tell you, overconfidence is dangerous. It loses lots of battles. Napoleon was overconfident at Waterloo."

"I'm not afraid of Norman. I never was and I never will be. The way I feel about him, I could lick him with one fist tied behind my back."

"There you go. Overconfidence. You'll be like John L. Sullivan if you don't watch out."

"Rubber!" said Joe.

Rubber! A pause is necessary to say something about the widespread use of this word during the nineties. Never before nor since has anything in the way of slang equaled it in popularity. It could be used on any occasion and for any purpose. It crept into editorials, it fell from the lips of politicians, it even found its way into sermons. The chief users were, of course, boys. With them it became everything, the cement which bound talk together. It was used to denote any kind of feeling—amusement, friendliness, warning, scorn, surprise, defiance, derision. A boy who had "Rubber!" hurled in his teeth felt he had been besmirched unless he retorted at once with the accepted rejoinder, "Stretch it! Throw it up and ketch it!" That made everything right. The use of the word went on and on, year after year, and finally settled down to its place in the common tongue with one form, the now immortal "rubberneck."

Ludar had been doing some serious thinking. "There's nothing else to it, Joe," he said suddenly. "*You'll* have to go into training. While you two gabbled about it I was giving this situation some thought. You'll have to train seriously like that young doctor who beat the Croxley master. You know, in that story by Dr. Doyle."

"Say, that's an idea." Joe nodded enthusiastically. "If Norman can get himself ready for a scrap, I guess I can too."

"How about letting your dad know?" asked Ludar. "Perhaps he would hire a trainer for you. He'll want you to win."

Joe shook his head at once. "That wouldn't do. Dad might be willing, but I know my mother would raise Ned if she heard about it. She's against fighting."

"Say, this is serious," declared Arthur. "We can't let Norman beat you, Joe, and yet what can we do if your parents won't help you?"

"We can take the whole business on ourselves," said Ludar. "Look. We'll make the barn our training headquarters. We'll have to be pretty slick about it and not let anyone else know. I guess, Arthur, you'll have to be sparring partner. I'll be trainer and handle the water bucket and sponges."

"And you'll be my manager." Joe was emphatic about this.

Ludar began to take his responsibilities seriously at once. "What will we do about boxing gloves and a punching bag? They cost money. There's one easy way. We could charge admission like the champions do. But if we did that your dad and mother would soon find out. They would get pretty curious if there were crowds in the barn each day, and what with yelling and cheering all the time."

"I can think up something to tell 'em," said Joe confidently. It was clear he liked the idea of money being paid to watch him train. "What would we charge, Ludar?"

"Five cents. I figure we might take in as much as a dollar a day. All the fellows at the Collegiate will want to be in on it. But, listen, your parents would soon know what's going on. Perhaps we could find an empty lot somewhere instead."

Joe refused to consider this. "How could we make 'em pay if we used an open lot? They'd just come pouring in and give us the laugh if we asked for nickels. It's just got to be the barn. I can talk the old folks into believing anything, but I don't know about that young Meg. She'll find us out."

"Let's try it anyway," said Arthur. He had picked up a new word during the summer and liked to use it on all occasions. "I think it's a spiff idea."

Ludar had some further doubts to express before the matter could be settled. "Look here, Joe, this isn't going to be easy for you. There'll be a diet to go on, you know."

"Hey!" cried Joe in alarm. "Do you mean I'll have to stop eating pie and ice cream and candy and all the things I like?"

"Probably. All the champions do. I've read about it. They eat a lot of beefsteak and they let their whiskers grow. I don't suppose you'll begin to grow a beard in time, but you can get plenty of steak and roast beef. Listen, Joe. You'll have to train so hard that you'll be able to fight twenty rounds if necessary. Your muscles must be developed until they're as hard as whipcord. Your stomach must be flat, but your arms and shoulders and your chest must bulge like Sandow's. You'll be the perfect specimen of manhood when we're through training you."

This picture was so enticing that Joe withdrew all objections. He cried, "Spiff!" The trio threw dignity to the winds and ran all the way to the Craven house. Not waiting to waste time in opening the gate, they climbed over the wall and rushed pell-mell to the barn. Here they started to clear out a space for the ring.

When Langley Craven reached home that evening he went at once to his wife's bedroom on the second floor, taking the stairs two at a time. He had a bulky newspaper tucked under one arm.

Mrs. Craven had the best apartment in the house, for it combined the functions of bedroom and boudoir. The ceiling was unusually high, an effect heightened by mauve-colored hangings of velvet with corded valances at each of the four windows. The room contained a huge walnut bed with back carved to represent the story of the Sleeping Beauty, in which, incongruously, Joseph Norman Craven himself had slept for twenty years. There was a chaise longue also, commodious and luxurious, with mauve velvet covering and pillows.

Mrs. Craven was dressing when she heard his quick step on the stair. Throwing a robe of regence, a rich silk with ribbed satin face, over her accordion-pleated petticoat, she went to the door to meet him.

He engulfed her in his arms, murmured, "A most seductive perfume for the mother of two children to be using!" and then drew her under the chandelier. Taking the newspaper from under his arm, he spread it out triumphantly. "Last Sunday's issue of the New York *Herald*," he said. "In it you will find information about all the plays in New York, the opera, the ballet, the restaurants. The ads are full of information about the latest styles. Read it carefully, Mrs. Langley Craven. You'll need to, because we're going to New York, you and I. In two days, moreover. The tickets are bought, a telegram has been sent, reserving rooms at the Waldorf." He leaned over and kissed her enthusiastically. "It's going to be like old times, my fair one, when we went to New York and Europe every summer. We'll see everything and we'll spend money like mad. You'll come home with a wardrobe like the trousseau of a princess." He gathered up the newspaper and waved it excitedly above his head.

Mrs. Craven's eyes were sparkling. "I can't tell you, Langley dear, how much I've longed for a trip to New York. I didn't say anything because I was afraid we couldn't afford it yet. O-oh, what a shopping spree I'll have!"

Her husband nodded. "The more you spend, the better I'll like it. I'll have you know, me proud beauty, that it was a group of bankers who suggested the trip. Harvey Twitty of Toronto himself was the guiding spirit. They want me to look into all these latest developments, to talk to manufacturers, and most particularly to attend the automobile show at Madison Square Garden. My love, anyone with a keen ear today can hear the ice of the *fin de siècle* cracking and heaving and getting ready for the breakup. Everything's going to change. Life is going to be faster, more complex, more exciting. But my great concern is with the horseless carriage. I want to make them for thousands of people. For millions. Let who will be the King of Steel. Let me be the Autocrat of the Automobile!"

His wife began to waltz around the room, the moreen in her petticoat rustling with each move. Then suddenly her mood changed. She stopped and seated herself on the chaise longue with her hands lying limply in her lap. She looked at him with grave eyes.

"Langley, this means—doesn't it?—that we're out of trouble at last? That we've come through the valley of the shadow, or whatever you call it?"

His gravity now equaled hers. "Yes, my dear, it means that. I'm prosperous again. A little better than prosperous, in fact. I'm a rich man. I have a conviction, Pauline, that in the course of time I'm going to become a very wealthy man, and in my own right."

He seated himself beside her and put an arm around her shoulders. "I realized for the first time today how much I've changed. I played golf and had a neat seventy-six. Even if we have no water holes and the traps are not too obstructive, that's good golf. I heard George Milton say to his wife and that broad-beamed guest of theirs, 'Lang Craven could be amateur champion of Canada if he tried hard enough.' And I wasn't particularly interested. Do you know why? Because I had a booklet in my pocket on automobile engines and I was thinking about that. I was thinking particularly of what it said about a new kind of engine—a sleeve-valve type." He turned and looked down the full length of his high-bridged nose at her. "I seem, my dear, to be turning into a businessman at last."

His wife said nothing for several moments. Then she patted the hand resting on her shoulder. "Do you know, Langley, I'm sorry to hear you say that. I liked you as you were."

"Even if we had gone on getting poorer and poorer?"

"Even if we had gone on getting poorer and poorer."

He got to his feet and shoved the newspaper back into his pocket. "What would you say if I told you I enjoy being a good businessman? I do. I'm tickled about it. I'm going downstairs now to finish reading about engines and I'm going to enjoy it."

At noon on the day following the departure of the Langley Cravens, Arthur fell into step with Ludar as they started home. He had a worried frown on his face. "Say, young Joe didn't lose any time in spreading the news," he said. "It's all over the place. I thought we were to keep it quiet for a while until he had a chance to catch up on Norman. Now he's gone and given the show away."

Ludar considered the matter carefully. "We'll have to change our plans to meet the new situation," he said. "That's what all great generals do. They improvise. We'll have to improvise too. Say, that's a good word, isn't it, Arthur?"

"Spiff," said Arthur.

As a result of the improvising of the trio they walked to school together the next day, Joe in the center wearing a heavy turtle-necked sweater (which he would have to discard before going in to classes) and carrying weights in each hand. They had a tail like a comet, made up of boys of all sizes who had fallen into line behind them as they came along. Already the need to get Joe in good shape for the impending fight had become a school responsibility. No softy who went to a private school could be allowed to beat a B.C.I. man!

Crossing the small playground between the school building and the gymnasium, Ludar drew a bottle from his pocket and held it to Joe's lips. The latter took a deep swig. Ludar held up the bottle to see how much had been consumed, nodded with satisfaction, and put it in Joe's back pants pocket. "Every hour, mind you," he said. "And don't let any of the teachers catch you at it. We don't want 'em finding out what's in it."

The whole school seemed to be swarming around them. "What *is* in it?" they wanted to know. Ludar looked at them with scorn. "Aw, go away back and sit down! Do you suppose we're going to let anyone widdle our training secrets out of us?" The bottle, as it happened, contained water with a small injection of licorice to give it color.

A schedule was drawn up shortly thereafter, and this was shown around with great stealth, although such precautions had become useless now. It read:

Mondays:	4:30 to 5:00, rope skipping and shadow boxing. 5:00 to 5:30, sparring with partner. Admission five cents. No credit. Don't ask.
Tuesdays:	Secret training. No one admitted. It is hoped the men of the B.C.I. will respect the need for such precautions.
Wednesdays:	4:30 to 5:00, limbering-up exercises. 5:00 to 5:30, three stiff rounds with sparring partner. Anyone desiring to contribute more than the admission fee will be regarded as an example of true school spirit.
Thursdays:	Road work. Routes to be kept a secret. Again we plead for the co-operation of our supporters.
Fridays:	Boxing with volunteers. Those selected to go a round must pay double admission.

The school talked of nothing else. Even the Thanksgiving prom, which was designed to introduce the new boys and girls into the social life of the institution, became a matter of decidedly secondary importance. Much concern was felt when Ecclefechan, the English teacher—so called because his name happened to be Thomas Carlyle—made a reference in class to what was going on. He had set the first form the task of writing something on Oliver Cromwell. The papers were lying on his desk in a neat pile, and he glanced down at them with an expression which seemed filled with loathing.

"Oliver Cromwell, Old Noll, the Savior of Liberty, the greatest of Englishmen," he intoned. He lifted the paper on top of the pile. "Here we have his story told in exactly one hundred and thirty-two words. And such words! Here is another, one hundred and eleven words, a sample of condensation carried to the point of evaporation. And this one—*this!*—a fossilized bone dried in the sun, when what I asked from you was a living, breathing portrait of a great man—ninety-two words!" He leaned across his desk and with an accusing eye surveyed the half of the room occupied by boys. "I am of the opinion that all the interest and enthusiasm of this class is going into *certain fistic activities*."

The training sessions had been in one respect a complete success. Already enough had been taken in to buy two pairs of boxing gloves, and soon there would be sufficient in the purse, which Ludar carried, to invest in a punching bag. As manager of all these activities, Ludar felt concern one afternoon when he noticed Meg at the door of the barn, which had by some mistake been left open. Her eyes were round with wonder as she watched Joe and Arthur, stripped to the waist inside a roped square, punching, jabbing, and feinting—and, it must be said, stumbling and making wild swings and falling into awkward clinches—while more than a dozen spectators watched.

Ludar hurried over to her. "No one allowed in, Short-for-Margaret," he said, trying to interpose himself between her and the proceedings in the ring.

Meg gave her head an indignant toss. "Don't call me that silly name, *if* you please. It's fit only for a child. You ought to know I'm much too old for it now."

It was true that she was growing up rapidly. Her eyes seemed almost level with Ludar's own. He was surprised to see that, whereas they had once been unmistakably blue, they had all of a sudden become a sparkling green.

"Well, Meg——"

"And I won't be called Meg. It makes me sound like the daughter of a washerwoman."

"Then what do you want to be called?"

"Just by my name. That's all." She treated him to a long and superior stare. "I have a good name, as it happens, little boy. Margaret. Is it asking too much that you use it?" Her indignation got the better of her at this point, and all traces of dignity deserted her. "And whose barn is this, anyway? How dare you order me out of it?"

"Please, Meg—Margaret, I mean——"

"And you're collecting money," charged the angry daughter of the house. "I've heard all about it. You're pigs, you three. You're growing rich and you haven't given me a single solitary cent. Now see here, Lude, I'll go away and not bother you any more and not say a word to anyone, if you'll give me a share."

Ludar knew that both Joe and Arthur would be angry if he dipped into the purse to satisfy this outrageous demand. Slowly and reluctantly he produced a quarter from his pocket. It was his own, the only money he had, and all that he could count on having for a long time. He held it up in front of her.

"Here! It's mine. But I'll give it to you if you'll go away and stop bothering about us."

She held out her hand, and he dropped the precious quarter into it. She gave him a flirtatious flicker of her eyelashes—already much too long for any good purpose—and said: "I know something about you, Master Ludar. Tony Milner likes you. And you like Tony. It's true, isn't it? Tony likes you and you like——"

Ludar's face was red with embarrassment. "You have your money. *Go away!*"

She laughed delightedly and went skipping down the path, chanting, "Lude the prude! Lude the dude!" She stopped suddenly. "But you'd be more of a dude if your long pants fitted you better." Then she added, "Rubber!"

Outraged in feelings, robbed in pocket, Ludar returned to his job as timekeeper. The round had already lasted seven seconds too long.

Late one afternoon Pete Work dropped in to watch. He was employed at Cauling's Stove Foundry, and he went home each night with a face as black as ebony. Something had happened at the plant to necessitate an early shutdown, thus making it possible for him to come by at this early hour. He watched what was going on in the ring for several minutes and then let out a roar of disapproval.

"Get out!" he said, waving one black fist at the ring of spectators. "Skiddoo, the whole lot of you. You're cluttering things up here. Twenty-three!"

A command from Pete Work was still something to be obeyed. The watchers left quickly and quietly, taking the few objectors with them. There were no demands for money back.

Pete took off his coat, called for a cake of soap, and started to wash himself in the bucket. The water was cold, however, and he could not make a thorough job of it. Streaks of black were left along the hairline and on his neck. He grinned and said, "Just a lick and a promise. I'll finish the job when I get home." Then he motioned to the three boys to gather around him.

"Joe," he said, "you're going to catch a licking if you don't watch out. I had a talk the other day with the English cop who trained your cousin this summer. This Tommy Berry—that's the bloke's name—says he taught Norman how to use his dukes. He says the kid's good enough now to give any greenhorn a licking. Get this into your head, Joe: you're a greenhorn. Someday soon this cousin of yours is going to pick on you in public, and if you try to stand up to him he'll whale the daylights out of you."

The three boys looked at one another. "It's what I'm afraid of," said Arthur. "We don't know anything about boxing."

Joe was looking much cast down. "I've always been dead sure I could lick Norman," he said.

Pete put himself in a position of defense, left arm doubled in front of him, his right lowered. "Now you, Joe, you come and hit me," he commanded.

For several minutes Joe punched and jabbed and lunged, and never succeeded in touching anything but Pete's arms. Arthur took a turn then and had no better success. Pete laughed loudly. "All right, boys," he said, "now you see how it is. As long as I take up the right position you wouldn't hit me in a week of Sundays. So all you boys got to do is hold your arms just the way I do, and I'll not be able to hit *you*. That's right, ain't it?"

"I guess so."

"All right then, young Gentleman Jim. Put 'em up."

Joe placed himself in the attitude of defense Pete had assumed. The latter then proceeded to go all over him like a boilermaker, tapping him on the face, the neck, the shoulders, the ribs. He would say, "Now, then, on the right cheek this time," and on the right cheek it would be, no matter what Joe might do to cover up.

Finally Pete dropped his arms and puffed loudly with his exertions. "By grab, I'm out of condition myself. Well, young Joseph, I guess you'll agree now that you need some lessons. And you need 'em quick. Do you know what I'm going to do? I'm going to be your boxing instructor myself. We were friends at school, all of us here. I don't want you licked and, what's more, I want to see this stuck-up cousin of yours all polished up and buffed and put into a chemical bath for finishing. I want to see him so well licked that his own mother won't know him. So when do we start on some real lessons?"

There was time for three only before Mr. and Mrs. Craven returned from their trip to New York. They had wired ahead to Creek and Stapley's to be met at the station, and it should go without saying that they were allotted the Creams. The Creams were a beautifully matched team of yellowish light brown and very much admired in town. Anyone hiring a team from Creek and Stapley would invariably ask in hesitant tones, "Will it be the Creams?" and almost always the answer would be an emphatic negative. The Creams were kept for special occasions, for weddings, for important visitors. To be given them was a real feather in the cap, a species of social accolade.

The Cravens rode from the station, therefore, in considerable state. Jerry Barber drove with style, and the Creams seemed to know that they should be on their best behavior. They tossed their heads and shook their manes and responded smartly to Jerry's urging. But what most people who saw the equipage on this occasion remembered about it was the hat Pauline Craven was wearing. It was of black velvet, a small and stylish thing with an impudent white wing on one side. Perhaps the best idea of it can be conveyed by quoting Mrs. Elihu Grant, who was one of the onlookers. "It made me feel murderous," reported Mrs. Grant, who was on the dowdy order herself. "I wanted to climb right into that carriage, stab her with a hatpin if necessary, and then tear the hat right off her head and run away with it." Everyone noted also that the vehicle was piled high with luggage and parcels of all shapes and sizes, mute evidence that the Langley Cravens had been shopping in New York.

Perhaps a pause is advisable to tell about some of the things Pauline Craven had picked up along Fifth Avenue and in the expensive specialty stores clustering like a star-studded constellation from Twenty-third to Thirty-fourth streets. She had two other hats, one an alpine, rather daring and high-steppish, with a spotted veil and designed quite certainly to be worn only by a handsome woman, the other a small gray toque which fitted the head as snugly as the folded wings of a swan and was trimmed with the head and crest of a goura. She had many dresses, for the house, for the street, for evening wear, formal and otherwise. There was in particular a twilled foulard calculated to send Mrs. Grant into a green frenzy because it was of the latest color, called Josie, a light yellowish olive. A street dress of Genoa plush would have captured the masculine vote because it looked much like velvet and had a gay sash. This was of Metternich green, another deliriously new (as the dressmakers said) shade. Most important of all, however, was a Tudor gown for evening wear, a ravishing thing (when Pauline Craven wore it, at any rate) of bengaline, with a well-fitted bodice, kilted frills on the hips, and sleeves with slashed puffs at the shoulders.

She had acquired as well all manner of accessories: high-buttoned and embroidered shoes, satin slippers, silk stockings with openwork and lace motif on the instep, and enough lovely underwear to fill a trunk—chemises of cambric and batiste and surra, flounced petticoats of silk and satin, stays of brocade or coutil which looked pretty and fragile but were as strong and durable as steel, nightgowns, yoked and frilled and pleated, and garters (this was talked about in whispers) in red and lavender and pink!

It was no wonder that all Balfour talked about the clothes Mrs. Craven had brought back from New York and that a deep envy took possession of every other feminine bosom in town.

A week after their return Langley Craven (who had been too busy to notice certain mysterious activities in the barn) said to his wife, "You'll have to change the date of that dinner, my love."

His wife was wearing one of her foreign acquisitions, a tea gown of gray crépon with full sleeves of ivory surra, and looking most exceptionally handsome. She glanced up from her embroidery frame. "Why?"

"I heard today that our much-esteemed sister-in-law is having an evening party on the same date."

Pauline nodded casually. "I knew about it. And that, Langley dear, was the reason I selected this particular evening for our dinner party."

Her husband raised his brows. "Hmhh! A test of strength, eh? Are you quite sure you're ready for such a direct challenge?"

Pauline smiled and put her embroidery frame to one side. "No, not quite sure. But I want to try. *Toujours l'audace,* you know."

Langley nodded his instant approval. "That's all I wanted to know—that you were aware of the situation. If you want to do a little toujours l'audacing, I am with you, my love. 'I will stand at thy right hand, and keep the bridge with thee.' "

The dinner proved a great success. All who had been invited came. In addition to the regular courses there was a platter of cold ptarmigan (brought at great expense from Toronto) over which everyone exclaimed with gusto, knowing that it was the favorite dish of Edward, Prince of Wales, and was invariably served in all fashionable homes in England. There was much talk of plays and the opera and concerts, and there were some of the latest records for the much-improved gramophone, one of which, a recording of Maggie Kline singing "Ta-ra-ra-boom-de-ay," set the whole company to humming and singing. All of the guests had been invited as well to Mrs. Tanner Craven's evening party, but only two of them—a pair of weak-spined lickspittles, as Langley said later—left early enough to put in an appearance there also. The rest remained until eleven o'clock, and then, needless to state, they went straight home.

At eleven-thirty Langley put his head in at the door of his wife's bedroom. She said, "Langley!" in a tone of reproof, because she was putting on her nightgown at that exact moment, a very charming trifle in nainsook.

"I just wanted to remark, my dear," he said, "that you seem to have won the first test of strength."

There has been no previous need to explain that an empty lot occupied the space between 138 Wilson Street and the corner. It was large enough for a baseball diamond or cricket crease, although seldom used for games. At the rear stood a frame barn which looked insignificant beside the splendor of the Milner stables. It did not contain as much as a wisp of hay, and no one seemed to know who had built it or for what purpose. A diagonal path cut across the space in the direction of Grand Avenue, and a wealth of burdock grew on all four sides.

Always on Halloween the lot came into its own. It had been accepted as the base of operations from which parties sallied out to ring doorbells, tear down fences, and topple over outhouses. Joe and Arthur arrived early this year, and quite a few other Collegiate beginners came with them, thus swelling the neighborhood quota. There had been quite a blow a few nights before, and a large limb had been torn from a tree on the Grand Avenue side. This was dragged to a spot in front of the barn, and a fire was started which in a few minutes was roaring high. One of the neighbors sent out some platters of taffy just right for pulling; another contributed a basket of snow apples. Chestnuts came from a third quarter, and a three-pound bag of mixed candy from someone else. With such ample supplies the celebration settled down to a happy circle well removed from the heat of the fire.

An interruption occurred when a group of boys of assorted sizes came skimming on bicycles around the Grand Avenue corner. The one in the lead, a strapping fellow of perhaps fifteen or sixteen, raised his hand as a signal, and the party came to a stop. They watched the proceedings for several moments in silence. Three of them, including the leader, were wearing caps of blue flannel.

"We'll honor this little gathering with our presence," announced the leader. "The natives seem to be celebrating. Perhaps, encouraged by an intelligent audience, they will play some rustic games and even intone their wood notes wild."

His voice carried such a note of conscious patronizing that the activities about the fire came to a stop. The boys gathered there (the girls had gone home on parental insistence long before) fell into an uneasy silence. The newcomers were grandly attired in bicycling outfits with belted coats and stockings, in all the colors of the rainbow, which folded back just below the knee. They laid their machines in the burdock bushes beside the road and seated themselves cross-legged some distance back from the fire.

One of the neighborhood boys called out: "What d'ye want? Go on and peddle your papers!"

This loosened other tongues, and there was an immediate clamor of "Dudes!" "Rubbernecks!" "Forty-love, forty-love!" (a favorite expression of contempt which rose from the belief that all dudes played tennis). One boy picked up a piece of blazing wood and tossed it at the intruders.

"This seems a very rough neighborhood," declared the leader of the newcomers. "Can it be that we've stumbled into the slums of Balfour? Perhaps we had better go on to more congenial surroundings."

"Not yet, Derek," protested one of the other wearers of the blue cap. "Let's study the antics of the *hoi polloi.* They may repay scientific observation."

Ludar was sitting back of the circle. He never felt much at ease in crowds and was willing to be a silent spectator at all games and gatherings. He happened to be in a depressed mood, having observed Aunt Tilly bending over her array of black books before supper and shaking her head in a doleful manner. A feeling of guilt always took possession of him when he thought how much loss and anxiety he had brought to 138 Wilson Street.

Joe came over and squatted down beside him. "Ludar," he whispered, "have you noticed that Norman's with this gang?"

Ludar sat up and looked closely at the group sitting just outside the area illuminated by the fire. He recognized the redheaded son of the richest man in town as one of the wearers of the school cap.

"I heard he had some fellows up from St. Paul's for the week end," said Joe. "I know all the others in the gang and I don't like any of them. I guess, Ludar, this is the night of the big fight."

It became evident at once that he was right. Norman Craven suddenly piped up, addressing the leader of the visiting party: "There's one fellow here, Derek, I've got something to say to. He's a cousin of mine, though I don't like to mention it. We had a fight once."

A complete silence fell. Everyone on the lot sensed that this was a signal for action.

Ludar looked frantically about him. He had endured the long trip to Canada with fortitude and had not hesitated to walk alone over the pitch-dark Hudson Road, but he had no stomach at the moment for tribal warfare. It was not fear for himself in this case but a dread of what might happen to his friend. He hoped that Pete Work, who had become quite a "fusser," would come along, dressed up in his natty brown suit and high brown fedora dented evenly on all four sides. It was always of Pete that he thought in moments of stress.

"Very interesting, Norman," said the leader of the intruders. "Perhaps this cousin of yours would like to renew the contest. What better time could there be for it than the night when witches ride on broomsticks and the powers of evil are loose?"

Joe got to his feet. "Come on, then," he said. "Might as well get this over with."

The invaders, who numbered eight, moved in toward the fire. Norman Craven, who had grown tall and heavy of body although still somewhat gangling of leg, was in the center. Joe stepped up to meet them. He drew a line in the dirt with one foot.

"If you want a fight," he announced, "all you have to do is step over that line."

Norman stepped across the line in the sand. The two boys squared off at once, and a close ring formed about them. The spectators crowded in so close, in fact, that the fight immediately degenerated into a tussle.

"Hold on here, hold on!" A man who had been standing under the lamp on the corner crossed the road and began to make his way into the crowd of boys. It was Ab Jack. He shoved in until he stood beside the two antagonists.

"Now you fellows listen to me," he said in a loud voice. "The best thing for you to do is to scatter and go home. You'll have a cop here if you don't, and then you'll be in trouble, the whole pack of you."

"We've got to have this fight sometime, Mr. Jack," said Joe, still holding the position of defense Pete Work had taught him. "It might as well be now."

"That's it," said Norman. "If I don't give him a licking tonight I'll do it the next time I see him, no matter where it is."

"Well"—Ab Jack's tone of voice suggested that he was not at all averse to having it go on—"if you're both set on fighting, I guess you've got to fight. But you've got to do it right. All you kids stand back. Now, then, form a ring. I won't have any interference from the crowd, do you understand? What's more, I'll hold my watch and call off two-minute rounds. And another thing. This must be a clean fight. There will be no kicking, gouging, or biting."

"Kicking, gouging, and biting!" The voice of Derek expressed the utmost contempt. "Is it necessary to warn against such ungentlemanly tactics? I don't know by what right you're intruding here, Mr. Busybody, but I'll tell you this: *our* man doesn't need to be told to fight like a gentleman, even if these hoodlums do."

"Stand back, you," said Jack, shoving Master Derek out of the front line. He looked about him happily, welcoming this chance to stand in the limelight. "Now, then, we must do this proper. In this corner we have Norman Craven—Kid Craven, if he must have a fighting name—who may be termed the challenger, as he lost the first fight between these two young gladiators. Over here in the corner we have Joseph Craven, the Norval of the North Ward." He raised his hand in the air. "At the word, come out, and see that you fight clean. Ready? In there, then!"

Ludar, who had managed to worm his way into the front row, discovered at once that what Pete Work had taught his friend was not going to be enough. Norman, handling himself with workmanlike ease, was able to block all of his opponent's blows and to land some himself whenever he desired. Joe was the recipient of an assortment of jabs and swings which reached his head and shoulders as well as some solid blows in the ribs and chest. All he could do seemingly was to retreat and dodge and wait for the chance to reach his opponent with his still inactive right fist. This chance never seemed to come.

At the end of the first round it was clear to everyone that Joe was getting the worst of it. A silence had fallen over the neighborhood boys, who occupied three sides of the squared circle. The fourth side, which belonged to the invaders, was alive with shouts of encouragement. The third wearer of the blue cap, who had a high-pitched voice, gave vent to the school yell, which was all about the storied halls and ivied walls of good old St. Paul's. Derek patted Norman on the back and said, "Stout fellow!"

Arthur, who had been standing behind Joe's corner, came around the ring in search of Ludar. "Say," he whispered, "Joe's going to take a beating if this goes on. What can we do about it?"

"Nothing," answered Ludar. "If only Sergeant Feeny would come along!"

"Perhaps I could pick a fight with one of the others and turn it into a free-for-all," said Arthur gloomily.

The second round had started. Norman, who was now grinning broadly and confidently, continued to reach Joe's face every time they came together. There was malice in every blow. "There's one for beating me the first time," he seemed to be saying. "This one's because your father beat my father. This one is because we don't like you or any member of your family." Joe was responding as well as he knew how, but for every blow he gave he received half a dozen back.

Arthur had returned to his post and at the end of the round he made a knee for Joe to sit upon. The latter, it was clear, was discouraged. "I can't get near him," he muttered. "He pokes me in the eye every time I try it."

"There's nothing else to it, Joe," whispered Arthur. "You've got to quit boxing and start to fight. Go right in. Forget all about science and swing at him with both arms."

"He fights like a machine," added Ludar, who had come around behind them. "Get him off his balance and he may go to pieces."

Joe was rubbing his nose, which had suffered considerable damage. "All right," he said, "I'll try it. He certainly has been lacing it to me, but I still think he's got a yellow streak in him."

He obeyed Ab Jack's summons for the third round by throwing caution to the wind. He bored right in with both arms flailing. Norman blocked the first blows. They were coming too fast for him, however, and one landed finally on the point of his nose. He let out an "Aouh!" of pain and stepped back. His guard was down, and Joe rushed in to take advantage of this, landing another below on his cousin's face and one on his ribs.

Arthur shouted jubilantly, "That does it, Joe! Keep right after him. Now watch that streak come out!"

Norman did not succeed in pulling himself together. He lost his head under the sudden punishment and began to retreat about the ring. Joe, seeing a chance to win back all he had lost before, kept after him and got in so many telling blows that when Ab Jack looked at his watch and shouted "Break!" it was apparent to everyone that the tide had turned decisively.

Norman had no stomach left for more of it. He shook his head at his supporters. "I guess that's enough for both of us," he said. "I'm not stuck on fighting any more."

"I am!" cried Joe. "I'm just getting started."

Whether Norman's friends would succeed in getting him back into the ring was much in doubt. Before the issue could be settled a voice sounded over the uproar, an Irish voice with plenty of authority in it. Sergeant Feeny, making his way in from the Wilson Street side, said, "And is it fighting ye are, ye young spalpeens?"

The crowd thinned with amazing speed, the first to vanish being Ab Jack. That amateur devotee of sock and buskin disappeared behind the barn and made the quickest exit of his career by scaling the fence in the rear. By the time the representative of the law reached the bonfire he had the lot to himself. The moon, coming out from behind a cloud, shone down on empty space.

"And a good thing," said Feeny to himself. "They're decent boys, and it's not in trouble I'm wanting to get them."

It was said around town next day that Mr. Tanner Craven had watched the fight from behind a fence on the cross street and that, moreover, he had sent the message for Feeny when he realized his son was getting the worst of it. The sergeant denied this, and so perhaps there was nothing in it. The people of the town, however, enjoyed this new chapter in the long story of the Craven feud, the first since that well-remembered day when Langley moved from the Homestead and Tanner tried to carry off the family possessions.

4

Crises might come and crises might go in the world around 138 Wilson Street, but life inside had become one long crisis since the Great Loss. Tilly never spoke of the shrinkage in her principal in any other way, and she spoke of it continuously. She had developed such an unpredictability of mood that William and the boy lived on the edge of trouble. No meal passed without some display of her ruffled feelings.

The head of the house said once to Ludar, giving his head a slow shake:

"Whatever you do, my boy, never get to be a holder-againster. It's an unpleasant sort of thing to be, because the things a holder-againster picks to—to hold against other people are things which aren't fair in the first place."

"Yes, Uncle Billy."

"Being a holder-againster isn't as bad as some things, of course. But it can—well, it can destroy the peace of a home just the same."

Ludar first felt the full weight of her new moodiness in the matter of his cat. Skeek was now an old fellow, a split-eared and somewhat mangy veteran whose one thought seemed to be to stay in a warm corner. Tilly suddenly decided he should no longer be allowed in the house at night. "He isn't reliable any more," she said. "I won't have an unreliable cat in the house."

Ludar hurried to defend his pet by saying that Skeek always went to the door and mewed when he needed to go out.

Tilly snapped him off with, "When you've lived as long as I have, young man, you'll know that old cats are never reliable."

There was no answer he could give to that.

It did not matter as long as the weather continued warm. Skeek would come to the boy's window and look in at his master lying comfortably in bed, as though saying, "Hey, have you forgotten all about me?" When the cold weather came, however, it was a different matter. Tilly agreed that the cat could be kept in the summer kitchen but would not move another inch, not even when William pointed out that the place became as cold as Greenland in winter. It was necessary, therefore, to fix up a special bed for the exile. Ludar made it himself, although he was awkward with tools, a box lined with the remains of a very old buffalo robe and with a flap over the entrance.

One night after Tilly had gone to bed—she was always the first to retire—Ludar appealed to William to follow him to the summer kitchen. The thermometer outside had dropped close to zero, and in the frame addition the air was already so frigid that they could see their breaths.

"He'll be frozen stiff by morning, Uncle Billy," said the boy.

"He's got a snug bed there." Then William remembered something in favor of Ludar's contention. "We used to take Gyp in when it got real cold. Of course he didn't have as warm a coat as this old nuisance."

"Please, Uncle Billy," said the boy earnestly, "let me take him to my room. I'll wake up when you do in the morning and get him out here. Aunt Tilly won't know a thing about it."

William looked at him sternly. "Don't you think that would be deceitful?" he asked. "There's one thing I don't like in young people, and that's deceitfulness."

"Deceitfulness isn't as bad as cruelty," Ludar pointed out. "How would we feel if we found poor old Skeek stiff and dead tomorrow morning and knew it was our fault?"

William gave in then, and the cat spent a comfortable night at the foot of Ludar's bed.

The next evening William did not come in at the side door. Ludar heard his steps continue on to the summer kitchen, and he went out to see what had caused this change in routine. He found the head of the house stowing something away inside the washstand.

"Shish!" said William, raising a finger to his lips. "I've got an idea. I'll tell you about it later. After your aunt Tilly has gone to bed."

By nine o'clock Tilly had closed her door, and William motioned the boy to follow him out. He produced from its hiding place a rusty and battered oil stove. This he placed close to the cat's box and set to burning.

"This will do it," he said. He seemed pleased at his own ingenuity and at the same time resentful of the need. "You've got what you wanted. On cold nights you can come out here and set this stove going, and it will soon warm the place up. But don't you let your aunt Tilly catch you at it or she'll warm you. I found it behind one of the boilers at the shop and I made a deal with Mr. Alloway about it. I paid him ten cents for the use of it. And get this through your head, young fellow, you'll have to buy the oil with your own money." He paused and looked anxiously at the boy. "Have you any?"

"No," said Ludar. "I haven't a cent."

William reached reluctantly into his pants pocket and brought out a small assortment of change. He handed fifteen cents to the boy. "That will keep you supplied with oil." Then his manner became stern again. "This is what comes of doing underhanded things. Your aunt Tilly says no cats in the house at night. You say your cat has to be in—and this is where we land, knee-deep in deceitfulness. And that's not the worst. We're throwing money around. Twenty-five cents spent already, and that's only the beginning."

"We'll keep the poor old fellow from freezing," said Ludar.

William pointed to the stove. "Look at it. Do you see how old and rickety it is? Suppose it toppled over or the cat got out and knocked it over. What would happen then?"

"It might start a fire, sir."

"That's exactly what it would do. And we might all be burned to death. How do you propose to deal with this situation?"

Ludar looked worried. "I don't know, sir."

"Then I'll tell you. Whenever this stove is going you'll sit out here beside it. You'll get into your overcoat and you'll stay here until the place is warm enough to turn the stove out."

Ludar looked startled and a little unhappy. This new duty would consume the time he usually gave to reading beside the heater in the living room. He had a library card now and had discovered S. R. Crockett and was at the moment deep in *The Red Axe*. Before going to bed he usually went out for a walk. Bundled up to the ears, he would stride briskly about the almost deserted streets, his mind full of dreams of the future, of the books he would write, of the lovely ladies whose portraits he would paint, of the great speeches he would make. His invariable rule was to pass the Milner house at least twice during his rambles, hoping for a glimpse of Tony at one of the windows. He rarely saw her now and, when he did, it was no more than an accidental glimpse.

The need to tend the oil stove would make these walks impossible. Nevertheless, he nodded and said: "Yes, Uncle Billy. I understand."

As things worked out, however, William did more of the watching than the boy. He would look at Ludar curled up with his book beside the fire and would give him a double wink, first with one eye and then with the other, which meant, "I'll take the cat watch tonight." This happened so often that Ludar finally became conscience-stricken. When he returned from one of his walks to find his guardian sitting crouched over the oil stove and looking very weary, he realized how selfish he had been.

"Uncle Billy," he said, "I ought to be ashamed of myself. You have to get up so early and you need your rest—and yet I've let you do this seven times out of the last nine."

William rubbed his eyes. "I was nearly asleep." He yawned and turned out the lamp. "Don't worry about it, Ludar. I've nothing else to do. I sit here and think of inventions and schemes. I tell you, if I worked out all the new things I've got in my mind we would soon be pretty rich people. Well, off to bed now. We won't have to do this much longer. Another month of it and April will be here."

The evening vigils came to an end in a most unexpected way. One evening early in March, William came home from work in a jubilant frame of mind. He was walking briskly and humming to himself.

"I've always said it pays to look out for Number One," he exclaimed.

Drawing a paper from an inside pocket, he waved it in the air and then planked it down on the table with a triumphant manner. "Well, I've done it this time. Do you see this paper? It may mean a million dollars for us. It may mean servant girls and gas fixtures and horses and ponies, and roast chicken twice a week. It's an agreement with Mr. Alloway. He's to have a half interest in the roller boat."

Tilly was making an omelet for supper. She laid down the spoon with which she had been beating the eggs and looked at him blankly. "Do you mean you've up and handed him a half right to it?"

"I've handed him nothing. I should say not! I drove a hard bargain with him." William tapped the paper with an excited hand. "He pays for everything. The expenses of all experiments and the materials for the model. It's going to cost him a pretty penny, I can tell you. There will be about fifty pounds of iron castings alone. Then, when we've got something to apply for a patent on, he's to pay the cost of *that*. It's all down here in black and white. I never drove a harder bargain in my life."

Tilly went back to beating the eggs. "Well, I *am* relieved. I thought at first you had given it away. It would be just like you."

As the work on the boat had to be started at once and the workshop was too small for the purpose, William announced he would use the summer kitchen. He bought a new oil stove out of the partnership funds. This he lighted as soon as he returned from work, and by the time supper was over it would be warm enough in the summer kitchen for work to begin. Ludar fell into the habit of finishing his homework before supper so he would have the whole evening free for literary pursuits. He would take his pad out and join William, using the seat of a wooden chair as a desk. The cat, delighted at having so much company, would emerge from his box and curl up at the boy's feet.

The contentedly busy couple would compare notes on the progress they were making. William would say, "Four months more and this flimsy contraption you see now will be a small leviathan"—this was a curious figure of speech which pleased him so much that he used it always in speaking of his invention—"capable of venturing out in the roughest weather and riding the billowy waves." Ludar's reports were more specific. "Just finished a battle scene," he would say. "I think it's good, Uncle Billy. And I killed off one of the villains. The hero rammed a pike into his midriff—that's a grand word, isn't it?—and he died before he could explain what he'd done with the gold." His pencil on this occasion did not return at once to its task but remained suspended over the pad. "When I sell this story, the first thing I'm going to do is pay back the money Aunt

Tilly took out of her principal. Then I'm going to get myself a pair of pants that fit."

"Better get the pants first," advised William. "Give your aunt Tilly a look at the money first and she'll have it out of you and into the bank before you can say Jack Robinson."

"But," said the boy, "we'll have to pay off the people who put money in the fund."

William lifted his plane and freed the blade of shavings. "Leave that to me," he said. "I've got a scheme. I'm going to give each one of them a share of stock in my roller-boat company. Those who put in twice will get two shares. Miss Craven will get fifty." He nodded as though the burden of this obligation had already been lifted from his shoulders. "Someday, when we're making money hand over fist, we'll double up the stock or give them a bonus or some such thing as that. All of these people who keep coming to me and nagging and complaining are going to be rich, if they only knew it." He stopped work again and looked at the boy with an expression of warning. "Of course, Ludar, we mustn't start counting our chickens like this. We've got to use common sense. We must wait until the money's rolling uphill before we spend it."

Going back to his task, he filled the room with the music of his smoothly propelled plane. He went on talking in fits and starts as he kept his attention on what he was doing. "I've been thinking a lot about the future. I'm against this idea of going to the old country to live. We wouldn't like it over there. People would look down their noses and say: 'Canadians! And just made their pile too. We'll put *them* in their places.' Now my notion is to build a house right here in Balfour. Is there a better place anywhere? If there is, I don't know it. I've found the very spot, on those bluffs along the river beyond Eagle Place. And, Ludar"—he turned toward the boy and indulged in a rather sheepish smile—"I've been working on a plan for the house. Three stories, red brick, two wooden pillars in front, a side veranda with a balcony on top, stained glass around the front door and in windows at both sides. Inside there's double parlors, a dining room, a library—yes, a library; you'll need plenty of books to read—six bedrooms, two baths, gas fixtures throughout. And woodwork! It will be the best and fanciest in town. Of course you mustn't say anything about this to your aunt Tilly. Not yet. She's so full of this English idea she'd get mad at me if she knew what I'm working on. But I'll show you the plan after I've thought out a few more details." He was assailed suddenly by doubts. "You don't want to go back to the old country to live, do you?"

Ludar's memory had almost relinquished its hold on his early years. His recollections of England were now confined to occasional flashes such as the visit of the old lawyer and of the great crowds in the purser's office. He could still hear voices clearly, although what they said was disconnected and with no meaning for him. He could still see the clock tower and an ancient man in a smock with a pitchfork in his hands. This, no doubt, was Old Baffle.

The boy shook his head. "No, sir, I'd rather stay here. I have so many friends. I want to stay with you, Uncle Billy."

William looked very much relieved. "I thought you felt that way about it, but it's good to hear you say so. Suppose there *is* an inheritance for you. They can send it out, can't they? And if there's a title you can use it better here. You'd be the only lord in Canada, and over there they must be a shilling a dozen."

By the time spring came the work on the boat had outgrown the summer kitchen, and the back yard became the new base of operations. Already two wheels had been made, over six feet high, with a cogged iron rim on the outside of one of them. The next step was to connect the pair with an axle of smoothly turned wood, eight feet in length. This accomplished, William began to nail narrow strips of wood from the circumference of one wheel to the other, fitting them so closely that they formed a solid outer surface. The sound of his continual hammering brought heads over the tops of fences all the way up the block and a constant stream of visitors to ask what in thunderation he was doing. His answer was always the same. "A new contraption. Too soon to tell you about it. Drop in again in a few weeks—or months—and I'll have something to show you."

He was very proud of his newly acquired sense of caution and would say to Tilly, "They don't get anything out of me, these rubbernecks."

By the twenty-fourth of May, which was the Queen's birthday and was celebrated by a callithumpian parade in the morning and a sham fight at the fairgrounds in the afternoon, William had completed the shell and was engaged on the stationary platform. This was to be suspended inside on the axle and entailed a great deal of careful fitting. By Dominion Day the platform was hanging in place and was in perfect alignment and balance. A month later all the iron cogwheels connecting the operating handle with the sprockets on the outer rim were in. Ludar could now stand on the platform and send the eight-foot cylinder rolling about the yard.

By this time the interest in William's achievement was mounting to great heights. Ludar's friends came over continually and begged to be allowed to run it. They began to use nautical phrases, such as "Keel-haul her, mates!" and "Abaft there, lubbers!" Mr. Alloway's first call was an important event because of his half interest. William acted as pilot and gave a convincing demonstration of his invention's maneuverability. There was an avid gleam in the Alloway eye at the finish, and he exclaimed, "William, you've done it!"

William climbed out. "There you are, Mr. Alloway," he said. "It works. All we have to do now is get the patent. We don't want any of these sharp lawyers cutting in ahead of us."

"We'll try it out first. At Tuscarora Lake. Some Saturday night, late, I'll send Dan Camp with the biggest lorry and we'll take the boat down to the park. Then we'll test it early Sunday morning when no one's about." He leaned over and slapped his employee on the back. "William, you've done it. I'm proud of you, man. We'll make a dollar or two out of this."

"Getting it out of here is going to be hard. You see, the boat's eight feet wide and there's only seven feet ten inches between the side of the house and the fence."

"Great Scott!" The Alloway mood lost its mellowness. "How did you come to make such a mistake?"

"Don't you remember we talked about the size and decided it ought to be eight feet?"

"Yes, yes, we did. But, ginger, didn't you measure the width of the lane before getting to work?"

William answered: "If we were taking it out the lane we'd have to remove the gatepost anyway. It will be easier to cut a couple of sections out of the side fence. I can do it and put them back so no one will know what happened."

Mr. Alloway was not to be placated easily. "Never heard of such a thing!" he declared. "It's like making something in a cellar and not being able to get it out of the house, or building a boat in a bottle. You'd better not let anyone else know of this or you'll be the laughing-stock of the town."

The whole town heard about the roller boat in the course of the next two days. William Pitt Milner came to the fence and asked Ludar to give him a demonstration. He seemed both surprised and impressed. "Damned if it doesn't work!" he said. "I'll have someone come up from the paper and do a story about it."

Sloppy Bates, looking more down-at-heel than ever because he had not yet qualified for a raise in salary, arrived early the next morning. He snuffled with the shock of seeing the boat in action. "The boss was right. It *does* work." The news story he wrote for the *Star* began, "Has Balfour contributed a great new invention to the world?" and it was given a prominent position on the front page. Mr. Alloway decided then that the test had better not be delayed, and the following Saturday Dan Camp arrived late in the evening with a large lorry and a team of horses. William had already removed a section of the fence, and the lorry was backed up to the gap. It took the efforts of four men and much tugging and shouting to get the boat in its place and securely lashed. The lot was filled with people by the time they were ready to leave.

William walked over to Tilly, who was watching from the side door, to say he was going to the park with them and that he would stay there all night.

His wife suspected that the excitement had caused him to lose his wits. "What kind of crazy notion is this? That precious boat of yours isn't a sick baby that you have to sit up with it."

"Well," said William, finding it hard to explain just how he felt, "it's this way. I don't suppose there's any chance of thieves coming along and stealing the castings, or some slick lawyer looking it over and getting plans in to the Patent Office before we can. And yet—well, I'm afraid to let it out of my sight. I couldn't sleep a wink anyway."

"Let me go too, Uncle Billy," said Ludar.

William tried to put his foot down, but, surprisingly enough, Tilly favored the idea. "Better a pair of galoots than just one," she said. "He'll keep you company, and I'll feel a mite easier myself."

Ludar, accordingly, took his place in the lorry, his overcoat under one arm and a couple of pillows under the other.

When the boat had been rolled close to the shore in a narrow inlet on the far side of the lake and the lorry had departed, William arranged all the cushions they had brought on the center platform and told the boy to make himself comfortable there. "I guess I'm too excited to try it yet," he said. "I'm going to walk up and down and do some thinking. This is a great event for me, you know." It was so warm that he took off his tie and loosened the neck of his shirt. "That piece in the *Star* got me all stirred up. Perhaps I *have* given a great invention to the world. Just think of that! Just think of all ships being built like this small leviathan of mine and rolling spang across the ocean in record time! Just think if there were never any more wrecks! I tell you, Ludar, I've got a lot to ponder about tonight." He started to move off down the road but came back to say, "You better lie down now and get as much sleep as you can."

The boy did not need any urging. He looked at the lights cast by the moon on the surface of the lake and at the thick fringe of trees on the other side. He could make out the roof of the pavilion and, well off to the right, the grandstand and track where the bicycle races were held. "Tomorrow," he said to himself sleepily, "we'll roll over there in a few minutes. We'll feel just like Columbus discovering America. Uncle Billy's going to be a famous man." Then he dropped off to sleep.

Dawn was breaking when he wakened. He raised himself on an elbow and watched the sun bring color to the massed trees of the park and light up the tin roof of the pavilion. Now he could see something which looked like a huge red umbrella but which, he was sure, was the merry-go-round. It was going to be a fine warm day. He realized that sleeping in the open air had made him hungry, and he remembered gratefully that Aunt Tilly had insisted on making up sandwiches for them.

He became aware that William was standing on the other side of the inlet, watching the sun come up over the horizon with absorbed interest.

"Hello!" called the boy. "The great day is starting, Uncle Billy. In a couple of hours you'll be the most famous man in Balfour."

William turned, grinned, and began to walk around the edge of the water with a step which resembled a skip. It was clear that he was in the highest of spirits.

Ludar asked, "Didn't you get any sleep?"

William laughed scornfully. "Sleep? Not much. I've been too excited. I've been walking about and thinking and planning." His mood changed and he nodded with a suggestion of gravity. "Yes, Ludar, this is the big day. Pretty soon we'll know if this small leviathan is worth anything—or not."

Ludar gave a confident laugh. "I'm going to take notes. I'm going to watch and listen and write everything down so I won't forget. Then someday I'll write an article for one of the big magazines about the day when the boat first took to the water. I guess they'll pay me a hundred dollars for it."

William smiled at him. "When's this important article of yours going to be published?"

"Oh, not for years. Not until you're an old man with a long white beard and they've put up a monument to you downtown. I think I'll begin it with a description of day breaking over Tuscarora Park. It would make kind of a literary start."

"Well," said William, gazing across the water, "we mustn't be too sure. How do we know what's going to happen?" The lake had brightened to gray with streaks of gold. A sound of hammering reached them from the park, where attendants were already up and about. "But, right or wrong, Ludar, this is the biggest day of my life."

It was shortly after seven o'clock that Mr. Alloway arrived in a buggy with Dan Camp and another of his teamsters, Ben Durston. They sprang out briskly, and the boss said: "A fine day for it. Have you been saying your prayers, Billy?"

They lost no time in propelling the boat down to the surface of the water. It rocked dangerously for a moment. Ludar, who had been skipping about excitedly on the edge of things, was afraid it was going to sink and carry all their hopes with it to the bottom of the lake. It gave itself a shake, however, and then settled easily on the surface. It began to turn slowly with the movement of the water, and the two teamsters had to rope one end of the shaft to keep it from drifting away.

"In you get, Billy, and show us what she'll do," said the boss. His joviality had left him, and it was clear that he was nervous. Undoubtedly he was thinking of the money he had sunk in this venture and wondering if he would get it back.

Dan Camp and Ben Durston held the boat well in to shore while William climbed through the spokes of the wheel. The inventor settled himself on one knee and put his hand on the crank.

Ludar said afterward that it was the longest moment of his life while they waited for the boat to begin its performance by rolling off across the water. He had closed his eyes, intending to keep them shut until he heard the shouts of triumph which would announce success. All was quiet save for the grind of the cogwheels and the deep breathing of the three men on shore. Finally the suspense became too great for him and he opened his eyes.

The first thing he saw was the long profile of Mr. Alloway. The latter had been chewing on a cud of tobacco, but now his jaws had ceased their rhythmic motion. It was quite clear that he was disturbed. Then the boy perceived that the boat had not moved and that William, his face red with his exertions, was still striving to turn the handle.

"I guess you didn't have enough breakfast, Billy," said the boss after more futile moments had sped. "Get out and let Dan try his hand. He's the strongest of the lot of us."

But Dan had no more success. He pushed and tugged and wheezed and sweated. The wheel refused to turn, and the boat did not move.

Finally Mr. Alloway gave his head a reluctant and angry shake. "It's over the dam," he said. "Forty-one dollars and seventy-five cents clean gone over the dam. I knew it was a gamble. I'm not complaining." He might not be putting complaints into words, but it was clear from the cold detachment of his eyes that inside he was burning with resentment over the failure. And then he smacked one hand against the other. "I know what's wrong with the contraption. It's something we overlooked, Christian. We didn't figure on the resistance of the water."

"That's it," affirmed Dan Camp. "There isn't a man in the world strong enough to force those paddles through the water."

"But——" began William. He was about to say that they could reduce the size of the paddles and thus get around the difficulty, but it occurred to him in time that he knew nothing whatever of water resistance. He had realized at the start that the boat would merely slosh about in the water if there were not paddles on the surface to take hold and provide propulsion. But how far could they be reduced? He looked anxiously at his employer. Mr. Alloway's face had taken on the set look his staff knew so well.

"All it needs is an engine to supply the driving power," said the boss. He indulged in a completely mirthless cackle of laughter. "Yep, that's all it needs. Not another thing."

William was at the stage where he was ready to clutch at any kind of straw. "I can rebuild it large enough to put in an engine," he said.

Alloway made an impatient gesture. "It's over the dam, I tell you. Have you any idea," turning fiercely on the downcast inventor, "of what it would mean to rebuild this thing so an engine could be put in? Great Scott, man, you'd have to start all over again. The whole design would have to be changed. It would mean metal construction, in part, at least. The cost might run into thousands! I tell you, it's a goner. It's over the dam."

He spat out his cud of tobacco and walked back to the buggy. "Let's get going, Dan. I don't want to set eyes on the thing again. I told my wife I wouldn't be going to church, but we'll be back in time, after all. Not that I'll have anything today to return thanks for!"

The red flush caused by his exertions faded gradually from William's face as he watched the buggy disappear down the road until nothing was left but a cloud of dust at the Stanger Mill turn. A sag had come in the line of his back.

"Will we have to walk home?" asked Ludar, who was feeling hollow inside with the disappointment.

William did not answer. He turned and walked slowly down the bank and inspected the knotting of the rope which held the boat from drifting away. "Well, Ludar," he said finally in a voice which did not sound natural at all, "you heard. It's over the dam." His composure left him abruptly. "If only it was over a dam and all smashed up! And me with it!" He sank down on the bank and rested his elbows on his knees. "The great day! The great inventor!"

Ludar felt it had devolved on him to take charge of things. "When the lorry comes back——" he began.

"The lorry won't come back."

"Why—why, it must. How else can we get the boat back home?"

"Mr. Alloway won't give that a thought. Here it is, and here it can stay for all of him."

"Then I suppose we'll have to send for a carter and get it taken back."

After what seemed a long time William said in a listless tone: "Yes, that's what I must do. I'll get to the nearest phone and call one of them." It was even longer before he rose to his feet. He was getting control of himself, however, for he was able to shake his head at the boy and smile. "It's a hard blow, sonny. Well, that's the way life is. One disappointment after another."

The boy remembered the sandwiches. "You'd better eat something before you go to find the telephone." He opened the paper parcel. "They look good, Uncle Billy. Corned beef."

"Ludar," said William, "there's a lot of things I don't want right now. I don't want to bother any more about this boat. I don't want to meet anybody I know and have to look them in the eye. I don't want to go home and tell Tilly I've failed again. But most of all, sonny, I don't want a corned-beef sandwich."

There was a long pause. Then William indulged in a deep sigh. "I don't suppose any of the carters have a wagon large enough to carry it home. We'll have to attach it to the back and drag it through town. What's more, we'll have to take the shortest route, right up the main street. Sonny, there's nothing you can do about it. I don't need you. Why don't you run along home now?"

"You need someone to go with you." The boy knew what staying meant, the humiliation there would be in such a trip through the town. It would be a victory procession in reverse. Amused crowds would gather to watch them, even to follow them. They would have to endure loud laughter and much joking at their expense. All his high school friends would see them returning home in defeat. He would have liked to accept the chance to escape. It was in a voice which did not express any high degree of resolution that he said, "No, I'll stay and go along with you."

William had lapsed back into a mood of bitter reflection. "I can see now why I was wrong." He seemed to be speaking for his own benefit and to have forgotten that the boy was with him. "I shouldn't try big things. I've had no education. What do I know about the laws of nature—electricity and strains and stresses and water resistance? Nothing at all. I should stick to what I know—improved clothespins and better breadboards. That's all I'm good for. I suppose it's just as well this happened to me. It's taken me down a peg. Yes, I can see it now. I'm just meant for little things, and I'm a failure even there."

Then the boy noticed there was more in William's face than disillusionment. There was fear. In a low voice which seemed to be forced from him the disappointed man said, "I wish it was tomorrow morning and I knew if Mr. Alloway was going to fire me."

Book Two

Chapter I

I

Half the world believed that the nineteenth century would come to an end at twelve o'clock on December thirty-first of the year 1899. The other half vociferously contended that the new century would not commence until a year later. In Balfour the first date was officially celebrated and, as it happened to be a Sunday night, the more frivolous part of the proceedings—the *fin de siècle* ball in the armory, the Board of Trade banquet, and the private parties at which drinking would be done—were advanced a day. At ten minutes to twelve on New Year's Eve a choir appeared on the Market Square and sang "Onward, Christian Soldiers" and, as the bells tolled and the factory whistles blew at midnight, the Doxology. Then they sang "After the Ball" and Tosti's "Good-bye" and, finally, "Auld Lang Syne." By this time there were several thousand people on the square and in the streets immediately adjoining it, and most of them joined in the singing.

There is solemnity to the passing of an ordinary year; the passing of a century is an event which can never be forgotten. It stirs up the deepest memories and most profound regrets and excites earnest speculation as to what the Deity has in store for the future. When the ringing of the bells announced that this century, the marvelous, unparalleled nineteenth, had finally consumed its last second of existence, it seemed as though for a tense moment the world stood still.

Andrew Cale, the president of an implement manufacturing company, the tails of his dress coat showing beneath the short fur jacket he had donned over it, said to those about him, "I solemnly swear I shall drive a buggy all the rest of my life and nothing else." The Reverend Elmore Nash, who had married the soloist of his church and was now the father of twins, propounded the question, "Should we all go home and destroy our gramophones as instruments of the devil?" Mayor Frank Sifton, in a speech he delivered from the steps of the Town Hall, expressed the thought which was in most minds. "If I could be sure," he declaimed, "that the century which stretches ahead of us will be as fine as the one which has just died, I could go home and sleep in great peace and content."

But the world was already changing, and it was as noticeable in Balfour as elsewhere. There were more smokestacks against the sky line, more red brick houses with towers and blue-glassed conservatories, more private carriages and expensive fur coats on the streets. The city had been growing. And with growth was coming a shifting of values and a change of thought trends. Life was becoming different—freer, perhaps; louder, certainly. There was a new feeling in the air, a new sense of freedom, and this was both good and bad; for, whereas it cleared minds of inhibitions and scattered bogies to the winds, which was good, it swept aside at the same time the restraints of moderation and good taste and paved the way for open indulgences and coarseness and the first manifestations of rowdiness and violence, which was very bad.

Change comes about gradually, and it is not possible to put the finger on a specific year or month or day and assert that here a new trend began. If it were ever possible, however, it was in the minutes which elapsed between the first sound of the bells and the last stroke on this particular night. Certainly men looked at each other and laughed (more loudly than before) and seemed to say, "Now we start over." Start over, most decidedly, they did.

2

The next day was raw, with a wind which swept down between the houses and leaped fences like a hurdle racer. Ludar, who had been working for the past two years after school and on Saturdays in James Haverstraw's book- and stationery store, was to spend the day in taking stock. He was up early and out in the yard for a look at the weather. He said to himself, "It will be a cold night for the skating carnival." This should have been a small matter, because he could not afford to pay for admission to the rink. He knew, however, that Joe was going as a circus clown and he wanted the carnival to be a success for his friend's sake.

William followed him out, and for a moment the pair of them stood together and stared with regretful eyes at what was left of the roller boat. The metal parts had been removed and sold for old iron (the proceeds going to Mr. Alloway, of course), and the wooden frame was showing a tendency to fall apart. "Your aunt Tilly wants me to split it up for kindling wood," said the inventor of the boat. "I think she wants to see the end of it because it's a reminder of failure."

"My story hasn't been sent back," said Ludar hopefully. "If it's accepted there will be enough to buy an engine and everything you need for a new boat."

The novel had been completed three months before. It had then been parceled up neatly and securely in a cardboard box and sent off to a publishing house. For many weeks thereafter Ludar's first thought on waking had been, "Perhaps there will be word today." He had been confident, when the manuscript went off, that it would be snapped up and published quickly. It was full of hair-raising adventures, and the heroine was a princess and very lovely. What more could they ask? But lately he had been entertaining doubts. Surely if the publishers wanted it they would have let him know by this time. Perhaps it had been lost in the mails and had never reached them. Perhaps it had gone astray on its way back.

Lockie McGregor came to the side fence. "William!" he called. Then, not bothering with the usual amenities, he plunged into the subject which concerned him at the moment. "You've wondered perhaps why I didn't put up a new sign after I took the old one down? I had my reasons and, you may be sure, they were good ones. I knew I would have to take in a partner sooner or later. Well, I've brought myself around to it. There will be a new sign one of these days."

"You work too hard, Lockie. It will be fine for you to have a partner. It's Clyde, of course."

The merchant nodded. "I've been doing well by him in the matter of salary. I've stepped him up a dollar each year. And I'll say this for him, he's never whispered the word partnership to me. Catherine has—many times." He had been employing a tone which suggested some distaste for the prospect, but now he brightened up visibly. "Saturday brought things to a head. He had a couple of invitations. He could have gone to the Board of Trade banquet, for one thing. But he stayed home instead and spent the evening with Catherine. I didn't go out either and I could hear them talking and laughing upstairs. I said to myself, 'This is the proof I've been waiting for, that he isn't going to change.' I made up my mind there and then. This morning I went to Catherine's room and told her what I had decided. Man, you should have seen her eyes! But she only let me start and then shook her head at me. She said, 'I shouldn't hear it from you. Please, Father, tell Clyde first, and then he'll have the pleasure of breaking the news to me.' I said to him at breakfast, 'You're working for yourself now, my boy.' He lifted his head with such a funny look that I knew he thought he was being discharged. Then the truth began to dawn on him. I winked and he stammered and said 'Thank you, Father-in-law.' He rushed upstairs and into Catherine's room, and I could hear them jabbering away to beat the band. They're still at it."

"I'm sure you'll never regret it, Lockie."

The merchant stroked his upper lip reflectively. "So now, William, I must decide about the new sign."

"McGregor and Carson will look fine up there over the front entrance."

The Remarkable Man frowned indignantly. "Tophet, man!" he cried. "Do you think I would ever let the firm name be anything but McGregor and Company? I started the business, and mine it must be as long as I live, and perhaps ever after." He paused and gave his head a doleful shake. "I've always regretted the old sign. There was something about it which warmed my heart every time I looked at it. I might have it copied, figure and all, and have the sign read:

"LOCKIE MCGREGOR
THE REMARKABLE MAN
AND PARTNER

It would be a business-bringer again." He was speaking now in an almost wheedling tone. "Give me your honest opinion, man. Don't you think it would be a canny stroke?"

Ludar had been following this conversation with interest, but at this point he returned to the house. The mailman made one round on holidays, and he was due to come on his crisscross course down Wilson Street at any moment now. "Perhaps I'll hear from them today," Ludar said to himself.

He was right: this was the day he was to hear from his publishers. While combing his hair in the bedroom—it had developed a distinct curl, and he took great pride in it—he heard the front door open and a loud cry of "Mail!" and then something landed on the hall floor with a thud which seemed to shake the house. Ludar knew this could mean one thing only: the manuscript had come back. He stood without moving for several moments, his eyes on the brush in his hand but seeing nothing. His disappointment was so great that he was incapable of coherent thought. He felt numb and sick. Life had become hollow; all prospects for the future had been taken away. He was sure he would never amount to anything.

It was some time before he summoned the resolution to walk into the sitting room. William and Tilly were standing in the kitchen and whispering. When they saw him they turned away and pretended to be interested in something else.

"My story's come back," he said.

William cleared his throat. "I'm afraid it has, sonny."

"First Aunt Tilly loses her money," said Ludar, speaking with some effort. "Then the roller boat doesn't work. And now they throw my story back at me. I—I guess we're an unlucky family."

When he reached the hall he discovered that the cycle of ill luck had indeed been completed. The manuscript lay on the floor. It had broken open in the fall, and the pages were scattered all over. There did not seem to be any letter with it, but he saw a slip of paper with printing on it.

Ludar looked down at what was left of three years' work and tried to wink back his tears. "This means I'll never be a success," he thought. "I couldn't do a better story, no matter how hard I tried, and yet they don't want it. There's no use trying again."

After several minutes had passed without a sound from the hall William tiptoed to the sitting-room door and looked through. He came back at once. "He's sitting on the floor," he whispered, "and gathering up the pages. He's as white as a sheet."

"He must be feeling bad, he built on it so much." Tilly sighed. "And it took so much postage to send it!"

They waited fully ten minutes, saying nothing. Then they heard the boy scramble to his feet. He came out through the sitting room and into the kitchen. The manuscript was under his arm, and to their surprise they found he was looking almost normal again; at any rate, he was keeping his feelings well under control.

"I'm going to hide this away at the bottom of my bureau," he said. "I'm never going to look at it again, but I think it had better be kept as—as a sort of lesson. It"—his voice became apologetic in tone—"it shouldn't have been sent. I began to read the start of it and—well, it's pretty bad. I was young when I wrote it. I don't suppose the editors got very far into it. But"—a look of resolution had taken possession of his face—"the last part was much better. Do you see what that means? I'd been learning as I went along. I can go on learning, can't I? If I learn a great deal I may have a story accepted someday." He managed to smile at this point. "It means I'll have to keep working at it, and working hard. I'm going to start on another one. Right away."

Later in the day William said to his wife: "His spirits bounced back like a rubber ball. I tell you, Tilly, that's a mighty good sign. He's a quiet young tyke, but this shows he'll be a fighter about things which count."

3

Ludar paid a call at the McGregor home that evening. "My story came back," he said to Catherine, who was in the back parlor, which had been converted into a sitting room for the young couple. She was in her wheel chair, of course, and looking a little thinner, he thought. She had been able during the past year, however, to dispense with the practical nurse who had been with her since the time of the accident, a cheerful and brawny woman who had been capable of lifting chair and occupant and carrying them out to the garden when desired. "I read some of it over," he hurried to explain, "and I'm not surprised they didn't want it. I'm ready to start on a new one and I hope it will be a lot better."

"Of course it will!" said Catherine stoutly. "And it will be published too. It will sell millions of copies."

Clyde, who had a touch of gray over his ears, came into the room in carpet slippers. Catherine cried in an apologetic voice, "Oh, my dear, your footstool has been shoved out of place!" Before anyone could make a move to rectify this domestic error she had swung the chair expertly into a corner of the room and brought back the misplaced stool. She had become so quick in the manipulation of the chair, in which all her waking hours were spent, that she got around the house much faster than anyone on foot. She might be heard at the front door talking to the mailman, and in the very next instant she would be in the kitchen discussing dinner with Margaret. Her face had become thin; but this, fortunately, enhanced rather than detracted from her looks. She was always cheerful and, if she suffered much pain, she kept the fact to herself.

"Now that Clyde's a partner—you've heard about it, I'm sure, Ludar—I'm hoping we can settle this matter of adopting a child. Clyde and I went to the orphanage this morning and we found the jolliest little boy. We would have taken him on the spot, but his name was Bert Hawkins and his parents had come straight out from England. We thought Father had better be consulted. He put his foot down. He wouldn't have any scion of an effete southern race as a grandson of his. That settled it, of course. This same thing has happened half a dozen times, and I'm beginning to think Father will be against all of them until a boy from Scotland turns up."

"I reached that conclusion long ago," declared Clyde. "He'll not only have to be from Scotland, but he'll have to sit up in his cradle and repeat the multiplication table before he'll be acceptable."

At this point there was an interruption. The front doorbell rang, and Margaret ushered in a young lady in a beautifully tailored coat with black fur trimmings and a big black muff. Ludar took one look at the visitor and felt a desire to disappear. It was Tony. He had not spoken to her for more than a year and he was acutely conscious that his coat had reached the stage which came sooner or later with every piece of clothing he had ever owned: when the sleeves were too short for his arms and allowed too much of the wrists to show.

This was a new Tony. She had become quite suddenly a young lady, and he was sure she was more beautiful than anyone he had ever seen before. Her nose had become a little longer, but its delicate chiseling had not suffered in the process and, whereas it had once been no more than pretty, it was now decidedly lovely. Her hair was no longer a mere brown; it had acquired luster and warmth. Her eyes, he decided, had changed the least. They were still as grave as those of the little girl who had stared at him through the pickets of the fence.

Her skirts rustled under the handsome coat as she came into the room. She said: "Hello, Cath. Hello, Clyde." Then she recognized the third occupant of the room. "Why, it's Ludar. It's—it's been a very long time since I've seen you, Ludar."

She took the chair which Clyde pushed forward for her and unbuttoned her coat, letting it fall back on her shoulders.

"I've seen you three times in the last year," said Ludar.

"When and where was it?"

"Once when you were getting into the carriage with your father and aunt. It was a Sunday afternoon, last summer. You were in a gray dress and you had a straw hat with pink ribbons."

Tony identified the occasion at once. "We were going to the Childers'. They had an artist from Hudson visiting them."

"Another time you were walking on Grand Avenue with some fellow from out of town. He wore a uniform."

"Ross Browning. He's R.M.C." After a moment she added: "He's very conceited. I don't like him much."

"He acted as though he liked you. The other time I was going past your house and I saw you come close to one of your front windows."

"Was it in the evening?"

"Yes, and very cold. I remember I was holding one hand over my nose because I was afraid it would freeze."

"What were you doing out on such a night?"

"I always take a walk before turning in. I've passed your house thousands of times—and I've only seen you the once." Fearing that he was making the conversation too personal, he asked, "Are you going to a party, Tony?"

"Yes, at Myrtle Wood's. It's such a short distance that I thought I could have a visit with Catherine first."

"You had better slip off your coat and overshoes, Tony," suggested Catherine.

Ludar had been sitting with his feet tucked uncomfortably under the chair in the hope of concealing their worn and scuffed condition. He got up at once and went down on one knee to remove the overshoes. The smallness of the feet which emerged in red satin slippers made his hands tremble at their task.

The removal of the coat revealed that the once small Tony had grown into a slender girl of slightly above the average height. She was wearing a party dress of red velvet with widely puffed sleeves and a gored skirt, that great and popular invention of the nineties. An antique necklace of garnets set in old gold was the only thing she wore in the way of jewelry.

"The last time I saw you two young people together," said Catherine, "was when you were mere children and had made a snow man. I climbed over the fence to inspect it."

"And Ludar recited a poem he had made up about you!" said Tony. "It was only two lines, but I thought it wonderful."

"And it came true. I *did* marry Clyde. You brought me good luck, Ludar."

It was clear from the light in Catherine's eyes that she honestly considered herself lucky. No one would have agreed with her, seeing how fragile she was and knowing that she was confined all day long to the wheel chair. The excitement of unexpected company was telling on her already, for her cheeks were flushed and her hands showed a tendency to tremble.

"It was a beautiful snow man," said Tony with a sigh. "I always hoped you would make another one, Ludar—but you never did. You were always so very busy. You've been very busy at the Collegiate, haven't you? I hear stories about all the things you do, serving on committees and making speeches. It makes me a little jealous." She added quickly, "I mean, I realize how much fun I miss by having to go away to a private school." She picked up her evening bag and opened the clasp. "I had a purpose in dropping in, Cathy. My contribution hasn't been paid to the Consumptive Fund and, as you're one of the officers——"

"A vice-president," said Catherine. "Strictly by courtesy, because I can't do a bit of work. But I can accept contributions."

Tony opened her purse. Ludar, watching intently, was delighted and awed by the graceful movements of her hands. His own hands were tucked deep in his pockets to conceal the inadequacy of the sleeves. He watched her take two dollars from a jumble of bills, and a hurried calculation convinced him that she had more money in her purse at that moment than he had possessed in the whole of his life.

The talk went on busily for half an hour, and then Tony announced that she must be on her way. She was calling for Grace Karns, who lived in the next block, and they were going together.

"Why not telephone and say you've made other arrangements?" Ludar suggested. "And then let me see you to the Woods' house."

Tony said promptly, "May I use your phone, Cath?"

She made a confession when they reached the street. "I hoped you would suggest this. We've so much to talk about. Why have we become such complete strangers?"

Ludar could have told her why. She had withdrawn, perhaps not willingly, from all circles where they might have continued to be friends. Most of the time she was away, at Molson's School or on trips with her father, and when she was home she was caught up in social activities of which he heard no more than the faintest echoes.

The way to the Wood house lay south, but by mutual though unexpressed consent they turned north. Then they wandered without purpose, save to continue their talk, finding themselves finally in the far reaches of the East Ward by way of streets which neither of them had ever seen (and which they did not see now), before Tony came to a halt.

"Whatever am I going to say to Myrtle?" she asked. "It's rude to be so late at a party."

"Must you go?" asked Ludar.

They had stopped beside a small corner grocery which was still lighted. The proprietor was busy behind the counter with his ledger, and there was a telephone on the wall behind him.

"Why not go in and telephone her that you can't come?"

Tony thought for a moment. "All right," she said. "But Myrtle will think it mean of me. What excuse am I to give? Shall I tell her the truth, that a selfish young man is insisting on monopolizing me for the evening?"

They walked for an hour and a half and were too concerned with what they had to say to each other to know where they went, or why, or to feel any fatigue. They talked with all the eagerness that the renewal of an old acquaintance can evoke, of books and music, of human relationships, of virtue and wickedness, of themselves, and of life. There were no pauses. If a moment's hiatus threatened, one of them would rush in with, "But don't you think——" and they would be off again. Once she called him "Ludie," and he felt himself lifted up to the heights because this meant she remembered the affection they had shared in their childhood.

Tony had been rather quiet as a small girl, but she now took her full share of the talk and he realized, whenever they passed under a street light, that her eyes were sparkling with animation. She confided her hopes and fears and ambitions, telling him that during the past year she had wanted above everything else to be a tennis champion. She had become, however, a believer in the new style of hard stroking, and this had led to a startling discovery, that it would broaden her knuckles if persisted in. "It was too high a price to pay," she said with a sigh. Ludar became pontifical on the subject of literature and laid it down that J. M. Barrie was the greatest living author, although Anthony Hope was giving him competition for the honor.

Mostly, however, they talked about themselves, and they reached one conclusion: that never again, come what may, in spite of interfering aunts and silly notions of pride, in spite of every adverse condition, even if Halley's Comet (which was expected with dread in a few years) should come ahead of time and upset the universe, would they fail to see each other regularly.

It was in every way a wonderful evening. There was a special vitality in the cold air which made them seem to be treading clouds. There was kindliness about all the houses, a special warmth to the lights in store windows, a graciousness to the churches. The sound of their footsteps became translated into music; there was magic in the stars, and even a touch of poetry about the lampposts.

But when they came to the Milner house Tony gave him her hand and said simply, "Good night, Ludar." He took it and let it go at once, reluctantly, however, saying no more than, "Good night, Tony."

He walked in a state of exultation around the corner to Wilson Street. What did it matter that so much of Aunt Tilly's capital had been squandered, that Uncle Billy's roller boat had been a failure, that the editors had unceremoniously rejected his story? Nothing of the kind mattered. He had fallen in love!

Chapter II

I

As soon as the close of the term brought Ludar's schooling to an end he went into full-time employment at Haverstraw's. His employer was a pleasant and bookish man who nevertheless figured costs with infinite care and a deep reverence for his own profits. He reluctantly agreed to pay the boy three dollars a week. It was a very poor wage, but nothing could induce him to raise it, although Aunt Tilly made a call on him to discuss the point.

The hours were long and the work sufficiently tiring to send the boy home at night in an exhausted condition. He never failed to settle down to his manuscript after supper, but under the circumstances he made small progress. He would struggle through a few pages with lead in his fingers and then take the paper and tear it into pieces. An editorial instinct, it seemed, was already showing itself.

One Saturday evening, when he had the house to himself, he proceeded to take stock and reached the unwelcome decision that he was wasting his time. How could he hope to write a story of biblical days (he was calling the new effort *The Bad Samaritan*) when he knew so little about the times? To escape from this dismal reflection he let his mind go back to the now very dim memories he retained of his first years. They had become so shadowy that Aunt Callie and Mr. Collison and Old Mr. Cyril had almost lost reality. The only tangible link with the past which was left to him was the sign he had worn on his back. On an impulse he got it out and spread it before him on the weak-legged little table he used as a desk.

The lettering was crude and the oilcloth was beginning to crack, making it difficult to decipher some of the words. He studied it intently, thinking how sharp Aunt Callie had been to devise such a scheme in order to keep the money for herself. Then he turned the sign over and discovered that she had been more than sharp. She had been sly as well.

In a tiny hand she had written two words in different corners of the oilcloth: "red" and "craft."

When Ludar saw them it was like opening a door and letting the light pour into a darkened room. He heard a voice, which was at the same time both familiar and strange and which he believed to be that of his mother, saying, "This, sir, is Ludar Redcraft." Memory flooded back, and he knew that he had rediscovered his real name. Strangely enough, he had created at the same time a clearer picture of his mother than he had ever enjoyed before. She had eyes which crinkled at the corners when she smiled and fluffy golden hair which she wore in a high pompadour. It seemed to him that she was poorly dressed, at least as far as the material was concerned. The words were addressed to an elderly man with a beard and a glass in his eye. Old Mr. Cyril, of course! The strange part of it was that this recollection was a new one. It had never been among his memories, and yet the picture was clear in every detail. He could see even that his mother had a small scar on one cheek.

The picture vanished and, as though his consciousness of the past had been stimulated by the discovery, others took its place. He saw Aunt Callie writing one of her interminable letters, desisting long enough to shake her fist and say, "These Redcrafts think they're so much better than us!" He recalled also that, when the old lawyer had made him repeat so many times, "My name is Ludar Prentice," it had been hard for him not to say instead, "My name is Ludar Redcraft."

The boy got to his feet and put the sign away in the bottom drawer of his bureau. His hands were trembling with excitement.

He must share this wonderful revelation with someone at once. Uncle Billy was at a meeting, and Aunt Tilly was calling on a relative. His thoughts, therefore, turned to Tony, and he wondered if he would dare go to the Milner house and ask to see her. He had never been inside it and he realized that it had been an undeclared prohibition which had kept him out. Would the circumstances warrant his breaking through the barriers? In a mood to make the attempt he put on his cap and walked out in that direction.

He was in luck. As he turned onto Grand Avenue he saw Tony going into Blaywood's grocery store, which occupied the corner on the other side. He called, "Tony!" and crossed the street on the run. She stopped and turned.

"Hello, Ludar."

"Something important has happened. I want to tell you about it." An inner sense of the dramatic prevented him from blurting out what it was. He did not want to spoil his story without first explaining how he had come to remember the name.

He was too excited to take anything else in. Otherwise he would have noticed how smart she was looking in a brown corduroy suit with a box-pleated skirt, her feet in the trimmest of brown oxfords. He did notice, however, that she seemed uncertain about carrying on a conversation so publicly.

"We can't talk here very well," she said. "I'm going to the library tomorrow night to return a book. I'll be there at seven-thirty. Do you mind?"

He minded, of course, for the secret was clamoring to be shared at once. But, realizing that she was right, that the entrance to the Blaywood store was in full view of the Milner windows, he nodded and said with a smile, "I'm going to the library, too, as it happens."

They were both on time next evening. Meeting at the delivery counter, Tony said, "Hello," in a casual tone and then walked to one of the periodical tables in the reading room. Ludar turned in a slip, received the book he had requested, and then strolled in the same direction.

In discreet whispers he told her his news. Tony said excitedly: "How wonderful! Now you'll be able to find out everything. Oh, Ludar, I can't tell you how glad I am!" Then she repeated the name: "Redcraft! It's quite different, isn't it? And distinguished too. I think it must be a very old name." Ludar had also reached this conclusion. For some reason which he could not have explained he was sure his family belonged to the North of England. It suggested to him the forests of Northumberland, and Hotspur riding at the head of his knights.

After they had discussed the matter as long as they dared Tony felt the need to explain her reason for suggesting that they meet here. "Ludar," she whispered, "you must have wondered why I haven't asked you to come to the house. It's—it's very embarrassing. Aunt Mona May has never believed you come of a good family. She puts so much importance in things like that. That was why she didn't invite you to my party that time. Won't it be wonderful when you find out that you *are!*" She busied herself with her library card and her gloves. "I'm afraid I ought to go now. Ludar, would you be interested if I said—well, that you might have a surprise if you happened to be at Catherine's tomorrow at eight o'clock sharp?"

The surprise consisted of a very fashionable-looking young lady in a fawn dress with a three-tiered skirt, each tier edged with velvet of a darker shade, and a smart brown hat, who said gaily, "Why, Ludar, what a very nice coincidence!" Clyde, who was curled up in a deep chair with a book, got to his feet and said: "All the conspiracies of history were no more than a series of coincidences of exactly this kind. Come on, Cathy, we had better make ourselves scarce so they can take full advantage of this most unexpected meeting."

Catherine smiled delightedly and wheeled herself out to the dining room. They could hear the click of the elevator door—Mr. McGregor had sacrificed his study for the purpose of putting in a shaft so she could get up and down stairs without difficulty—and they knew they would have uninterrupted possession of the living room for as long as they cared.

Tony seated herself on the sofa and said, "Now we can talk." They talked for a full half-hour, entirely about the possibility of continuing the friendship which had flared up again between them. The great obstacle to this, she confessed, was her aunt.

"In another year," she said, "I'll be old enough to come out and then I'll refuse to let her interfere in my life. But what are we to do now? It would be nice to go on seeing you, but to meet like this very much—well, it would be noticed in time and we would be talked about the way Mrs. Gadsby was when she used to run around with that terrible Mr. Corliss. No matter how sensible we were, it would be kind of underhanded."

"I suppose so," said Ludar gloomily. "But it's better than not seeing you."

Tony had been in a most serious mood, but at this she smiled. "I'm glad you want to see me. I had become certain that you didn't care about me at all—— Oh, I know, of course, that things have been difficult." She gave the problem some thought, her brow wrinkling in a way which had been familiar to him in the early days. "Perhaps we should see each other just once in a while. Here or at the library. Don't you think that would be sensible?"

"Yes," he said. "But I don't like being sensible."

She gave a sudden gasp of dismay. "I've just remembered something. We're going to Europe, Father and I. We'll be gone all summer."

"When do you leave?"

"In two weeks. We sail from New York."

As soon as she had left Ludar heard the elevator doors click and the rattle of the wheel chair. "Well," said Catherine, "I hope the conspirators had a good time. You don't look as though you had."

Ludar shook his head. "It's terrible to be young when a crisis arises in your life."

"Yes, I know it is. And when you're young life is just one crisis after another."

"I can't do a thing about it. I just have to wait and hope—and dream of accomplishing great things. It's not very pleasant to be in such a fix."

Catherine leaned across the arm of her chair. "I can tell you this," she whispered. "Tony likes you. She likes you very much."

2

There was a visitor at Number 138 this evening, none other than William Pitt Milner. The irascible publisher seemed surprised and a little dismayed at the bareness of the rooms and the shabbiness of the furniture. After one quick glance around him, however, he seated himself beside William in the sitting room.

"I've news for you," he said. "Bad news in one sense. But good in another, because it serves to clear the air."

He brought out a sheet of flimsy paper from a pocket and handed it to his old school friend. "Willing brought this to me just now. It came over the wire from Toronto."

It was a news dispatch from a small town in eastern Ontario, and it told of the sudden disappearance of a printer named Lamb, an Englishman. He had been instrumental in putting some townspeople in touch with a lawyer in England who claimed to be in a position to forward their interests in a large estate. After collecting considerable money from them without any results, the printer had cleared out. Lamb, it had then been discovered, had acted in a similar way in other towns. He would get a job on a local newspaper and keep his ears open for word of residents with any kind of expectations in the mother country. There were always plenty such. A letter would come from a lawyer saying that Uncle Robert's will had contained provisions for heirs abroad, or that the title to a great estate was in doubt and were they descended from one Miles Morton Mainwaring, etc. This never failed to send them into ecstatic willingness to invest money for further information. The dispatch concluded with a description of Bert Lamb and the information that the police in half a dozen towns were now on the lookout for him.

William read the story through and then handed the sheet back to the publisher. "It calls the lawyer Headstone," he said, his face red and unhappy. "I suppose he used a different name each time. I guess we were his most profitable suckers."

"Probably. He certainly got a lot of money out of this town." Milner seemed quite unhappy about it also. "We'll have to publish something tomorrow. I warn you, Billy, it's going to cause a lot of angry buzzing and clacking of tongues. All these bighearted people who contributed a dollar or two are going to be furious. They'll blame the *Star* for starting it."

"It will make things bad for the boy."

Milner nodded. "I'm afraid it will. He seems to have grown up into quite a nice young fellow."

Tilly had been stirring about in the kitchen and keeping her ears open. William could tell from the stricken look she now wore that she knew the worst. He half expected her to go into one of her violent exhibitions of grief, but nothing happened, perhaps because of the presence of a visitor. Instead she seated herself near one of the windows and stared out into the darkness. "Poor Tilly!" he thought. "This will go hard with her."

"What a simpleton you were, Billy," said the publisher, "to throw away your savings like that! And yet at the time I must confess that I had some admiration for you. I said to myself, 'He may not be looking out for Number One, but he's showing that he has a stout heart.'" The rockers of the chair he sat in were widening a hole in the straw matting on the floor, but he was not aware of it. "Your money may not be lost at that. My daughter tells me the boy has remembered his real name at last. Well, I'm sailing for a tour of England and the Continent before long. It's partly a business trip, partly for my daughter's sake. I'm taking her with me so she can see something of the world. Now that we know his name we ought to be able to ferret things out and get in touch with his father's people. If you're willing to have me do it I'll take this business in hand while I'm over there."

William was quick to assent to this plan. He had such faith in his old friend's capacity for getting things done that he now considered the mystery as good as solved. One consideration only caused him some hesitation. "I'm afraid," he said, "it will take a lot of money."

The publisher brushed this aside. "We sat in the same seat at school, Billy, and we were always the best of friends. I'm only too happy to be of help. And to be perfectly honest about it, I rather fancy myself as a sleuth. This business will be a pleasant relief from all the visiting of churches and museums I'll have to do."

Chapter III

I

The most-talked-about event of the year was the arrival of the first automobile in Balfour. It was driven from Toronto by a mechanic named James Blarnish and was delivered at the Langley Craven door on the afternoon of the first Monday in July. Ludar knew nothing about it until he was ready to leave work for the day, and then his first intimation was in the form of a great uproar in the street which suggested mobs and riots and the breaking of heads.

He hurried out, buttoning up his coat and slamming the door, to find drawn up at the curb a small red automobile with a front seat for two and a curious arrangement behind which would be called in later years a rumble seat. A man in a peaked cap was at the wheel and Joe was beside him, and all Balfour, or so it seemed, had assembled on the street and the market square to watch and laugh and jeer. Men going home from work with their empty dinner pails had lined up along the sidewalks; dozens of rigs had been hauled to a stop; the members of the Fin de Siècle Bicycling Club, which was set for a run out the Hudson Road, had dismounted and taken possession of the sidewalk opposite and were laughing and exchanging jokes about "the coffee grinder on wheels."

Even the Creams were in evidence, driven by a traveling salesman with a cigar in his mouth and the youngest of the Tawney sisters beside him, the dark and impudent one. Considerable surprise was being manifested over this, as it had always been understood that the Creams were reserved for ceremonial purposes. Taking one of the Tawney sisters out for a drive in the country might be considered a ceremony, but hardly of the right kind.

Joe called excitedly: "Hop in, Ludar. Aunt Lish wants to see you, and I'm going to drive you there. This is the new car. A beaut, isn't it?"

Ludar stepped into the front seat in a mental daze. There had been much talk in the Craven household of the car they intended to get, but it had always seemed hazy and very much in the future. And yet here it was, bright and shiny and a symbol of the changing times; and he, Ludar, was to be one of the first to enjoy the privilege of a ride in it.

Blarnish was in front of the car, turning a crank. There was a sudden splutter, the car shook from stem to stern, and then something inside it began to rumble and purr. The driver climbed into the rear seat, grinned, and said, "How's 'at, Joe?" and Joe answered, "Good work, Jamesie. You got her going first crack out of the box."

The car was actually moving, and the gaping crowds were hurrying to get out of the way. Horses reared and showed a tendency to bolt. Even the dignified Creams tossed their yellow manes and snorted and made it necessary for the drummer to spit out his cigar and hang onto the reins with both hands. This did not seem to bother the dark and impudent Tawney girl. She smiled at them as they passed and said, "You must come and take me for a ride in that thing, Joe." Joe smiled back and said, "Sure thing, Bella."

When the car turned northward and the streets were no longer crowded with jeering spectators Joe started to talk. "Now, young skullurk"—a favorite word of his own concocting—"I've got news. And, holy man, Ludar, it's great!"

The car buckled and threatened to stop. Then the purr of the motor was resumed. "I'm not going to college," said Joe. In sheer exuberance of spirits he let out a piercing whistle and then leaned over and sounded the horn three times. It had a high shrill note which sounded like "Pleep!" Ludar was equally excited. It had seemed certain that the happy triumvirate would soon be a thing of the past. Arthur was going to Toronto University in the fall to take an engineering course, and Joe's parents had been determined that he also was to go. Ludar had been certain that the winter would be a time of loneliness for him. "The well-known firm of Pop and Mop, Inc. have given in. They agreed that since I wasn't interested in more education—boy, how interested I *wasn't!*—it was foolish to make me go. Dad was upset about it. He said all gentlemen went to college, and I said there were grave doubts as to whether I was a gentleman or not. Then he said you got culture at college, and I said I wouldn't recognize culture if I met it face to face on Holbrook Street. He laughed at that and said 'no, culture and I weren't even fifth cousins, God help you, Son.' Then he gave in."

"Are you going into the business with him?"

"Yep. He asked me what I would like to do, and I said I wanted a job which made it necessary to drive around town a lot in an automobile. He said nix—you know how he picks up our talk sometimes—that I would do all my driving behind a desk. Anyhow, I'm not going to college. Hooray!"

They were approaching Miss Craven's tall and unprepossessing house. "Look, Ludar," said Joe, "I've got to go right on to the station and pick up Mother, who's been in Toronto. I don't want to stop the car and take chances that it won't start again. We'll slow up and you jump out, will you?" He was still thinking of his good luck. "No more lessons," he said. "No more silly examination papers, no more sour teachers telling me I'm ignorant. Holy man, it's a great feeling!" And he leaned over and sounded the horn louder than ever.

The car spluttered almost to a stop, and Ludar hopped out. "I certainly am tickled that you're not going to college, Joe," he called. He was so happy about it, in fact, that he put a hand on the side gate and vaulted over it instead of opening the latch and walking in.

Al Hanley was in the kitchen and took him right up to the top floor. "Listen to that," he said as the hoist went into a paroxysm of shaking and rumbling. "I know the meaning of every sound this old lady makes. No one else in the world can get her to start. Miss Craven knows it and valleys me accordingly, but she's the only one does. Here we are, young Master Duke, Your Grace. Last station."

Miss Craven's long room was deserted when he stepped out of the cage, but the door into the inner room was open. He was very much surprised to see that it contained an English billiard table. The owner appeared in the door with a cue in her hand.

"Good evening, young man," she said. She chalked the cue and deftly sent the red ball into one of the corner pockets. "This was my father's. He was one of the best billiard players in the world, I guess. Neither of my brothers took much of a shine to the game, so he used to play with me. We played every afternoon at five-fifteen sharp, and he taught me how to make all the shots. I'm good at it, if I say so myself." She continued to make shots as she talked, and it seemed to the onlooker that she was amazingly expert. "Mum's the word about this, Ludar. People would think it very unladylike of me to play this game, so I keep it a dark secret. I'll tell you something, my boy. I could take on any stuck-up squirt of a man in this town and stand him right on his head. But I didn't send for you to tell you that." She straightened up and regarded him accusingly. "Why didn't you go to the closing prom at the Collegiate?"

"Well," said the boy, "I was too busy."

"You needn't think you can fool me like that. You didn't go because you didn't have suitable clothes."

Ludar's face turned red. "Perhaps that had something to do with it," he acknowledged. "We haven't much money for clothes."

She let her eyes rove over the shabbiness of his coat and the worn condition of his tie. "Now see here, young man, you know I want to help you in every way I can, and you probably know also what a careless old thing I am. I should have been keeping an eye on you and not letting things get to such a pass. Why didn't you come to me?"

"Why, I—I couldn't do that."

"And why not? What kind of silly pride is this? From now on I want no more of it." She desisted long enough to execute a difficult massé shot. "Now, Ludar, before I tell you what I've decided I want to say a word about that Joe nephew of mine. You and he have been such good friends, and I want you to keep right on that way. But this Joe nephew has found another great interest in life. Girls!" She gave the ball a vicious ride around the table. "Why, that boy doesn't seem to care what they're like as long as they wear skirts. He's becoming a regular philanderer! It's going to get worse now that his father has bought this horseless carriage and the boy will have all the dark country roads as his territory. I'm afraid he'll be getting himself into trouble and having to marry some giggling chipmunk of a girl. I tell you, Ludar, I'm frightened.

"Do you see what I'm leading up to?" she went on. "You two have got to be still better friends and have all sorts of interests between you so his mind can be kept off girls. You go in tomorrow to Huban and Hill's and get yourself a new outfit of clothes. Ask for Hugh Hill, and he'll have his tape measure out before another word can be said. He knows exactly what he's to do for you. All you'll have to do is choose the materials and the colors you want."

She replaced her cue in the rack and turned off the green-shaded light over the table. "I'm going away next week. To California, to a little place called St. Angelus or something like that. You may find yourself in need of money while I'm away, and all you'll have to do is come here to the Chief. She'll keep you supplied with enough so you can travel right along with this Joe nephew. I can see by the way you've got all flushed up you don't like this. You're an independent young coot. I admire it in you, but I don't want my plans upset because of it. Here's the way we'll leave it. You know that the money is here if you need it, and I hope you'll come in often."

2

Norman Craven stepped off the train with a friend from school and saw the car drawn up at the station platform. He turned an angry red when he recognized Joe sitting in it. Norman had grown quite tall, being at this stage close to six feet, and he seemed to think that an air of gravity suited best his importance as son and heir of the richest man in town. He decided that he must control his feelings.

"Kenneth," he said, "I'm very sorry, but there's going to be a peach of a row when I get home."

The friend, being Kenneth Cleary, the son of a very rich man in Toronto and consequently the most-looked-up-to boy at St. Paul's, did not seem much concerned. He had been looking about him and had spotted Mrs. Langley Craven stepping off the train. "Say!" he exclaimed. "There's a pip for you. I admire a handsome mature woman like that more than these little snips of girls we have to endure at parties."

"That woman is my aunt," said Norman. "And I hate her."

"Must be hard to hate a woman as handsome as that," said his man-of-the-world friend.

"It's just as I expected," declared Norman darkly. "My parents have let my uncle get ahead of them. That must be his car over there, the first in town, because it's my cousin sitting in it. I was sure it would be this way."

His friend had nothing but contempt for the little red roadster. "We have a Mercedes at home. You ought to see *that.*" He became unhappy in turn over the conduct of parents. "I'm not allowed to drive it yet."

"I'm not going to give my father any peace until he's bought a Mercedes. He can afford it easy enough."

As they got into Barney Grim's cab Norman entered on a necessary explanation. "I must warn you, Kenneth, that my mother doesn't allow cards or dancing in the house. Or smoking. She's a perfectly terrific Baptist."

"Say!" The visitor looked shocked. "This is going to make it tough, isn't it?"

Some of the younger set had been invited in for dinner. It was served buffet fashion, but it was a substantial meal with a roast of beef, a cold salad in aspic, and a truly magnificent charlotte russe for dessert. Some of the girls were pretty enough to catch the blasé eye of the young house guest. Watching them over the tallest and stiffest linen collar which had ever been seen in Balfour, he decided that Tony was the best of the lot. He waited on her assiduously throughout the meal.

"I've been told a couple of things which worry me," he said to her while coffee and chocolate were being served. "First—and this one is a blow from which I may never recover—that you and Norman are—ahem, how shall I put it?"

"Oh no," said Tony hastily. "That's not true."

"It makes me very happy to hear *that,*" said Kenneth. "The other thing was that no dancing's allowed in the house."

"That *is* true."

"Is flirting also banned? This, you know, is positively maddening." Then a doubt assailed him. "See here, you're not against it, are you?"

"Dancing? Oh no." There was a hint of amusement in Tony's eyes. "I like to dance."

"Then we'll have to manage it somehow. I claim the first one and as many more as I can manage to beg, borrow, or steal—and I'm good in all three departments." He lowered his voice. "I happen to know that Norman is quite crazy about you. Well, from now on I am dedicating myself to blasting his hopes, to trampling on his aspirations. In short, to carrying you off right under his nose."

Tony laughed and began to sing in a low voice:

"Try again, Johnny, try again do.
Fair village maids are the right sort for you."

The visitor looked surprised. "You saw that show? I didn't know it came this far out."

When Mrs. Craven finally left the young people to themselves the house guest suggested a plan for the circumvention of the obnoxious house rules. They would dance in the library, which was in a corner of the house, and someone would always stand in the hall as a lookout. Norman, who gave in with some doubts, volunteered to stand watch. It was lucky that they had taken this precaution, because Mrs. Craven came sweeping down the stairs very soon after the dance records began to sound through the house. She discovered nothing incriminating in the library, where an earnest discussion of books and plays was being carried on. Perhaps, however, she detected a slight strain in the air, and perhaps also a flush on some of their cheeks. At any rate, she said on leaving, "As you're all *here,* you'll hardly need the music going in the living room," and she went in there and turned the gramophone off.

After a moment's silence Kenneth said, "I suspect, Norman, that your mother knew what was going on."

"Of course she did," said the son of the house.

The guest of honor sat for another moment in gloomy silence. "What do we do now?" he asked. "Have you any conversation lozenges to hand around?"

The party ended at eleven sharp, and the visitor went promptly and grumpily to bed. Seeing a thin line of light under the door of his mother's bedroom, Norman knocked and went in.

For some time past Mrs. Craven had seemed like a woman who had given up. She had become almost massive in frame, and her face had broadened in proportion. Her bedroom, however, did its level best to contradict this. There was a frivolous note to the appointments. The bed was French, a thing of slender legs and rose-colored satin panels. The dressing table was covered with cosmetics in expensive bottles, and the hangings were of the costliest materials.

"Have your young friends gone?" she asked, dropping in her lap the Bible she had been reading.

"Yes." Norman spoke in a glum voice.

"They were dancing. I said nothing because I didn't want to create a scene. But, Norman, it must never happen again." She frowned. "I'm sure it was the work of that boy you brought home with you. He must never be invited here again."

"Hey, wait a minute!" Norman was shocked. "Kenneth was the most important fellow at St. Paul's. And he's going into varsity with me."

"I don't care who he is. He's not to be allowed in this house."

Norman seated himself beside the bed, stretching out his long legs. "Mater," he said abruptly, "I didn't come here to talk about Kenneth. There's something much more important. Look here! I've got to have one too. I tell you, I must. It's been a terrible mistake to let him get ahead of me this way."

"Whatever are you talking about?"

"I'm talking about Joe Craven. And that red car he has."

Mrs. Craven laughed. "Oh, *that*. Why are you so excited about it?"

"Because, Mater, it's a mistake for him to have the only automobile in town. I tell you, it's an unhealthy state of affairs and I must have one too, right away. If I don't get one I'll spend my vacation somewhere else. I couldn't face people as long as he has a car and I haven't."

Mrs. Craven took off her reading glasses and folded them into their case slowly. "My dear son," she said, "it's going to be necessary for me to disappoint you. There will never be an automobile in this family. Every time I open the Bible, Norman, I become more certain that they are instruments of the devil. They go against nature. Against the teachings of God."

Norman stared at her belligerently. "You can't have any idea how silly and old-fashioned that sounds."

"I may be silly and old-fashioned," declared his mother sharply, "but there's one thing you seem to be forgetting. The Craven money was made out of carriages. What would happen to us if everyone started to use these new contraptions?"

"There you go! It's clear you haven't studied this matter the way I have. No one with any sense thinks everyone will own automobiles. They'll always be too expensive for that. Only the best people will have them. The rest will keep right on walking, as they do now." He pulled back the sleeve of his coat and consulted his wrist watch. "It's getting late, but we might as well get this settled now. We ought to have two cars, one an electric for your special use. I saw plenty of them on the streets of Toronto when I went down for that dance. They were all Waverleys, and I want to tell you they were slick. It was something to see them breezing along at ten or twelve miles an hour. Certainly the leader of society in Balfour should have one of them."

Mrs. Craven drew a deep breath. Her face had taken on a tallowy look, and the roll of flesh under her chin shook with the intensity of her feeling. "My dear son!" she said. "My fine handsome big son! It breaks my heart to do this, but—things have changed, Norman. I'm no longer the leader of society in Balfour."

So great was Norman's surprise that the one foot resting on the other fell to the floor. He pulled his hands out of his pants pockets and sat up straight in his chair. He looked at his mother as though he suspected her of a temporary loss of sanity.

"What foolishness is this?" he demanded.

Mrs. Craven spoke in a resigned tone of voice. "I'm telling you nothing but the truth. Things have changed a great deal in the last six months. I don't suppose there really is a leader any more, but if there was it would be the wife of your uncle Langley. She's in charge of charity drives now and she's the first patroness of the dances. Your uncle is president of the golf club. Oh, they're cutting a fat pig again! Their house is always full of people. They're worldly, you know. Your father and I are no longer in tune with things, I guess. We're too serious, too concerned with the *real* things."

Norman had not taken his eyes from his mother's face as she talked. His cheeks had become almost as red as his hair, and he was breathing hard. It was clear that he was not only surprised but angry.

"Isn't Father still the richest man in town? And the most influential?"

His mother nodded her head. "Yes, Son. But things have changed. The city is so much larger. There are new factories. And new stores and banks, and new people with plenty of money. Your father hasn't the influence he used to have." She sat up in bed and gathered her jacket about her. She began to speak in a more decided tone. "You mustn't think, Son, that I'm giving up. It's the swing of the pendulum. Right now it's swinging away from us. But it will come back. Make no mistake about that, it will swing back! Not that I care for myself——"

"What have pendulums to do with it? We're the Tanner Cravens, aren't we? We're the richest and the best family in town, aren't we? If what you say is true, we're not going to stand around and wait for pendulums to swing back. We're going to step in and set things right ourselves."

"It's not as simple as you think, Son. There's nothing we can do——"

"There are lots of things we can do!" declared Norman in a savage tone. "Father can—well, he can crack the whip, can't he?"

"It would have no effect on the people who count. They're not dependent on us at all. I mean the younger set—the people who dance and play cards, like this new game of bridge, and go to the country club and play golf. They think we're dull and old-fashioned."

"And you are! I want to tell you this, Mater, that if you keep on being against cars you'll be left behind in the parade. I can see that clearly enough."

"Norman, there's no use talking." His mother's voice had taken on a sharper edge. "Your father and I agree. We will *not* descend so low as to own one of these things."

Norman's face wore a sulky look now. "What's the matter with Father, anyway? I haven't heard him say half a dozen words since I came into the house. All he did was grunt when I introduced Kenneth to him. That's no way to treat a guest whose father's as important as Kenneth's."

His mother's mood had achieved another complete change. She shook her head with an air of despondency. "Your poor father is not in good health. He has headaches a good deal and he complains of ringing in his ears. Dr. Clarke is worried about him. He says he's too young a man to be like this but that something will happen if he doesn't take better care of himself. But he goes right on. Nothing I can say has any effect."

"He goes right on! What do you mean, Mater? What does he go right on doing?"

Mrs. Craven turned in bed so she could look squarely at her son. "He goes right on smoking—and *drinking*. He buys two bottles of whisky a week and drinks it all himself. He thinks no one knows about it, but it's become the talk of the town. How can he expect *not* to have a ringing in his ears when he keeps himself intoxicated like this?"

Norman was finding it hard to accept these unpalatable facts. He got up and began to pace about the room. Every few steps he would raise his arms and give them an impatient wave. "I come home," he said bitterly, "expecting to find things as they were. And what *do* I find? I find relatives usurping—yes, usurping—our position in society. I find my father shirking his job. That's what he's doing. He may be in poor health, but that's no excuse for—for abdicating in this way. I don't understand how such things can happen." He stopped pacing about and looked down at his mother as though he felt like breaking down and crying. "Don't you think I have any rights? Don't you know that when you let things go like this you're letting my part of it go too?"

3

The Craven automobile was more than a nine-day wonder. It was a topic of conversation in town for weeks, and the adventures of the little red car were retailed at full length. Mort Rattigan, for instance, kept tab on the different girls seen in it with Joe. He worked in the tobacco shop at the corner of Wilson and Abercrombie and was in a position to observe everything. "Twenty-two to date," he announced one night at the Yukon Club. "And tidy little bundles, all of them." Wesley Albert, the auctioneer, who never got out of the habit of calling it an "ordermorebill," was always telling jokes about the car. "The ordermorebill got seventeen punctures going to Hudson and back last night. Why don't they just sit in the back yard and pump up the tires as they blow out, and save all the wear and tear?"

Once when this had happened—at least the car had gone dead in the yard and refused to move—Peter Work came along on his way home from work and saw that Joe was in difficulties. He hopped over the fence and labored with the son of the house for half an hour. By that time Joe was as black of face as his companion, but the engine was running sweetly again. Mr. Craven, who had come out to watch, was very much impressed.

"Peter," he said, "it won't be long before there will be lots of cars here in Balfour, enough to make a public garage and repair station necessary. I'm inclined to think you are the man to start it."

Peter's eyes gleamed in the black expanse of his face. "That is what I'd like to do, Mr. Craven. I've a way with machinery. By grab, I can make things run."

"This is a deal between us," declared the owner of the first car. "When the time comes I'll see to it that you're set up in business."

Joe fell into the habit of calling to take Ludar home from work. Sometimes Meg would be in the car. She would say, "Hello, Lude," in an indifferent voice and pay no more attention to him. Occasionally Mrs. Craven would be sitting beside her son, being driven home from some afternoon function. She was almost as fond of the car as Joe and had all manner of motoring clothes, in which, as was to be expected, she looked stunning.

On the day when Tony and her father were leaving on their trip abroad it was arranged that Ludar would ask for an hour off at the store and that Joe would call for him and drive him up to the station. Ludar hoped that his new suit would be ready for this important occasion, and he called in at Huban and Hill's several times to urge that it be finished in time. They gave him solemn promises, but at noon of the momentous day it had not been delivered. Ludar was so bitterly disappointed that when Joe came by for him he felt like telling him to go to the station alone. It did not help that Joe was looking his very best. He had grown into quite a handsome fellow, not as tall as Ludar but well-proportioned, and he wore his clothes with an air. He had just been at the barber's, and his hair was neatly trimmed and plastered down with bay rum. He looked spick and span and decidedly debonair, with a flower in his buttonhole and a new watch fob.

Tony was standing on the platform when they arrived, surrounded by a group of friends. It was a very gay party. The girls were wearing light summer coats, and they had brightly colored parasols. The boys wore straw sailors or bowlers, and they were carrying gloves and canes and looked stylish and, to use their own word, doggy. It was impossible to see much of Tony because her arms were full of flowers and parcels.

Ludar's heart went down into his new boots which, at least, had been ready. "Is it customary, Joe," he asked, "to give presents to people when they're going away?"

Joe nodded indifferently. "Yep," he said. "It's an expensive and silly custom, to my way of thinking."

Ludar had never heard of this. Sitting in the car without making any move to get out, he began to realize that there were undoubtedly many such things of which he was ignorant.

"Joe," he said, "I'm feeling kind of sick. Without my new suit I'd look like a country bumpkin in that crowd. I think I'll stay here if you don't mind. You go over and say good-by to Tony for the two of us."

"Heck!" said Joe. "What are you afraid of all of a sudden? That gang? Why, you know them all."

Ludar swallowed hard. "I used to know them," he answered. "But I'm not sure that I do now. They've changed and they may have forgotten. In fact, I'm sure they have."

"Tony has spotted us. Won't she think it funny if you don't go over?"

Ludar thought for a moment. "I think, Joe, she'll understand if I don't."

Joe sat back and watched the proceedings, making comments on the participants. "Say, that Linda Dorr is pretty snappy in that yellow dress.

Have I been overlooking something? And will you take a gander at little Billee Brown? What color is she wearing?"

"Mandarin," answered Ludar, who had an accurate eye for color.

"She's a plump little grig, isn't she? I kind of like 'em plump, Ludar." He exploded suddenly: "That Norman is certainly making a play for your girl, Ludar. I hate to see you sitting back and giving him a clear field."

"This isn't the time to do anything about it, Joe. I'm—I'm outclassed today."

Tony, standing beside her father, her left arm around a large bouquet of flowers (Ludar was certain it had come from Norman), was as gay as the occasion demanded. She looked a little distrait at moments, however, and her eyes kept turning to where the red car was backed up against the platform. Standing on the steps as the train for the junction began to puff and shake, she called, "Good-by, everyone!" and kissed her hand. It seemed to Ludar that her eyes were on him when she did this.

"Damn that Norman!" said Joe as they drove away. He seldom swore, and so it was certain that he was feeling deeply. "He acted as though he owned the girl. You'll have to be ready to stand up and fight when Tony comes back, or he'll cut you out."

Ludar made no comment. He had been realizing that the lack of his new suit had been a small matter after all. The lack of other things had held him back. The possession of new clothes alone would not have fitted him to play a part on an occasion like this.

"I'm taking a new girl out for a drive this evening," announced Joe.

Ludar looked worried. "Not Bella Tawney, I hope."

"You keep harping on that. No, it's not the Tawney piece." He looked at his friend and grinned. "I've got a little sense—not much, but enough not to take her out in this car."

The new suit was at the house, of course, when Ludar arrived home for supper.

He took his troubles to Catherine after supper. "I've just found out something," he said. "There are so many things—social rules, I guess you would call them—which I don't know a thing about. Like sending presents to people when they go away. Is it—is it proof of bad breeding when you don't know about them? Should you know about them instinctively, like telling the truth and trying to be decent?"

Catherine smiled at him with full understanding and sympathy. "I wondered about that once myself. You see, my mother died when I was quite young, and so I had no one to tell me what was right and proper.

I didn't know about these rules either and I had a bitter experience or two. No, Ludar, it hasn't anything to do with breeding or instinct. The rules of politeness change, and you have to be in a position to know what is being done and what isn't. It's all a matter of being in the swim."

He looked very much relieved. "I was afraid it was something lacking in me. From now on I'll have to keep my eyes open and ask lots of questions."

Clyde had been reading in a corner. He got to his feet, laid his book aside, and walked to the door. "All your life you find things you can't do or don't know about," he said in a carefully restrained voice. "Look at me. I discovered I didn't know how to manage a livery horse."

Chapter IV

I

Langley Craven said to the man across the table from him, "May I play hearts, partner?" and the partner responded in the bridge ritual, "Pray do."

Bridge had taken such a hold on what was called the upper crust of the city that two rooms in the Balfour Club had been given over entirely to it. Both Langley (who had joined again) and Tanner Craven were regular players, although never in the same room. By an undeclared arrangement Langley sat in the small cardroom which looked out over the lawn-bowling greens. Tanner clumped his way upstairs, slowly and rather resentfully, to the larger one where tournaments had been held in the days of whist.

"Gentlemen," said Langley when the opening lead had been made and his partner's hand had been exposed on the table, "it gets clearer to me all the time that this game has one serious defect. Why do we use only the red cards for trumps? It's going to keep me hustling to make one heart out of this cranky assortment of cards, but if Andy had been allowed to bid his spades we could make five or six odd."

Dr. Harley Clarke, one of the opponents, took advantage of this opening to ride his conversational hobby. "What is this, Langley?" he asked. "Are you saying that the conditions and rules of life, which so many of you regard as sacred and unchangeable, should not be reflected in the game of bridge? Why, the game as it stands is a perfect picture of social conditions. In the first place, only the dealer or his partner is allowed to bid. Their adversaries have no control over the manner in which the hand is to be played. The red suits are the aristocrats; the black suits are the lower orders, because only the red suits may hope to become trumps. Well, what's wrong with that? Isn't the same principle employed in everything else? Isn't——"

"Shut up, Harley," said Langley, who was trying to concentrate on the problem of making the odd trick in hearts. He succeeded in doing so, being an expert cardplayer, and was triumphantly sweeping in the last trick when a club attendant appeared in the door with an agitated face.

"Dr. Clarke is needed on the floor above," he said. Then he turned to Langley. "If you please, sir, your brother has been taken ill. I thought you would want to know, sir."

Dr. Clarke walked swiftly to the stairs with Langley Craven at his heels. "I've been warning him," said the physician with a worried frown. "But, after all, what could he have done but use common sense and moderation in all things? Tanner has refused to be moderate in anything. He eats too much and, as you must know, he's been drinking heavily."

Langley was amazed at the change in his feeling for his brother. The hatred was gone, and he realized with a start that he was sorry for Tanner. If there had been time for him to analyze his state of mind he would have come to understand that his antagonism had been lessening gradually. Affluence had worn away the sharpness of his ill feeling.

As he took the steps two at a time he found himself thinking of the days when he and Tanner had slid down the banister at the Homestead. They would make the swoop downward together, Tanner behind him and hanging onto his shoulders for dear life. Tanner had been pudgy as a boy (he did not begin to shoot up until he reached his teens) and he had been kind of jolly and amusing. He, Langley, had been very fond of his little brother. The first hint of hard feelings between them had come when he married Pauline. The antagonism had grown into hatred in one black moment when Tanner had produced the proxies which gave him control of the Craven interests. The violence of Langley's reaction had been partially the result of surprise, partially because it had been his younger brother who had taken advantage of him and thrown him out so unceremoniously.

The other players had succeeded in carrying Tanner to a couch beside the wall. He was lying there, motionless, his face mottled, his breath coming in stertorous puffs. Langley, who knew a little of everything, was well enough versed in medical matters to recognize at least some of the symptoms of apoplexy. "He's young for that," he thought grimly.

He was not surprised to hear the doctor say: "Hand me a pillow, please, someone. His head must be kept higher." Then Clarke added in a sharp tone, as though ill pleased with the stricken man's condition: "Phone for the ambulance, Henry."

The other players, feeling their presence an intrusion, were beginning to leave the room. The one who had been Tanner's partner said in a rueful voice, "We'd been having the most frightful luck and then he picked up a cornwhacker." The sick man's cards were lying on the floor, many of them face up. Langley saw both red aces and an unusual share of other honor cards. "Too bad he hadn't the fun of playing this hand," said the partner.

Orders from Dr. Clarke had produced an ice bag, which he proceeded to fit down over the head of the unconscious man. A few drops of medicine were forced between the lips, and then the medical man straightened up. He said to Langley, who was now the only one left, "Everything else will have to wait until I get him home."

"I've sent for a carriage," said Langley. "Why don't you ride out with me?"

"Very well. The ambulance is a small one, and there's nothing to be done on the way." Clarke turned suddenly and looked with curiosity at the older brother. "This does something to you, doesn't it? I can see that the ill feeling you've kept up so long is blowing away like—like fog before a wind."

Langley nodded. "You're right. It seems to be gone, completely gone. All I can think of is when we were boys together. I was very fond of the little coot."

"You know he's had a stroke?"

"I was afraid so. There are degrees in apoplexy, aren't there? What would you say about this?"

"Too early to tell. If he recovers consciousness quickly it will be a good sign, of course. If the stroke's not too severe he may have many years ahead of him. But that is—well, it's expecting a lot."

The ambulance arrived and the carriage for Langley came a few moments later. The doctor indulged in a rather wry smile when he saw that Creek and Stapley had sent the Creams. He patted one of them on the flank. "These fine fellows must be getting some surprises," he said. "They're being sent out on all kinds of trips."

They drove in silence for several blocks, and then Langley began to talk. "I suppose a crisis like this brings people together. God knows I've been so bitter about Tanner that there were times when I felt I would be glad if—if something of this kind happened. I don't think it was entirely real or I wouldn't be feeling as I do now. I can't conceive of a family feud being carried on after a death. The shock of it, the irrevocability, should be enough to sponge out the past."

"With normal people, yes. But I've seen the other side of the slate pretty often: hates continuing to smolder over a coffin and blazing up furiously over a grave. That's generally the way it goes when there's something to be divided up. How much bitterness there can be over a few hundred dollars!"

The Creams were brought to a stop on the drive behind the ambulance. After what the doctor had said Langley was wondering whether he should walk in as he had intended. He was no longer certain that Tanner's wife would be as ready as he was to forget the past. "Harley, will you tell Effie I'm here and that I'd like to come in?" he asked. "If she would rather I didn't, you can send one of the servants out to tell me."

Several minutes passed. The ambulance drove away. Langley, who became more worried over his brother's state as the waiting continued, found it necessary to wipe the perspiration from his brow.

Delia Connor, who had made an unfortunate marriage and had returned to service shorn of much of her bloom, opened the front door and looked out. "Please, sir," she said in a flurried manner, "the doctor says you had better not wait."

The older brother had more than half expected to be dismissed and so he did not particularly care. His concern was all for his brother's condition. "What does the doctor say about Mr. Craven?" he asked.

"Nothing, sir," answered Delia. "Not a word, sir."

Before dinner Langley telephoned the doctor's residence. "He's conscious," reported Clarke in a cheerful tone. "He rallied half an hour ago."

"You consider that a hopeful sign?"

"Yes, it's good. He has no headache, and that's better. But at the best there will be plenty of effects, you know. One side will be paralyzed. I'm afraid poor Tanner's active days are over."

There was silence for a moment, and then Langley said, "I suppose we should be grateful it's no worse." He hesitated about reopening the other matter. "I take it that Effie had no desire to see me."

"All she said was no." There was another pause, and then the doctor added: "You had better count on me for reports. I'll keep you advised, Langley."

He kept his promise by telephoning each day. The reports became progressively more hopeful, and finally he said that the worst was over. If

Tanner would be sensible and take things easy he might reasonably count on many years of life. There would be drawbacks, of course: a partial paralysis of one side, an impediment in the speech, and perhaps a slowing up in his mental processes.

"I don't want to drag you into our family troubles," said Langley after hearing all the details, "but perhaps you'll give me a word of advice. I would like to see my brother. How does the barometer stand? Would there be a scene if I called?"

The doctor took some time to think. "That wife of his knows her own mind," he said, finally. "I don't believe you would be welcome as far as she's concerned. Tanner, of course, is different. I took it on myself to tell him you had gone out with me. It pleased him. It seems to me that standing on the brink has softened him up too. The first good chance I get I'll ask him if he wants you to go over."

2

When Tanner Craven was permitted to get out of bed it was found he would need a cane. He made a tour of the room slowly, his right leg dragging. He paused at the window to look out over the grounds and then, with a heavy sigh, settled down in an armchair.

"Perhaps," suggested his wife anxiously, "you had better go back to bed."

He looked up at her with a face which sagged conspicuously to one side. It was several moments before he brought himself to the point of replying, and he used one word only.

"No."

"Is there anything I can get for you, dear?"

Again a stare lacking in expression, and again a long delay in answering. "No," he said. "I—I am quite—all right."

When the doctor arrived an hour later he was still sitting in the armchair and staring out the window.

"Well!" said Clarke cheerfully. "So far so good. You've taken your first steps successfully, it seems. How are you feeling?"

"I feel—terrible."

"Did you expect to feel like jumping fences and shouting greetings to the sun? You're an unwieldy hunk of flesh, and for a long time still a part of your body isn't going to take orders from you." He had been holding the patient's wrist. Concluding the count, he gave his head a noncommittal shake. "Is there anything you feel the need of?"

"Yes," declared the sick man with a surprising degree of promptness. "Three fingers of whisky. Neat."

"Huh!" exclaimed the doctor with fitting scorn and disapproval. "Get this through your head, young fellow. You've had your last drink of whisky. One small swallow of the stuff might do for you."

"I don't care." Tanner nodded his head slowly and, it seemed, painfully. "I need a drink. Two fingers, then. Neat."

The doctor frowned still more fiercely. "Didn't you hear me? You must never take another drop as long as you live. Not a drop, not as much as a smell of it. This is an order, Tanner Craven, and one you must obey if you value your life."

There was a pause while the doctor replenished the mixture in one of the bottles beside the bed. "I've been telephoning Langley every day," he said. "He's been very worried about you. When you feel well enough for visitors he would like to come and see you."

There was not a vestige of expression on the face of the convalescent, but after a long pause he managed to say, "I—would like to see—Langley."

The doctor smiled and nodded. "I'm glad to hear it. Hoped you felt that way." He began to close and lock his medicine case. "Here's what you must do, Tanner. You must tell your wife you want to see Langley and ask her to let him know. The invitation must come through her. You see that, don't you?"

After the doctor had left, Mrs. Craven heard her husband walk down the upstairs hall, dragging his feet as he went. When she reached him he was trying without any success to get the telephone receiver off the hook. She took him by the arm and directed his steps back to the bedroom.

"You must get used to the idea, Tanner," she said, helping him into his chair, "that for a time you'll have to depend on others. You should have known you can't manage the telephone alone. Who were you trying to get?"

The sick man said in a low voice, "Langley."

His wife's manner changed instantly. She said, "Who?" in a voice which had become rather shrill.

He nodded his head slowly. "My brother. I—I want to see him."

"And may I ask why you want to see him, of all people?"

Tanner Craven thought for a long time. "He came. He—he asks about me."

"And what of that?" Mrs. Craven was finding it hard, quite apparently, to hold herself in check. "Tanner, I'm sure you don't mean this. I would have thought your brother the last person in the world you would want to see."

He answered in such a low tone that it was hard to follow what he said. "No use holding—grudges—forever. I want to see old Langley—again."

"I suppose," said his wife, studying him with angry intentness, "that you would like to see his wife too."

He stared at the window for several moments. Then he turned in her direction, slowly rubbing a hand along the stiffened cheek. "What—did you say?" he asked.

"I said," raising her voice, "that I supposed you would like to see your brother's wife also."

He nodded at once. "Yes. I would like to see—Pauline."

Mrs. Craven turned and left the room with swift steps. He did not see her again for several hours. No reference was made then to the matter, nor at any later time.

The Langley Cravens were not invited to pay the sick man a visit.

While Tanner Craven had been ill his affairs had been suffering from the need of his firm hand at the helm. As soon as it was believed that he had recovered sufficiently to transact business the manager of the carriage and furniture plant, Harrison Pollard, paid a call. After his talk with the convalescent he sought out Mrs. Craven with a worried look. "I don't think he took in what I was saying," he reported. "I couldn't get an opinion out of him."

From that time on she was always in the room when there were business visitors. She not only took a part in the discussions but prompted her husband's answers and in many cases dictated the decisions which were reached. She discovered, to her great alarm, that Tanner had little recollection afterward of what had been said and done. At first she kept this disturbing fact to herself.

One morning, when Mrs. Craven was out shopping, Delia Connor came to the bedroom and said that someone was on the telephone and that it seemed urgent. Wrapping his dressing gown about him and taking his cane, the master of the house shuffled unsteadily down the hall. His mind was less clear than usual this morning, and he had difficulty first in identifying the caller and then in reaching any understanding of the matter in hand. When it came to the point of deciding what was to be done he thought so long about it, and then struggled so hard to find words to use in answering, that the man at the other end thought he had left the telephone. He called, "Hello, hello!" repeatedly, and jiggled the hook with impatience. Finally his stricken boss heard him say in a tone of intense annoyance, "The son of a bitch has hung up on me!" There was a loud click, and Tanner Craven realized that the receiver had been banged down hard at the other end.

This was one incident he did not forget. It rankled in his mind the rest of the morning, and when his wife returned just before lunch, he muttered angrily, "I—I was called names—over phone—I was called—son of a bitch."

His wife, who was wearing a rather indifferently designed summer dress of brown holland which any other woman would have recognized as a bad choice for her, stared at him as though she did not entirely believe what he was saying. "Who was it?" she asked.

A long pause followed. He had forgotten who it was! His wife waited as long as her patience lasted and then said, "Surely, Tanner, if you remember what the man said, you must remember who he was."

The sick man resented the reproof in her voice. He managed to say with an air of hurt dignity: "Should I be—snapped at—because I forget names? It will—come back."

His wife stared down at him for a full half minute without making any further comment. She was thinking, "My poor husband, what has he come to!" The incident served to bring her to a decision of the greatest importance. After lunch she again put on the brown sailor hat which went with the dowdy holland dress and sallied out on a series of calls which occupied the whole afternoon. She did not return, in fact, until her husband was having his evening meal on a tray beside his bed. She settled herself down in a chair and proceeded to draw off her gloves.

"I found out who called you," she said. "It was Gordon Simmons at the bank."

Tanner Craven went on eating a custard and did not seem very much interested.

"I would have demanded that he be discharged," she continued, "but I realized he didn't mean anything. He has a bad temper and he thought you had hung up. They told me he turned quite green when he heard you had been on the line all the time."

Her husband went right on eating. "I'll have something to say to *him,*" he declared.

Mrs. Craven rolled up her gloves thoughtfully. "Tanner, I've something to tell you. Something very important. You may not like it, but—what I've done is the only sensible thing. Until you're fully recovered again I am going to attend to all matters of business. Things can't go on any longer without a head. I've been to see everyone concerned, and they all agree it's the wise thing to do." It was evident that she was apprehensive of his reaction, for she went on in a hurried tone: "You're not strong enough yet to carry such a load. They all say that. Of course it's a temporary thing only. You'll soon be yourself again"—to herself she said, 'If only that were true!'—"and then you'll get right back into things. I'll consult you and tell you what's going on. In that way you'll be able to—to keep a finger on things, my dear."

She waited for the angry refusal she fully expected. Instead there was a long pause. "It tires me to talk business," he said finally. "I guess—I guess you're right." He was dabbing ineffectually with his fork at a leaf of lettuce on which there was some yellow salad dressing. Then he laughed. "I—I thought it was butter. Must be something wrong with my eyes."

3

Norman Craven had been in a camp in Muskoka with some of his school friends and had intended to stay all summer. A letter he received from his mother, however, caused him to cut his stay short. He returned home in a consuming fit of indignation.

He barked at Barney Grim, tossing his two bags into the hack, "Take me to 97 Grand Avenue."

"I know," said the cabman. "Your family moved a week ago. Your father certainly got a fine fat price for the Fife Avenue house."

Norman's face froze. "What do you know about it?"

The cabman laughed easily. "Why, boy, everybody in town knows what the figger was. Things like that get around. But no one sees how that Ernest Brake had the money to pay for it. Can't make any fancy profits out of that dinky foundry of his. Guess his wife wants to break into sussiety." Grim turned to wink over his shoulder. "And your old man bought the Alworthy proppity cheap enough. That's what I call good business—sell high, buy cheap. They say your dad is pooped out, but his noodle must be working just the same."

They had turned onto Grand Avenue, and Norman was startled to see a figure, slowly walking up the street with the use of a cane, who at first glance suggested his father. He decided that he must have been mistaken.

"Say, there's a girl waving at you," said the cabman. "Thought you ought to know."

Norman was in a fury by this time. "See here, Barney Grim, I hired you to drive me home and not to talk my ear off!"

"Won't charge you any more." The cabby grinned and said to himself, "He was a mean little puke as a boy, so what could you expect he'd grow into?"

They stopped before a three-storied house of white brick with red facings and a heavy wooden veranda on one side. It was squeezed in between the houses on each side, and the front lawn, on which an automatic sprayer was at work, consisted mostly of flower beds on each side of the cement walk in which unhealthy petunias and zinnias were planted.

When Delia Connor answered his impatient rap on the front door he said, " 'Lo, Delia," and brushed by her into the hall. "So," he said, "this is the pigsty we've moved into! What a state of affairs this is!"

He looked about him with such an air of hurt and mortification that it would not have been a surprise had he burst into tears. The hall was high enough and dark enough to suit the most ardent lover of the architecture of the period, but, remembering the grandeur of the entrance hall at Fife Avenue, he sniffed in a scornful rage. "A dismal rathole! Is Mother at home, Delia? I've got some things to say to her, and they won't keep."

"No, sir, Mrs. Craven is out," answered Delia. He had never been a favorite with her, but it was clear she was feeling sorry for him.

The parlors opened off to the right, and there was another door farther down the hall on the same side. The sound of a typewriter going at full speed drew him there. The room had once served as a study, for there were bookcases on three of the walls, but it was now being used as an office. An office in the home of the Tanner Cravens! He was so amazed and upset that he did not give a glance at first to a thin girl in a black dress who was seated at the machine and making her fingers fly.

"Have they gone mad?" he exclaimed aloud.

Then the girl turned and looked at him with cool gray eyes under really fine brows. He walked in and seated himself in the chair behind the huge desk which filled a large part of the room.

"I went to school with you," he said. "At the Central."

She went on with her work. "So you did," she answered.

"You were one of the quiet ones. You stuck to lessons and always got good marks and didn't have much to say."

"Not a flattering picture. But I'm afraid it's an accurate one."

He noticed that her arms moved with grace as well as speed.

"I used to sit across from you in Miss Reeve's room. I remember thinking you had a nice nose."

"My head will be completely turned by such praise." She looked up abruptly. "But you don't remember my name, Norman Craven."

He had been struggling to recall it, but all he had been able to bring to mind was that she was the daughter of the minister of a small church on the edge of town. The family had been in hard straits, and her clothes had always been plain. Her father had died recently. The Reverend Stephen Bennet! That was it.

"Your name," he declared in a tone of triumph, "is Lou Bennet."

"Goody!" She suspended work long enough to clap her hands. "I am remembered and identified. How wonderful!"

Norman liked the spirit she was showing. He grinned and asked, "What in Sam Hill are you doing here, Lou?"

"I'm your mother's secretary. And, fascinating though this conversation is, I must point out that it is now over. I have a dozen letters to finish."

She got to her feet and walked to a letter press in a corner to secure some data. "She's trim," he said to himself. "It's odd I never noticed she has a good ankle."

"Lou," he remarked, "I'm glad you're here. I'll be seeing a great deal of you."

She returned and seated herself at the machine without giving him a glance. "No, your mother would discharge me if you did," she said. "I'd rather keep my position, if it's all the same with you."

Norman had gone over the house from top to bottom by the time his mother returned. She came in the side door in a very great hurry, calling, "Lou, I've more letters, and they must catch the five o'clock mail." Then she saw her son standing with an accusing look beside the newel post which carried the bronze figure of a Roman soldier.

"Norman!" she exclaimed. "What's wrong? I thought you were staying longer."

"Do you think I could stay a minute longer after getting your letter? Great heavens, Mater, what's going on here? Have we lost all our money?"

Mrs. Craven gave her head a hurried shake. "Of course not." She kissed him and then said they had better go out to the conservatory where they could talk freely.

The conservatory opened off the dining room. It was built of dark blue glass, blown into tortured shapes, and so was dark and airless. A number of assorted chairs, for which there had not been room, obviously, in any other part of the house, had been dumped in here. The only concession made to the proper function of the place was the presence of a rather sickly cactus plant, bending patiently as though mourning for hot sands and bright sun.

Mrs. Craven took one of the chairs, which creaked under her weight, but Norman was too agitated to sit down. "Do you realize," he demanded, "how humiliating this is for me? Why, why did you move from our house, the best house in Balfour and one to be proud of, into a dismal wood box like this?"

"Sit down, my son, and I'll explain." Norman paid no attention to this suggestion but stood beside the cactus plant and regarded its eccentricities of shape without any real interest. "There were two reasons for moving. First, your father. He's not himself, my son, and you had better know now that he never will be the same. He can't concentrate and he takes no part in business any more. It seems to him that everything is going out and nothing coming in. He thinks we're going bankrupt, and it got so he couldn't stand to live in the old place. When we moved in here he went all over the house, counting the gas fixtures and measuring the rooms. He seemed much relieved."

"I thought I saw him on the street just now. Is he still quite lame?"

His mother nodded in somber assent. "He'll never recover the full use of his right side. He has to hobble, but he insists on going out twice a day. He's fallen into the habit of going as far as he can and then seating himself on the nearest doorstep. People are being very kind about it. They either bring him back or telephone us where he is."

Norman began to pace about between the chairs in a state of intense agitation. "This is frightful! Why, he's a town character! My father, the richest man Balfour ever produced. Mater, we've got to do something about it. Why can't he always go out in the carriage? Drives would do him as much good, and it would be more dignified."

"We've sold the carriage and the horses."

"What!" He stopped in front of her and glared. "This is beyond me. You say we haven't lost our money——"

"He has other habits which are harder to bear," she went on. "Because he thinks we have no money coming in he tries to help by collecting things. Every day he comes back from his walks with his pockets filled. Pieces of string, apple cores, old spools. I'm afraid it's getting to be a common joke. And, Norman, he still drinks. I don't know how or where he gets it, but he always has a bottle hidden away. I suspect Delia has something to do with it. They get along well, and she calls him 'Boss.' I'm watching her, and on the first bit of proof, out she goes!"

Norman moved a chair in front of her and sat down. "You said there were two reasons for selling our house. What, in God's name, is the other?"

His mother glanced at the open door and instinctively lowered her voice. "Do you remember that talk we had, Son? About your uncle's family and the swing of the pendulum? The pendulum is never going to swing back for your poor father and me. When it does, it will be for you. I'm forgetting about the present, my fine son, and making all my plans for the future—and for you. We must not only make more money, we must save every dollar we can."

"Has our position been—insecure?"

She raised her head proudly. "No, Norman, we're in better shape than ever." She reached out and took possession of one of his hands. "I'm only concerned about the time when you and Langley's son will be rivals. When it comes you must be rich, rich, rich. From what we hear, your cousin will carry on his father's concerns, and that means he'll be nothing but a small-town businessman. I don't want you to be in business at all. I want you to be far above him, a power in public affairs, perhaps a senator in time or Canadian Commissioner in London. That means you'll have to be one of the richest men in Canada."

Norman was finding his mother's argument quite persuasive. "Well, I can't say I've ever wanted to be a businessman particularly. I believe I'm better suited for the other kind of life. And I'm as anxious as you are to outshine Master Joseph Craven."

"This," said his mother, "is the revenge I'm planning. I'm ready to wait for years."

Norman's mood darkened again. "They must enjoy seeing us in this pigsty while they're planning to build a big place on Fife Avenue."

Footsteps came briskly down the hall, and Lou Bennet appeared in the door. "That call from Toronto is coming through, Mrs. Craven," she said. Her manner suggested that if she were aware of the room's having another occupant she was in no sense interested.

Mrs. Craven got to her feet briskly. Although she was growing almost massive in build she remained quick and light on her feet. Before going to take the call she laid one broad and tallowy cheek against his. "Son, trust me in this," she whispered. "I know what I'm doing. And I'm thinking of nothing but your interests."

Chapter V

I

One evening Ludar returned from the store to a realization that something was wrong. He read it in the set, grim face of Aunt Tilly, who was going about preparations for supper with an unusual clatter of dishes. Still more significant was the lack of a cheerful greeting from Uncle Billy.

The latter was in the sitting room, propped up in a chair with a blanket over his knees and one hand bandaged so thoroughly that it looked twice the size of the other. He was already in his flannel nightshirt.

"What's wrong?" asked the boy, walking to him.

"Well, it seems I've lost a finger and a small part of the hand as well." William spoke lightly. "Every carpenter loses a finger or two sooner or later. I've been lucky up to today. I'm not worrying now that it has happened. I can get along with one finger less."

"This is terrible, Uncle Billy." The boy was frightened in spite of the cheerful attitude of the injured man.

"All it means is that I'll be laid up for a few days while the wound heals."

Tilly came into the room. "A few days!" she said bitterly. "The doctor told me it would be weeks before you could go back to the shop. What are we going to live on in the meantime? The little bit the boy gets from that miser of a Haverstraw?"

William interposed mildly, "We could dip into the principal to carry things on."

"No!" cried Tilly. "Not another cent do we take out of the bank. Not as long as I live. I've had my fill of dipping into the principal, Billy Christian, and make no mistake about that. We'll live on bread and skimmed milk first."

The handwriting was on the wall. Ludar knew that he must take another job, a better-paying one, which meant finding employment in a factory. This prospect had been hanging over him since the end of the term, and it was one from which he shrank, knowing himself unfitted for that kind of work. He was certainly not prepared, however, for the announcement which Tilly proceeded to make.

"As soon as they brought you home I went in next door and called

Uncle Alfred. I asked him to make inquiries. He phoned back to Margaret just before six and said for Ludar to go tomorrow morning and speak to Mr. Bender at Cauling's Stove Foundry. There's a job open, and Mr. Bender has promised to wait until Ludar sees him."

William looked perturbed. "But, Tilly, not Cauling's. That's heavy work. And very dirty."

Tilly demanded fiercely, "And what's wrong with good, clean dirt?"

"Well," answered William, "I'm not certain it's either good *or* clean. I'm sure it's terribly hard to get off."

"He can earn six dollars a week at Cauling's." This seemed to Tilly the answer to all objections. "Will any other factory pay a green hand as much?"

William was not sure. He thought clean work could be found somewhere at about four dollars a week.

"We can't live on four dollars a week," declared Tilly. "I've got my figures to prove it. Six will barely serve. This is where my little black books come in handy."

Ludar could not bear to listen any longer. He went into his own room and got out the manuscript of his new novel. He had not progressed far with it. "If I hadn't been so lazy," he thought bitterly, "I might have had this finished and sold by this time. But here it is, and I've got to work in a stove foundry instead. It serves me right."

He sat for some time in the growing dusk, staring blindly at the side wall of the McGregor house. The thought of what lay ahead of him was so repugnant that he said to himself he would rather die. But, unfortunately for him, dying would do no good. He was not like the young English poet Chatterton, whose suicide had been discussed at school. The poet had had only himself to think of. "They've got to have six dollars a week," he thought. "And they can't get it any other way."

He was learning one of the bitterest lessons that life has to offer, that personal preferences count for nothing, that fears and dislikes have to be controlled when necessity dictates.

2

Mr. Bender's office was a glassed-in cubbyhole just inside the grimy entrance to the Cauling plant. He had been applying a silver toothpick to its proper function and now was using it to point out something on a balance sheet to one of the clerks. He faced around when Ludar was escorted in, a pudgy man with hard eyes and a high, whining voice.

"Well?"

"Mr. Alfred Hull told me to come and see you, sir," said Ludar, whose knees were near the shaking point.

"Oh yes. Oh yes. Calamity did speak to me." He looked the boy over. "Calamity didn't say much about you. It's a good thing the job's not in the molding. A day of *that* would crack your spine. Well, we'll go up and see George Dankin."

He led the way through the assembling department, where men were banging away at a fearful rate on partly erected heaters and ranges, most of them richly festooned with nickeled parts. After climbing two of the dirtiest and most rickety stairways in the whole world, surely, they entered a long room filled with a row of machines at which men were holding castings against revolving wheels and filling the air with a cloud of emery particles.

"Dankin!" shouted the superintendent.

A door opened at the far end of the department, allowing a glimpse of long tanks over which was suspended a most curious-looking apparatus filling the place with a strange green light. The liquid in the tanks had the same unearthly green tinge to it, and into this men were plunging the nickeled parts of stoves. The foreman came out, closing the door after him.

"Will this one do, George?" asked Bender, pointing at Ludar with his always useful toothpick.

Dankin, who had a high bald head and a mustache which suggested that a piece of porcupine skin had been grafted on his upper lip, looked critically at the boy.

"Not very blasted rugged," he said.

"Skinny," said the superintendent.

The foreman addressed Ludar. "Want to glue emery wheels?" he asked.

"Yes, sir," said Ludar. "I'm thin but I'm wiry, sir."

"You mustn't think I'm recommending him," said the superintendent.

"Shucks! Anything will be better than the Glue I've got now." Dankin raised his voice sufficiently to be heard over the noise of the machines. "Corns!"

A boy about Ludar's age had been keeping very busy (the whole room was electric with activity under the eyes of both superintendent and foreman) over a long wooden box which might have served as coffin for a circus thin man. He straightened up at this summons and came hobbling over.

"Corns, I'm taking you off the glue," said Dankin. "This one's going on it. You'll go into buffing, Corns."

In spite of his weak feet, the boy, whose face was murky and whose fingers for some reason were as large and black as those of a gorilla, did a sort of shuffling dance, his eyes gleaming excitedly. "Gee, boss! Gee, Mr. Dankin! That's great. I've always wanted to get into the buffing. Three years on the glue is enough for anyone. Gee, my old man will be pleased."

The foreman said to Ludar with a gloomy air, as though he already doubted the wisdom of the step he was taking, "Report for work on Monday, you."

3

It has always been agreed that a sunrise is the most beautiful thing in the world. The sky becomes shot with glorious light, the birds chirp, all nature stirs and rouses for the activities of the day, and the horizon becomes a fence which can be cleared at one bound with all Wonderland beyond. But the rising of the sun is a different matter to those who must go to work early. To them the orb is a jailer who shakes the bars and says, "Come now, up with you!" Certainly it affected Ludar that way on the following Monday. He rose early, feeling sick and nervous. No other footsteps echoed the fall of his feet as he hurried down Wilson Street. The cheerfulness of the sun made him feel all the more certain that he would be a complete failure and that he would suffer in the process.

The polishing department was empty, and it looked incredibly dark and dirty and sordid. The windows had never been washed, and little light got through them. The walls were black, the corners were filled with debris, and the machinery looked old enough to break down at a touch. Ludar had been reading *David Copperfield,* and the thought occurred to him now that the rotting factory on the water where David labeled bottles in company with Mick Walker and Mealy Potatoes must have been very much like this. He inspected the emery box and the sink, at which the wheels were glued, with increasing trepidation. He was certain now that he would not be capable of performing the duties of a glue.

The next arrival was a boy of his own age with a weak round face and unruly reddish hair. It was not until the newcomer spoke that Ludar realized this was his predecessor in office, Corns.

"Hello, Glue," said Corns in a lordly tone. His transfer to the buffing had already given him a sense of superiority. "I'd start learning you what there's to do, but it isn't allowed ahead of the bell. That's one thing you never do, start before the bell. Another thing, drop your tools when the quitting bell begins to ring, even if they land on your toes."

The men were coming in now, rolling up the sleeves of their flannel shirts, and looked about with loathing. It was clear that the beauty of the morning had left them untouched. One of the last to arrive said, "Oh, slush, what a life!" and took down a dirt-encrusted apron from a nail on the wall. Another one said: "God bless Papa and Mama and Mr. Cauling and Mr. Bender and that nice, kind Mr. Dankin, the sonofabitch."

At this point Mr. Dankin appeared at the inner door, waving them impatiently to work. Castings were pressed up to the whirling wheels and sparks began to fly.

"See these machines?" asked Corns. "They buff the parts that got to be nickeled. The emery wears off the wheels, and the buffer throws the wheel on the floor and hollers, 'Glue!' You go and collect the wheel and then you rub the edge with glue and you pound it in this box. It's filled with emery sand, see, and you got to be careful there's no lumps in it. Then you take the wheel back there," pointing to the closed door of another department, "and hang it up to dry. That's all you got to do. Easy, isn't it?" Corns threw back his head and cackled. "Just you wait. Just you wait until two or three of them yell 'Glue!' at once. You'll have to hump then. When I first came here I used to hear them hollering 'Glue!' in my dreams." He stopped and looked shortsightedly at his successor. "What's your name?"

"Ludar."

"Oh yes. You're *him*. They won't be kind to you here, not these guys. They wouldn't be kind to St. Peter or the Queen or Santa Claus. My name's Wilbert, but everyone calls me Corns because of my feet. They ache so bad at night I can hardly get to sleep."

Someone yelled, "Glue!" and Corns, through force of habit, sprang to get the discarded wheel. He handed it over to Ludar. "Get to work, Glue. I'll watch and see if you do anything wrong."

One of the men, who had a nose which jutted out from his face like the beak of an owl, came over to say a word to the new boy.

"Just want to set you straight, Glue," he said. "You'll like it here if you don't mind working under a man who was born crooked and grew up mean enough to rob a bird feeder. That is, of course, if you're willing to be worked to death, and jumped on, and harped at, and bawled out for nothing. When he comes at you with a look on his face like a dog spewing raisins and his voice is quiet, watch out! He's going to rip you right up the back." The owl-nosed man, whose name, Ludar learned later, was Windy Smith, nodded. "Mind you, Glue, outside of that, Mr. Dankin is a fine—square—lovely man."

"Glue!" cried two voices at once, and both boys went into action together.

After an hour or more of work Ludar made a startling discovery. His fingers had become as large and black as Corns's had been the day Ludar was hired. The reason was clear enough: some of the glue got on the hands, and the emery dust adhered to it. He held them up in front of him and gazed at them in dismay.

"It'll be hell to get off," said Corns. "Every night for three years I've had to rub mine in turpentine, and dig at 'em with a pick, and pry at 'em on the foot scraper. Ain't it queer how easy this emery sticks to your fingers and how hard you got to pound to make it stick to the wheels?"

By twelve o'clock the two boys had glued and rolled and pounded so many wheels that Ludar's back ached, and he wondered if he would be able to last through the afternoon. "Tomorrow," he thought in a panic, "I'll be alone on the job."

When the noon whistle blew, everyone dropped what he was doing instantly, grinned, stretched, and reached for his dinner pail. Corns came over to Ludar, an old shoe box under his arm.

"Come on, Glue," he said. "We got to go down to the yard."

"I'll eat my food up here," said Ludar. The polishing department had already taken on some of the qualities of sanctuary. The rest of the plant was foreign territory, inhabited by unfriendly men and boys and filled with strange terrors.

"Nix," said Corns. "You come on. It's your first day. You got to be in the yard."

Ludar's mind filled with dread. "Why?" he asked.

"When I first came on as the glue," explained Corns with great cheerfulness, "they yanked my pants off and glued newspapers all over me. Took me a full week to get rid of it."

It was many seconds before Ludar spoke, seconds packed with the most intense dread. "Will—will they do that to me?"

"Guess so. It's the usual thing for a new glue."

The men were making their way to the stair, with the intention clearly of being on hand for the fun. "Come on, Glue," they called. "Won't do you any good to hang back. Might as well get it over with."

Ludar followed the procession down the stairs, feeling like a condemned man on his way to the gallows. The yard was a large open space full of slag piles and heaps of broken castings and a surprisingly large number of wheelbarrows in all stages of use and disuse.

He was greeted with a roar of welcome from as many as a hundred men sitting on the ground in small groups and already munching on the contents of their dinner pails.

"Here he is! Get out the gluepot, boys!"

"Fix him up with a new suit. One he won't get off unless his skin goes with it."

"Off with his pants!"

Ludar was being led to the center of the yard (now that the terrible moment had arrived he did not try to fight or get away), when an interruption occurred. A familiar voice said: "Hey, what's this? Hold your horses now. I've got something to say about this." It was Pete Work. He had come out into the yard from the assembling department, where he worked, his lunch under his arm. It was clear that he was surprised to discover the identity of the new glue.

Ludar's spirits rose a trifle. Even if Pete's arrival did nothing to save him from the ordeal, there was some consolation in having a friend at hand.

"Duke!" said Pete, crossing the yard. "What in Sam Hill are you doing here?"

"I'm in the polishing department," answered Ludar.

"Holy crow, this is a queer one. I never thought——"

He stopped abruptly and looked around the circle of grinning and expectant faces. Then, his instinct for the dramatic coming to the fore, he cleared a wheelbarrow of its occupant and mounted it himself.

"I got something to say," he announced. "The Duke is an old friend of mine. Someday he's going to be pretty important. Now then, listen to me, the whole lot of you. There's going to be no initiation today. I won't let it happen to a friend of mine." He gave a long and challenging look about him. "Anybody who picks on the Duke picks on me. We're in the ring together. Who's the first with anything to say?"

A voice from somewhere in the yard called, "You're not the boss here the way you were at school, Pete Work."

Pete stepped down from the wheelbarrow and walked slowly in the direction of the voice. "I think," he said, "it was Mr. Loud-mouth McGinn said that. I've been keeping an eye on Mr. Loud-mouth McGinn for quite a while. First time I set eyes on him I said to myself, 'Someday you'll have to polish off this know-it-all guy.'"

It was like a slow-motion tableau. As he talked Pete was walking forward and Mr. Loud-mouth McGinn was stepping backward.

"I said to myself, 'Someday I'll have to settle his hash for him. This slick customer's in need of it.'" At this point he took a quick step forward and his antagonist almost lost his balance in trying to retreat with equal speed. A loud roar of laughter rose from the yard.

Pete stopped. "I guess that's all we need to do about *him*. Now, then, who else has anything to say?"

Ludar took a panicky look about him. It was clear that none of the men liked this interference. They glowered at Pete Work and muttered among themselves. But it was equally clear that none had any desire to start trouble. They knew all about Pete and the explosive power in his fists.

There was a long moment of silence. Pete made another survey of the sulky faces. "Come on, Duke," he said. "We'll eat lunch together."

They seated themselves in front of a slag pile. "Pete," whispered Ludar in an ecstasy of relief, "I was scared." He was thinking of what his champion had said, *Anybody who picks on the Duke picks on me.* It had sounded better to him than Magna Charta or the Bill of Rights or Habeas Corpus as a guarantee.

"Duke," said Pete, a bun in one hand and a roll of boloney in the other, "I can't make this out. What are you doing in a dump like this? I thought you were going to be a professor or a preacher, or something like that."

"Uncle Billy lost part of his hand in an accident. I've got to bring in the money while he's laid up." Ludar gave his head a shake. "I'm afraid it's going to be a long time."

Pete dissented emphatically. "Nah, you won't be here long. Not you, Duke. You got a head on you, and brains. Not me either. I'm going into business for myself. I'll tell you about it later."

Corns brought his lunch over and squatted down beside them. Gradually others followed his example. They formed a circle around the pair and started to eat with enormous appetites, their grimy jaws biting vigorously into sandwiches and pieces of cake and wedges of pie. Most of them ended the meal with an apple. One man peeled an orange and popped the slices into his mouth with more than a hint of ostentation. Oranges were the greatest of all luxuries, usually reserved for the Queen's birthday and Christmas.

By six o'clock the most tired body in all Balfour dragged itself away from the gluepots and the coffinlike box of emery dust and walked on legs as stiff as stilts to the ground floor. Pete Work was waiting for him at the timekeeper's door.

"How are you, Duke?"

"I ache all over. This has been the toughest day I ever remember."

"The first day's always like that," declared Pete, leading the way outside. "You get used to it in no time at all. And what's more, you needn't worry about what happened at noon. They won't bother you again. Once you get over the first day, you belong." They turned onto Wilson Street. "Duke, I'm leaving this place flat on its back the end of the week. Mr. Craven and a couple of other men are putting up the dough to start a general repair shop. I'm to run it and I'm getting a slice in it too. We'll repair everything that runs—buggies, wagons, bicycles, hobbyhorses. Pretty soon we'll have plenty of automobiles to look after. By grab, I'm a businessman now." His manner became less assured and he kept his eyes lowered as he went on talking. "And I'm going to be married. At least I hope I am. I'm crazy about a girl. Mary Slaney. Do you know her?"

"Doesn't she live in the East Ward?"

"That's it. She's pretty and little and I just love her. She doesn't know how I feel, but I think she kind of likes me. Once I've got my business running well I'm going to say to her—— Geez, Duke, I don't know how I'll say it or how I'll get up the nerve to say anything!"

A neighbor of the Christian family passed them at this moment, and Ludar raised his cap. She looked startled at first and then puzzled. "I didn't think she was stuck-up like that," said Ludar to himself when she brushed past them without any acknowledgment.

He discovered the reason when he reached home and started to wash in front of a small cracked mirror in the summer kitchen. She could not have recognized him because his face was pitch-black.

Curiously enough, he was surprised. Although he had seen the faces about him get blacker and blacker as the day progressed, it had not occurred to him that his own would be the same.

4

For the next few weeks Ludar existed in a state of fear: fear that he would get the glue too thick or too thin, that he would not pound hard enough, that he would hang the wheels on the wrong arms in the drying room and bring down on himself the ire of the buffers, who wanted to be able to reach up blindly and put a hand on the right kind. These fears never left him because he knew that, no matter how hard he tried, his work never reached the instinctive accuracy of good craftsmanship.

One day the wheels started to clatter on the floor and the air was filled with shouts of "Glue!" Ludar ran frantically back and forth, rubbing the wheels, pounding them, rushing them to the drying room. Corns left his machine and came over. He tested the glue and sniffed contemptuously.

"It won't hold the emery when you get it as piddling thin as this," he said. "Here, I'll get it right for you."

It happened that Mr. Dankin chose this moment to come out for a look over the room. When he saw the pile of wheels on the floor he went over and investigated the glue himself.

"You ought to soak your head!" he roared, the quills of his mustache bristling. "Corns, you stay with this until the stuff has some stick in it." He glared at Ludar as though the latter was no longer entitled to respect or to hold up his head in the company of his fellows. "Get this into your head, Glue. Next time this happens you'll go and get your time, you damned sissy!"

As usual Ludar strove hard to make himself presentable that evening, pouring turpentine on fingers as round as country sausages (but being careful of the turpentine, which cost money, as Tilly pointed out) and scrubbing his neck and shoulders. William watched him from the kitchen door for a few moments and then turned back to the kitchen, shaking his head sadly.

"I can't bear to see him in this condition," he said to Tilly. "Cauling's is no place for him."

Tilly slammed the loaf of bread down on the table. "There you go! Why are you always saying things to make him more dissatisfied than he is anyhow? And what's wrong with Cauling's, pray? The work's steady and the ghost always walks."

"He's not made for this kind of thing," protested William.

"Then why didn't he take the job Mr. Langley Craven offered him?" demanded Tilly. "He was to get the same money, and the work would have been clean."

"You know why he didn't. He knew Mr. Craven was just making a place for him and offering more than the job was worth. The boy was too proud to accept anything like that from friends. He preferred to go on this way." William gave vent to a deep sigh. "I sometimes wish he had taken it. There's something fine about working with wood. There's a sweet smell to it, and the dust doesn't get into your skin. I couldn't work at anything else."

"Don't go saying things like that before the boy," said Tilly. "Anyway, this is an iron town. Half the men come back from work with faces as black as Ludar's."

"That," declared William with another and wearier sigh, "doesn't seem to prove anything."

Ludar came in to supper, complaining that his hair was almost black again, although he had given it a thorough scrubbing the night before.

"Be sure and use that cotton duster on your pillow tonight," cautioned Tilly, handing him his cup of tea.

William had no appetite. He took a few bites of food but for the most part sat at the head of the table in an unhappy silence. When he got up and walked back into the sitting room Ludar was shocked to see how thin he had become. The cords in the back of his neck were standing out tautly, and his coat hung loosely on his shoulders. The boy followed him in and said, "You're not getting well as fast as you should."

William assumed an air of cheerfulness at once. "I'll be fine in no time," he said. "In fact, I've decided to start back at the shop on Monday. The hand is pretty well healed now. A man can't go on working for Street and Walker like this. There's no money in it."

He was so obviously unfit for work that Ludar paid no attention to this. He was accordingly very much surprised on Monday morning, as he hurried with his dressing in the semidarkness of his room (Tilly insisted that a lamp in the kitchen was sufficient to light all the bedrooms), to hear William call him urgently. Running in hurriedly, Ludar found him stretched out on the floor.

"Uncle Billy, what's happened?" he asked in a frightened voice.

Putting his arms around the shoulders of the recumbent man, he assisted him to a standing position. William promptly dropped over on the bed, breathing in quick gasps. "I'm afraid I haven't the strength yet," he said after a moment.

"Were you getting up for work?" asked the boy in amazement.

William nodded. "I was never so surprised in my life." His breathing became easier. "I thought all I had to do was to make up my mind I was going to the shop and that the strength I needed would come to me. I got up and then—suddenly—everything went black and I toppled over."

"This settles any idea of work for a long time," said the boy severely.

After a few moments of rest William began to speak in an apologetic tone of voice. "I've always known, Ludar, that you're cut out for different things than this. To see you coming home from such work is—well, I can hardly bear to see you. And yet—and yet it can't be helped, I'm afraid. I know you hate it, but can you—can you manage to hang on for a while longer, until I get some strength back? I don't seem able to tackle it yet."

"Were you going back to work so I could quit Cauling's and get an easier job?"

"Well—perhaps."

Ludar looked down at the thin figure on the bed. "You'll not try that again," he said. "Not until the doctor says you're strong enough. It will be a long time before he does. I'll keep on at the foundry, and you must stop worrying about me. I'm getting used to it."

The last statement was far from the truth. He was not getting used to it. Each morning he found himself increasingly unhappy. As he filed past the timekeeper's wicket and climbed the stairs to the dingy hole where black castings were turned into shiny nickeled parts he felt like a lost soul on the ferry crossing the Styx. The men he worked with were like the inhabitants of another planet. He did not understand the talk which went on around him and, which was worse, he had no desire to acquire understanding of it. He always sat alone over his lunch box. There was a second reason for that: he was ashamed of the extreme plainness of the food he brought. The men made it a rule to compare what was in their pails and to suggest trades. "I've got three chicken sandwiches today, Rube. What'll you give for one of them?" "A piece of my wife's layer cake. She makes the best in town, bar none." He could not very well boast of the superiority of his cheese sandwiches or offer to trade one of jelly.

At the beginning the men had decided he was a queer species of fish, and they were willing to leave him alone. As time went on, however, they began to resent his standoffishness. "Guess we're not good enough for His Lordship," they would say, or "Why doesn't he have his footman carry in his lunch box for him?" They became more critical of his work, more inclined to discard a wheel before it was scraped clean in order to give him more to do. The foreman seemed to make it a rule to reprimand him loudly and blasphemously at least once a day. Ludar would almost jump out of his skin when he heard the Dankin voice roar, "Glue! Come here! I got something to say to you!"

He lived in fear that they would decide to have the initiation ceremonies after all. Pete Work was no longer there to stand up for him. He had located his repair shop at the end of Holbrook Street, near the Iron Bridge, and was the busiest and happiest young man in town.

5

Ludar never went out in the evenings and he saw Joe only when the latter came to 138 Wilson Street. This was quite often, for Joe was not a fair-weather friend. He would rap at the side door, put his head in, and ask, "Anyone here think they can beat me at crokinole?" Ludar's unhappiness of mood left him as soon as his friend arrived. They would sit and talk and laugh boisterously. Ludar would make light of his work at the foundry, depicting Mr. Dankin as an amusing tyrant and the men as interesting character studies. "I'm getting material I'll use in a book someday," he said once.

Joe always made the same remark on leaving, "Hold your tail high, doggy, because better days are just around the corner."

Sometimes they would go out for drives in the red automobile. Joe would look quizzically at his friend and say: "Still Old Faithful, I suppose? No use mentioning that I could find two of the prettiest, gigglingest little chicks with the nicest round bustles without going more than two or three blocks from here?"

Sometimes Joe had reports of Tony for his friend. "Myrtle Wood got a letter today from You Know Who," he said one evening. "It came from Paris, and she gave me the stamp. My collection's getting pretty big."

"But what was in the letter?" asked Ludar, unable to control his eagerness.

"Well, it seems she's had a proposal from an English officer. She didn't like him because he talked of nothing but horses and dogs and thought Canada was the name of a town. She wrote pages about the dress shops in Paris. Have you noticed that she's batty about clothes?"

Ludar had noticed it. He had the most vivid memories, in fact: Tony in a golden velvet cloak with a cape in which her head nestled like a jewel in a box, Tony looking her loveliest in brocaded blue bengaline, Tony looking like a dancer in an accordion-pleated skirt with orange panels, Tony looking very young and gravely sweet in a straw hat with pink ribbons, Tony going to a party (he had seen her driving by in a carriage) in a dress of white silk with pin points of heliotrope.

"You know," went on Joe, "Norman is going to rush Tony for all he's worth when she gets back. I heard him give it out at a party the other night."

What chance would he have against the son of the richest family in town, particularly when she came back and found him working in a stove foundry and walking through the streets with black face and hands?

"Tony will cut quite a dash when she gets back from Paris," said Joe. "From what she wrote Myrtle, she's bringing back a lot of French clothes. She'll be the belle of the ball." He paused and then grinned. "But she's going to have competition before long. You haven't seen that young sister of mine lately. I want to tell you, she's getting to be a corker. She'll be having her claws out for you. She'll have 'em out for every man in town, that Meg."

"I'll bet she will." Ludar smiled, thinking of little Meg swishing her silk skirts through town and drawing all men after her like a feminine Pied Piper. Then he sobered and asked another question. "Did Tony say anything about me? I mean, whether they had found the Redcraft family?"

Joe shook his head. "She wouldn't dare say anything about it to a gossip like Myrtle."

He went to see Catherine that evening. "Had any letters from abroad?" he asked.

Catherine was holding her new Pekingese pup in her lap. It was biscuit-colored and very small, and both lively and affectionate. A bark of welcome greeted him.

"Yes," she said. "It came yesterday. Such a nice, chatty letter." A far-away look took possession of the invalid's eyes, one of regret for lost opportunities. It vanished quickly. "She seems to be having a wonderful time."

"Did she mention the officer who proposed to her?"

Catherine nodded. "Of course. She wouldn't neglect to tell me anything as important and exciting as that."

"I hear she turned him down."

"Yes." Catherine's smile became somewhat enigmatic. "She intimated that she had other plans. I wish I dared show you the letter, but—— Well, I'll start right in now and tell you everything you ought to know."

6

Ludar learned to dread the coming home from work more than the going. In the mornings he either had the streets to himself, always being early because of a dread of tardiness, or he fell into step with a little middle-aged grocery clerk who had to be at the store in time to open up and do some straightening of stock before seven. The clerk was a chatty fellow who confided many things to Ludar, even that he and his wife had decided years before not to have a child until they had saved up a hundred dollars. However, they had been saving ever since and still had no more than forty dollars in the bank, because his wages had not risen above six dollars and a quarter a week. They were beginning to wonder if forty dollars would be a safe margin for the raising of a family of one.

Provided, he always added in a devout tone, the Lord was kind enough to grant their wish.

But returning from the foundry at night had become a form of torture. Wilson Street was very lively just after six o'clock, and he never failed to pass many old school friends. Girls he had known quite well pretended not to see him when they met, and even some of the boys sidled by without speaking or gave him a very casual greeting. He suspected that they enjoyed a laugh at his expense as soon as his back was to them, but he lacked the courage to turn.

Finally he could bear it no longer and began to seek other ways of getting home. He thought of making a detour through the Hollow and then by open fields to the north end of Wilson. This, without a doubt, would have gained him the solitude he sought, but it involved a walk of two miles and was not practical. Instead he began to cut over to Diamond, which paralleled Wilson. It was a quiet and even humble street with white brick cottages so close together that there was barely room for fences between them. Here he could walk in the company of other men returning from work, with empty dinner pails swinging in their hands, and never be noticed. After several blocks the houses began to thin out. There were open meadows where horses and cows were pastured, and an abandoned lumber yard which occupied a full block. Ludar always turned in here, through the sagging gate, and walked in the desired isolation between piles of rotting lumber, never failing to look with curiosity at the remains of a small office. It had gaping holes for windows and was leaning drunkenly toward an inevitable collapse. The owner had failed and had been found here, dead of a broken heart, slumped down in his chair.

After emerging at the other end of the yard the boy would cross over through the rear of a monument maker's lot. This brought him to a meadow on which a few shanties backed, their clotheslines invariably displaying an assortment of red flannel underwear. Then came the willows, a long row of fine old trees standing well back from the street. Quite apart from the need which brought him on this crisscross journey homeward, he found pleasure in leaving the sidewalk and walking on the winding dirt path beneath the arching willows. It seemed to him that this must be like England with its lanes and hedges, and his mind never failed to race ahead to the probability that soon the mystery of his past would be solved.

The willows ended at Eugenie Street. He would hurry over the two blocks which led to Wilson and then, after a cautious survey to be sure the coast was clear (if not, he would loiter and tie a shoelace or even retrace his steps), he would race across the street and into the side gate of his home with the haste of a fugitive of the Middle Ages bolting into sanctuary.

7

He had followed this course one evening in mid-September and was on the last lap from the willows when he heard the warning toot of an automobile horn. It was far different from the thin "Pleep!" of the Craven car, for it said "Auk!" in a deep and commanding tone. Ludar was not surprised, therefore, when a long snout appeared in sight on Wilson Street and then turned onto Grand Avenue, but he was not prepared for the full impact of what he saw. It was a blue car, standing high off the ground on monstrous wheels, and it had seats for at least half a dozen people. It rolled along with the ease of hidden power.

Ludar whistled. "Must be a foreign car," he said to himself. "Now what is it doing here in Balfour?"

He had already noticed that it was driven by a man in a plain uniform and that it had two passengers, a man on the front seat and a lady in the back with her face swathed in veils. Then he discovered that the man on the front seat, who was holding onto his bowler hat with one hand, was William Pitt Milner.

People had rushed out to the street to see this amazing spectacle, and for once Ludar walked through them unnoticed. Joe had been right, he said to himself. The Milner family had come home in style, and it might be expected that Tony would indeed cut quite a dash.

Joe came over that evening with all the news. "It's a Mercedes," he said in an awed tone. "A German car and the best in the world, I guess. They had it shipped on ahead and picked it up in Toronto. They came right through without a stop, nearly seventy-five miles and over bad roads, in less than seven hours!"

William, whose eyes had a curiously strained look these days, was listening with intense interest. "They've started," he said, "the changes which are going to upset the world. I tell you, boys, life is going to be interesting from now on. Things are going to hum."

"Mr. Milner is going to sell his horses and keep the car in the stable," explained Joe. "That fellow who was driving is to stay with them. He's English and his name's Philip Beal. He used to be a chauffeur for some lord over there." He turned and grinned at Ludar. "Perhaps it was a relative of yours. Wouldn't that be funny?"

"I've noticed, Joe," said William, "that you don't seem to feel your nose is out of joint."

"Heck, no, Mr. Christian. Why, all this is exciting. I want to see lots of cars here in Balfour. Pete Work needs the business, for one thing. We'll have one as big as this someday soon. Perhaps a Daimler or a Panhard." Joe smile affectionately. "But I'll always love this little red rooster of a car the best."

He was so excited over the arrival of the Milner family in their Mercedes that it was not until he was ready to leave that he remembered his most important piece of news. "They passed Father on Holbrook Street," he said, "and they stopped. Mr. Milner told Father they had news about Ludar. He'll be over at eight o'clock to tell you what it is."

Chapter VI

I

Mr. Milner was horrified at his old friend's appearance when he came into the sitting room. "Great Scott, man!" he exclaimed. "What have you been doing to yourself? You've lost a lot of weight."

William held up his injured hand by way of explanation. "As you see, I'm minus a finger," he said. "It happened at the shop. Dr. Gibbon says the shock to my system brought on Graves' disease."

"I've never heard that one before. Is it—is it serious?"

"No," cheerfully, "it's uncomfortable, but it isn't serious. It's the name they have for a goiter."

Milner looked puzzled. "Only women have goiter. Billy, are you sure Gibbon knows what he's talking about?"

"It's a special kind, and men are liable to it as well as women." Like all patients, William was not averse to discussing his symptoms. "There's no swelling of the neck with this kind. Dr. Gibbon says it's a more acute form and he's filling me with medicine for it. Mostly iodine."

"Iodine!" The publisher now looked thoroughly alarmed. "The stuff you use on cuts? You—you swallow it? Billy, that stuff is strong enough to eat out your insides."

William smiled. "It hasn't done anything of the kind yet. I think we can trust Dr. Gibbon in this matter. He says what I need is a long rest—I've had a pretty long one already—and that I mustn't let myself get excited or worked up about things. It's bad for the kind of disease I've got."

"Well, then, you must promise to take things easy before I tell you what I found in England. It's a snarled-up situation." The publisher turned to Ludar, who had come into the room in a consuming state of anxiety. "Well, Master Ludar Redcraft, I have a great deal to tell you about yourself and your family in England."

2

"Redcraft?" said the manager of the hotel in London where the Milners stayed. "Indeed, yes, Mr. Milner. If it's Godfrey Redcraft you're interested in I can tell you a great deal about him. The failure of Godfrey Redcraft and Company was—well, it *was* a crash. I lost a bit myself in that one." From what the manager was able to tell them about the family of this financial titan who had gone down in such a spectacular way, it became reasonably certain that they were on the right track. Godfrey Redcraft himself had been dead for some years, but there was a son running a brokerage house, and quite successfully. "He's in Birmingham," said the manager. "Cyril Humphrey Redcraft. A rather difficult little man, I should say, from a few experiences I've had with him. But he has plenty of it. The brass, I mean. And he seems to know how to—well, conserve it, shall we say?"

So, on Tony's insistence, they went to Birmingham and found the offices of C. H. Redcraft and Company located in a shabby old house on the edge of the business section. They went in and gave their names to a gloomy clerk behind a wire partition.

They waited for half an hour, and Mr. Milner was on the point of a violent outburst when the clerk finally nodded to them. "Last door down the hall, sir," he said. "There's a brass plate."

Cyril Humphrey Redcraft had a long nose, which was bent close down to the sheet of paper on which he was writing, and shrewd dark eyes, which he slanted up at them without the rest of his head being raised.

"Balfour, Ontario," he said. "That means you've come about the boy."

"Yes," said Milner.

"Won't do you any good." The head of Redcraft and Company touched a paper on his desk. "The boy's growing up. Living with a respectable family. Said to be clever but a bit too partial to books and things like that."

The broker raised his head and indulged in a metallic laugh. "Any son of old Viv's *would* be queer that way. You see, Mr.—uh—Milner, I had a few inquiries made through a law firm in Toronto some years ago. Now, then, what's the point of this visit?"

The publisher started to explain about the poor circumstances of the Christian family and the money which had been raised in Ludar's interest, but the lawyer interrupted impatiently. "Fools!" he said. "Hiring a lawyer they knew nothing about. They deserved to be done in. And what did they expect, anyway? A special consignment of gold bullion on the first ship to Canada?"

"I'm not sure they expected anything," said Milner. He was finding it hard to control his choler over the offhand and even contemptuous attitude of the English broker. "The boy had been cut off from his relatives and was living under a name not his own. The fund was raised in the hope of getting things straightened out for him."

The broker sat up in his chair. He had small shoulders and a thin neck and he was fastidiously dressed in a dark gray morning coat. He had a black pearl stickpin in a red satin tie. A ring with an extremely fine toad's eye looked enormous in conjunction with his small, neat hands. "I've no scruples about telling you what you want to know," he said. "Viv's dead, and so are our parents, and no particular harm can come of it. But I see from your card that you're a publisher and I must be assured there won't be any use made of the facts. I won't have publicity, not even in a colonial newspaper."

"I give you my word of honor that not a line will be published."

After giving this assurance Mr. Milner thought he detected a suggestion of doubt in the eyes of the broker, as though the latter was wondering about the standards of honor in the colonies. This made him angry because he had already conceived a dislike for Cyril Humphrey Redcraft, both as a man and as the director of a business. He had noticed there was no telephone on the premises, that typewriters did not seem to have been heard of, and that all the work was being done in a truly Dickensian manner by humble clerks driving pens across paper in an almost frenzied hurry. The visitor's face showed a tendency to redness about the eyes, a sure sign that the Milner irascibility was rising.

"In any event," said Mr. Milner, "I know something of the story already. You see, I've been making inquiries too. Your brother Vivien made a disastrous speculation in the market, using as margin some stocks belonging to a client of your father's brokerage house. As a result he found it necessary to go to Canada under an assumed name."

"Quite so." The broker accepted this brief summary without any effort —or desire, obviously—to soften the facts. He toyed for a moment with the pearl in his tiepin. "I might as well tell you the rest. My father, Godfrey Redcraft, was a very ambitious man. He was on his way to making a big fortune. He wasn't a Rothschild, you understand, nor a Baring. But still he was doing something quite handsome. He wanted social position as well, which was quite proper. We're an old family. It must be clear to you from the name that we come of soldier stock." The little man puffed out his puny chest, and his visitors found it hard not to smile. "We've traced the line back to one of the Lords Marchers of Breconshire. Fourteenth century, no less. And, mind you, there's no guesswork or flummery about it. My father built a summer place above Brecknock Mere and he talked of retiring to that section someday when he had made his pile.

"It was a great blow to him when Viv, who had shown himself perfectly useless in the office, jilted a girl he was to marry. It's true she had been Father's choice and not his, but she was handsome in a bosomy way and her father was in the House of Lords. On top of that, he runs off and marries a pretty shopgirl named Harriette Railey, whose father was a gamekeeper. That was what you could always expect from that brother of mine. A left-handed honeymoon and a present of a hundred pounds or so would have been the usual thing, but Viv was determined to marry her. The old man went into a great rage and refused to give him a penny more than the small salary he was supposed to be earning in the business. That was what caused the trouble. The girl was going to have a child—have you noticed there's never any time lost about that in such cases?—and Viv didn't see how he was going to manage things. He appealed to Father again, got an emphatic no, and then went out and plunged in the market. As you've already pointed out, he used for collateral some stocks which— well, which *wasn't hisn*."

"If you had known Vivien," went on the broker with a shake of his head, "you would be certain that any stocks he would pick were sure to fall off. This one fell right through the bottom of the market. There was a great to-do, naturally. It happened that the owner was a peppery retired colonel who ranted and roared all over London. Viv *would* pick a man like that to steal from. My father made good on the stocks, but the colonel's sense of law and discipline had been outraged and he went to the police about it. Father had to get Vivien out of the country overnight."

Redcraft desisted from his narrative long enough to shake his head, as though the things he was telling were completely out of range of his sympathy and understanding. "There hadn't been time to collect the wife and send her with him," he continued. "Father asked her to come and see him, intending to pack her off after him. She couldn't be found. Actually, she had bolted. And what's more, we never did hear from her. A curious case of obduracy."

"Or pride," said Milner.

"Four years later a sister was heard from. The girl was dead and had left a son. The boy was with her, the sister, and she was determined not to be burdened with the expense of his keep. At that time the affairs of Redcraft and Company were getting involved—my father was never content with anything but a gallop ahead at full speed, even when business was bad—and he didn't do anything about it. The woman became so insistent finally that he sent Collison to see her. Old Colly made arrangements to ship the boy out to his father in Canada. I'm inclined to think something would have been done for Viv then. But, as you know, he elected to make an end of it."

"You waited too long to think of helping him."

The broker passed this over without comment. "Godfrey Redcraft and Company failed before we had any report as to what had happened to the boy. We heard about Viv's death and we knew that efforts were being made to locate the boy's family. Things were in such a mess that we did nothing about it. It was a complete smash, in fact, and nothing was saved. Father's houses went, his carriage and horses, his rings and gold studs. Mother had some money of her own, so they were able to get along after the crash. They're both dead now. This business, Mr. Newspaper Publisher from Canada, is mine. I got together a little capital and made a start for myself. I've built it up by my own efforts and I'm doing well. No one else has a penny in it and no one else is ever going to make a penny out of it. I'm an intensely acquisitive man, and proud of it. Is there anything more for us to discuss?"

"The memory of your brother——" began Milner.

Redcraft had been noticing Tony in a sly way as he talked. She was looking particularly well in a suit of brown Bedford cloth with Italian sleeves and collar turned down sufficiently to show the youthful roundness of her throat. Her eyes were shining with interest. He found her so pleasant to look at that there was a moment's delay before he made any answer.

"The memory of my brother is an unpleasant one," he declared then. "You may as well know that I hated him. He was such a soft, fumbling fellow and yet he had all the advantages at the start. My father would talk of the title he was going to get and hand on to Viv, but he had no plans for me. I would gladly have married the girl they picked out for him. As it is, I've never married." With a side glance at Tony he added: "I'm not lacking in admiration of the fair sex, but it's hardly likely that I'll marry now. I've come this far alone and I might as well go on that way." He hitched around in his chair, and his manner suggested that he wanted to close the discussion. "Well, Mr. Publisher, there you have it. I'm the only one in the family with a dib to my name. The girls married useless fellows and will always be poor. I'm under no obligations to my brother's son and I don't intend to do anything for him. Not a shilling will you screw out of me, not even if your pretty daughter goes down on her knees and begs me."

"We're not here to beg," declared Milner, getting to his feet. "Come, Antoinette, let's be on our way. But first I have a word to say to this most affectionate of brothers. You say your mother had enough property of her own to keep herself and your father in comfort. That leaves the matter of her will to be inquired into."

The surviving son said in a sharp tone, "There was no will!" He had straightened up in his chair again and he bristled like a fighting bantam cock. "And there was nothing left. She had her money in annuities, although I don't know why I bother to explain all this to you. As soon as I laid eyes on you I knew you were going to be troublesome."

Mr. Milner placed both hands on the edge of the desk and looked down at him. "Yes," he said. "Yes indeed. I intend to be very troublesome." Then he allowed his own irascibility to take control. "Let me tell you, sir, that with all the smartness you admire so much in yourself, you are hopelessly behind the times. Permit me to suggest that you study the methods in modern offices. If you don't, progress will sweep over you like a tidal wave. Someday you'll find yourself in a world of great business, a world of millions, and you'll be making shillings and saving pennies. It may be, of course, that this is the true measure of you."

3

It was again on Tony's insistence that they paid a call on Poppy, the sister described by Chief Jarvis as a "pudding-faced little girl" and who had written the affectionate letter found in Vivien Redcraft's possessions after his death. She and her husband, whose name was Tatterton, lived in Kent, and the Milners hired a big imported car and a small imported driver and were taken down without a stop. This, their first experience of motoring, was so pleasant that it resulted in the purchase of the Mercedes the very next day.

The Tatterton residence was an old brick house, rather large and obviously drafty and suffering from defective drains. Its windows (in the Byronic manner) looked down on a beautifully kept lawn, and the lawn looked down on a stream which rippled and meandered in the most lackadaisical way through the grounds. It was evident that a kennel was being maintained, for a dozen or more young dogs were racing about in a securely enclosed field, and the tooting of the horn started a loud chorus of assorted barks from somewhere back of the stables.

Vivien Redcraft's sister was still small, but no one observing her as she received them at the door could have described her now as pudding-faced. She was pert and pretty and had a reddish tinge to her hair, which had been cut short so that it curled closely about her head. She wore an old sweater and a skirt with unashamed patches and shoes abominably scuffed, and she looked completely charming. Tony, admiring her instantly and immensely, wondered what would be said in Balfour if she were to copy this style of hairdressing.

Their hostess showed them into a high-ceilinged room filled with furniture in a state of advanced dilapidation. "Out, Breezer, old fellow," she said. A heavy old dog removed himself painfully from one of the chairs and left the room in a mood of sorrow and indignation. "He's our old champion," she explained. "He has been ruining all the springs." Looking at the couches and the chairs, one might have assumed that strange mastodon animals had been enjoying the run of the place.

"So you've been to see Cyril," said Poppy when they had all perched themselves precariously on chairs. "Oh, dear! *What* a bad impression you'll have of us. I'm sure he wasn't nice to you, that *ghastly* brother of mine."

"He was of no help," admitted Milner.

"He wouldn't be if it had to do with Viv and the boy. Or if it concerned money. And"—she shook her head soberly—"I'm afraid I can be of no real help. I know little of what happened at first hand. You see, I was away on a visit and didn't even know about it until it was all over. As for helping the poor little nephew, my husband hasn't two brass farthings to rub together in his pockets. Our two boys are at their father's school, and there's hardly anything left over for us to live on."

"The people who've kept Ludar are very poor," said Tony. "You can have no idea, Mrs. Tatterton, how poor they are."

"In Canada that doesn't matter much," added her father. "Most people are poor. We had hoped, of course, there would be prospects for the boy. Your brother made it very clear that there wouldn't be."

"I'm sure he did," sighed Poppy. "I'm sure he was most explicit. He's such an explicit man." She looked at Tony and smiled. "But I can tell you about the boy's father and a little about his mother. That might be something to take back for the small nephew, at any rate."

"He's not small," said Tony, returning the smile. "He's seventeen now, you know, and quite tall. Very close to six feet, I think."

"Is he actually?" There was a suggestion of envy in the young mother's voice. "My two nice fellows are growing fast, but I'm sure they won't run to that. Would it do any good to speak to them about it, do you think? I'm surprised, because both Viv and his wife were rather on the smallish side. My father was called the Little Napoleon of the Stock Exchange, and he wore false heels in his shoes."

"Please tell us about Ludar's parents," said Tony.

The sister folded her hands in her lap and looked out through the nearest window, which afforded a glimpse of the stream and the old willow trees growing on the far side. "Poor old Viv!" she said. "I loved him dearly. And I was always a little sorry for him. He was so gentle and generous—and so helpless! He was a great reader, which was a sin in Father's eyes. Do you know what he wanted to be, that nice little sobersides of a brother? He wanted to be a teacher. Actually! And not a headmaster or a professor or anything important. He had no managing instinct in him at all. He would have been quite content with some obscure post." She sighed unhappily. "But Father, my Napoleonic father, was determined to make a businessman of him. It was absurd and it was pathetic—like trying to turn a household pet into a pit dog. Poor Viv!"

"Did you ever meet his wife?" asked Tony.

Poppy nodded. "Once. I had tea with them. She was very pretty, and I liked her. Well, I think I'd have liked her if I had come to know her better. I'm sure she would have suited Viv. They would probably have been quite happy if that dictator of a father hadn't been so determined to have his own way. He had his way—and Viv and his wife are both dead, and nothing came of Father's great plans anyway."

She went on talking and, out of the random stories she told, a clear picture of the Redcraft family began to emerge for the visitors. Godfrey himself occupied most of the canvas, and it was clear that, in spite of everything, his daughter had entertained for him a dazzled respect and a reluctant affection. "Cyril," she said, "is a puny, marked-down copy of

Father. When he fails—and of course he will someday—it will be a poor, fiddling affair, not the grand, resounding crash my father made. *He* pulled down some of the pillars of the temple first."

She was very curious about Ludar. Tony answered her questions as impartially as she could (knowing her father was watching and listening with special attention) and succeeded in making a puzzle of him. "He really has talents?" asked Poppy. "He can write things like poems and draw pictures? How *very* odd. None of the Redcrafts have ever had talents like that. You don't think it possible they came into the family through his mother? No, surely not. After all, she was a gamekeeper's daughter." She was pleased with the picture in the main, however. "He seems nice. I wish I could bring him back to live with us, but that would be thin rations for him. Worse than life in Canada, I'm afraid. And I wouldn't hear of him going to Nancy—even if she would have him, which is extremely doubtful. She and Viv didn't hit it off. Are you going to see Nancy?"

"Should we?" asked Tony.

"Perhaps you should look the whole family over, unrewarding though we are. Nancy is married to a lawyer and lives in London. I must warn you that she's difficult. She believes in speaking her mind."

"Can you give us the address of the aunt who kept him after his mother died?"

Poppy rose and went to a secretary-desk in a corner of the room. It sagged with age and hard usage and was so crammed with papers that to find anything in it was an obvious impossibility. "I was so interested in Viv and his troubles," she said, "that I did keep that woman's address. I couldn't have put it in here, though, because I would have realized it would be like sinking something in quicksand." Then she straightened up and said, "The safe, of course."

The safe, it developed, was a grandfather's clock in the hall, a very disreputable type of grandfather and no longer a clock in reality, because the works had disappeared from behind its face and the hands were broken off. Opening the door, she reached down into the shaft and brought out a handful of papers. She carried these back into the room and proceeded to go over them.

"We really ought to have a better place for the family papers," she said. "The will is here someplace, and the only bond we possess. Ah, here we are! The will may never be found again and the bond does not seem in evidence, but here we have the address of that terrible woman." She handed Tony a slip of paper. "That's in London, of course. A shady neighborhood, I suspect. A bad time you'll have if you go near *her*. They say she's a tartar."

They had tea on the lawn. The head of the house arrived home from some errand having to do with dogs just in time to join them. He was a silent man whose conversational efforts did not run too much more than an occasional "Oh, rather!" and his advent brought a canine avalanche upon them. They were the friendliest dogs in the world and the most voracious. Tony, with her arms wrapped defensively about her knees (there were no chairs on the lawn, so they all sat on the grass), saw six watercress sandwiches and as many squares of toast vanish from sight with sudden flicks of long red tongues.

Poppy's husband looked sufficiently piscatorial to warrant the nickname of Dear Old Fishface which his wife had once applied to him. It developed that his silence was due, in part at least, to the fascinated interest he was taking in the automobile. They all went over to see it when he could no longer keep his concern to himself, and he proceeded to look it over with a thoroughness which disturbed the imported driver. The upshot of the examination was, however, a prediction that it would never succeed in getting back to London.

"It's exactly the kind of thing my father would have had if he could have delayed his failure a little longer," said Poppy. "I'm sure he would have had two of the very largest money could buy."

She called to them as they drove off, "I'll give that—that *horror* of a brother of mine no peace until he does something for the nice little boy in Canada."

"The Tatterton family is well named," said Mr. Milner when they had reached a safe distance. He was busy brushing dog hairs off his clothes.

"I liked her very much," said Tony. "Didn't you, Father?"

He answered with an emphatic negative. "I can't like a woman who runs a house as badly as that," he said. "But she may stir the brother up to doing something. You can see she's determined."

4

They found themselves one rainy afternoon not so long thereafter in the parlor of a severe little house in a respectable district of London and talking to a severe woman who made no bones of the fact that they were not welcome. This was Nancy and, in spite of her disinclination to talk, she did contribute some information with reference to her mother's estate.

"It's true there was no will," she said, her lips showing a tendency to tighten. "Mother bought an annuity in order to increase their income and make it possible for Father to continue the indulgences he couldn't do without. He kept up his club memberships and he had his clothes made by his old tailors, the most expensive in London. And there was always a bottle of good wine for his dinner." She stopped, as though unwilling to give any more information to strangers. Her indignation getting the upper hand, however, she went on with her tale. "Cyril says she put all her money into the annuity. My sister and I know she didn't. There's a paper in existence somewhere—if we could only get our hands on it!—giving full details of what she possessed after the annuity was taken out. But when she died—Father went first, but only by a few months—there was nothing. Poppy and I even had to contribute to the expense of the funeral."

Mr. Milner broke the pause which ensued. "Your brother told me that he picked up a little capital and started his present business with it."

"I'm making no charges," declared the sister. "All I can tell you is this: neither my sister nor I received as much as a penny. My husband, being a lawyer, has tried to do something about it. He has had no success whatever."

"I have it in mind to place the boy's interests in legal hands before I return to Canada."

"I'm afraid it will do no good. But more power to you if it forces Cyril to do something for all of us. There's no doubt in my mind that he got his hands on what was left. Mother, the simpleton, may have given it to him. She always worshiped her two sons and didn't care what happened to Poppy and me. I can imagine her saying, 'I know the girls are hard up, but poor, dear Cyril needs every penny of it to get started.' She preferred Cyril to Vivien, I think."

This led to an anecdote. "Viv and I didn't get along. I take after my father and I couldn't stand such lackadaisical ways. He was always in trouble of some kind. When he went to Canada, Mother gave him a Bible and made him promise to read it every day. Pretty soon he wrote her saying that he had no money and was starving and that the only consolation he had was reading the Bible as he had promised. She knew this couldn't be true because she had stuck a five-pound note to the page telling the story of the prodigal son. She decided then he must be punished by not letting him know the money was there."

Tony, deeply concerned, cried, "And didn't he ever get the five pounds?"

Nancy sniffed scornfully. "Yes. That sentimental idiot of a sister of mine sent word to him. She couldn't even wait to write. She cabled."

5

Knowing that Miss Railey was employed in the daytime, they took a cab to her address after dinner. Mrs. Griffen said incredulously, "Her?" and then, "Oh, you're from Canada. I hope——" She checked herself and did not tell them what she hoped. Instead she went to the foot of the stairs and bawled out in a loud voice, "Visitors, Railey. *From Canada.*"

The parlor into which she showed them was patterned after the best models. There was a three-foot Buddha at one side of the fireplace, by which much store was set, obviously, because it was scrupulously dusted. At the other side was a copper coal scuttle, and over the mantel a purple drape which had not been dusted for a very long time. There was, of course, a large framed picture of Queen Victoria and some wax fruit under glass. The table, at which Mr. Milner seated himself, was covered with a fringed chenille cloth and, after tapping his fingers on it impatiently, he was dismayed to find the tips quite black.

Callie Railey had become shrunken and ancient-looking. She stood beside the table and faced Mr. Milner with a hostile air.

"You needn't ask any questions because you'll get no answers," she said, repeating the formula she had used with Ludar.

"I think," said the publisher, studying her intently, "you will find it hard to answer satisfactorily the questions I propose to ask."

"This one, first," he went on after a pause. "What became of the money Mr. Collison left with you for the boy's journey?"

Miss Railey made no response. She stared at the wall with no more expression than could be found on the clay vacancy of the Buddha.

"I have full information as to the amount left with you," continued the publisher. "The sign you sewed on his back has been kept. It will be an excellent piece of evidence."

Still no response.

"It would be a pleasure to assist in sending you to prison for theft," he declared.

A crimson wave passed over the face of Ludar's first guardian. She did not speak, however.

Tony joined in at this point. "Don't you think, Father, we might agree to say no more about the money if she would answer some questions for me?"

Milner hesitated and then said, "Well, you might try."

Tony asked, "Where did Ludar live with his mother before she died?"

There was a long pause, and then the woman said, with every evidence of a bitter grudging, "Pursley Castle."

"And where is Pursley Castle?"

An even longer delay before the stubborn mind could persuade itself to answer. "It is near Oxford."

On their way back to the hotel Mr. Milner began to talk. "It would be a real pleasure as well as a public service to send that woman over. She's the most dangerous of criminal types. She would put strychnine in the tea of anyone she disliked, with no hesitation whatever."

Tony's answer was, "Poor little fellow!" She was thinking of the time Ludar had spent in the gloomy boardinghouse in the care of his mother's formidable half sister.

6

The Milners acquired Philip Beal before they paid their visit to Pursley Castle. They had done a great deal of riding about in the blue Mercedes in the meantime (and a great deal of waiting for the engine to start and for Beal to repair punctured tires), and Tony had spent much time in the stores and had met the officer who proposed to her.

It was a beautiful morning when they finally rode out into the Thames country. The car was running with sweet efficiency, and in no time at all they came whizzing along the main street of Abbot's Pursley. Here they turned north, leaving a cloud of dust behind them, and in a few minutes Mr. Milner said as the gatehouse hove into view, "There's the clock tower, sure enough." In less than a minute more they came to a stop on a circular drive, which had been designed for carriages and not for long automobiles, and saw a thin old gentleman with a surprised look on his face. "And there we have, as big as life, Old Mr. Cyril himself."

Old Mr. Cyril it was: a courtly gentleman of what was still called the old school, with the brow of a scholar and the neatly trimmed beard of an Elizabethan dandy, carrying a cane to assist his not too steady legs. The reediness of age seemed accentuated by his Bond Street correctness of attire.

"Good morning," he said, coming down the steep stone steps to meet them.

"Good morning," said William Pitt Milner. "We are from Canada, sir, and we're in need of information which we have reason to believe you can supply."

"I shall strive to be of service," said the old gentleman. "My name, sir, is Cyril Anthony Tresselar."

"I am William Pitt Milner. And this, Mr. Tresselar, is my daughter Antoinette. We come from a place called Balfour in Ontario."

Tony had dressed with particular care this morning. She was wearing a suit of fine black cloth, with a plain skirt and a double-breasted Newmarket jacket piped with velvet. To compensate for the comparative staidness of the suit she had donned a wide hat of yellow straw which was scooped up daringly from the face and sported a wide black ribbon bound around the rim and hanging down the back to her shoulders.

"My dear young lady, permit me to congratulate you," said Mr. Tresselar, helping her to alight. "You are the best-dressed woman I've seen in many a long day. I wish my nieces, poor dowdy little creatures, could see you and benefit by your example. But that would be expecting overmuch."

He would not agree to any discussion of their errand until he had shown them about. Tony was delighted and, walking eagerly beside him, asked a steady stream of questions. The castle was a mixture of periods, but all of it was old. Some parts had their masonry sunk in Norman days, and the ivy twining on the most recent addition, the clock tower which Ludar had remembered to such good effect, had been growing for more than two hundred years. The large dark rooms which the boy had dimly recalled were filled with fine old furniture, some pieces going back to Tudor days, which is a very great age indeed.

Mr. Tresselar was particularly proud of a section in the picture gallery on which appeared a complete set of Rowlandson's Dr. Syntax plates. "To get this space, my dear," he said, "I was compelled to oust half a dozen ancestors. I did it gladly because they were a moth-eaten lot of old boys and girls and had served one purpose all my life, to put humility into my soul."

When they had made the rounds and had ensconced themselves in chairs on a comparatively modern terrace, they were served a dry sherry by a butler a shade more ancient than his master and in thin glasses which were older than either.

"Let me tell you the favorite anecdote of Pursley Castle," said its owner. "Do you see the knoll over there on the left? And then, on the other side of the valley, a bit of road winding around the base of the hill? Six hundred years ago an archer serving the master of Pursley of that day sent a shaft from the knoll and lodged it under the heaume—as they called then the iron pots they wore on their heads as helmets—of a knight riding around the bend. The knight was in the service of King Henry III, and so that very bad King had one knight less when he met the soldiers of the good, the great, the glorious Simon de Montfort at Lewes.

"It's an almost unbelievable distance to send an arrow," he went on. "For centuries the owners of Pursley had a standing offer of ten shillings—and that was a huge amount then—for any bowman who could match it. Hundreds of stout yeomen came to try their hands, but never once was the prize money earned. This isn't one of the pleasant legends which make up so much of history. I know it's true. As a boy I found a score or more of arrowheads sunk deep in the valley beneath the hill."

He placed his glass on the small table beside him and turned to Mr. Milner. "And now, sir, in what way may I be of use?"

"We're looking for information about a boy. His name was Ludar Redcraft, and we have reason to believe he was here with his mother. It would be twelve or fourteen years ago."

The old man looked puzzled. "They came here on a visit?"

"If so, it was a long one. My understanding is that they made their home here."

A look of understanding replaced the doubt in Mr. Tresselar's face. "Now I remember. Redcraft, of course. That is hardly a name to be forgotten, is it? The mother was employed here on the understanding she could bring her son. I recall them clearly now. The mother was delicate-looking, and the boy was a bright little codger. I sometimes took him with me when I inspected the rabbit snares." He straightened up with a sudden access of interest. "It's all coming back. The mother died while they were here. It was quite sudden, but I don't recall what she died of. An old sister, a most disagreeable person, came and took the boy away with her. I was sorry to see him go."

"Where was the mother buried?" asked Tony.

"I believe," he answered, wrinkling his brow in thought, "that she's in the village graveyard. That could be ascertained for you."

"We'll look for the grave when we drive back."

The owner of the castle glanced inquiringly at Mr. Milner. "Was that what you desired to know?"

"Yes, thank you," said the visitor. "May I ask further in what capacity the mother was employed here?"

Mr. Tresselar's brow wrinkled again as he cudgeled his uncertain memory. "I'm inclined to think," he said finally, "that she was an assistant to the cook. Yes, that *was* it. I remember praising a ragout and being told she had made it."

Assistant to the cook! Tony's face was a picture of dismay as she glanced at her father. This was not the kind of information they had ridden out to Pursley Castle to find, not the kind of story to take back to Balfour. How the townspeople would laugh!

On second thoughts, however, the story assumed a more favorable guise. Vivien Redcraft's wife, after his hurried exit to Canada, had been too proud to accept help from the family and had turned to the only form of employment in which she could support her young son.

"You could do me a great favor now," said the master of Pursley, not aware of the effect his statement had produced on his guests, "by remaining for lunch. If you don't I must sit down alone, and that is a dismal prospect. Breakfast is, quite properly, a solitary meal, for one is not yet disciplined for the day to the rigors of human companionship. There is something about dinner also which makes solitude a pleasure. You are tired, you have your wine, you are in a reflective mood and desirous of getting to bed early. But lunch is different. Lunch, served in the full sunlight, when all your faculties are active, clamors for companionship."

They stayed, of course, and it proved a bewilderingly delightful meal to Tony. Food in Balfour, which was so distant from the great markets and the sea, was plain and wholesome and, from force of habit, she and her father had been avoiding unknown dishes. Mr. Tresselar had a French cook, and much of the food which reached the table in the oak-paneled dining room was strange to her. There was a crab bisque to start, then a sole covered with a sauce containing sea food she had never heard of, thick rich mutton chops, a salad with a dressing which made her nostrils curl with delight, and finally a fresh black raspberry tart with the flakiest of French pastry. Tony had always been a healthy eater and she did extremely well, finishing even the tail of her mutton chop and her slice of the tart to the very last crumb.

Old Mr. Cyril, delighted, leaned over and patted her shoulder. "Splendid fellow!" he said. "*That* is what I like to see. My nieces, the anemic little sparrows, do no more than peck at food. And now, my dear young lady from Canada, is there anything more you would like to see?"

Tony nodded at once. "I would like very much," she said, "to see the place where Ludar fell into the pigpen."

Their host looked startled for a moment and then threw back his head and laughed. "As I have already made clear, I consider you a charming girl. I would go to endless effort to satisfy your merest whim. But I very much doubt my ability to help you in this curious concern over pigpens."

"Perhaps Old Baffle would know," said Tony. "If he is—if he is still about."

"Old Baffle is still about. I'm beginning to think he'll always be about. He claims a hundred, and I suppose he's ninety by this time. He dodders about the stables, pretending to be busy with a pitchfork. Shall we go down and consult him on your problem?"

Old Baffle proved to be so pink of face and so white of hair that he might have been mistaken sometime for the spirit of a new year (as artists like to depict them) if he had not been wearing a filthy smock under which his skinny old legs formed almost a complete bow. He remembered the incident at once, yus, zur. Leading the way to the rear of the stables, he indicated the exact spot where Ludar had tumbled inside after a hard upward climb. It developed that Old Baffle's exactness of memory was due to the fact that he himself had been responsible for the rescue. "I hooked 'un with pitchfork through belt," he said proudly. "And I lugged 'un out. Yus, zur."

"What kind of little boy was he, Mr. Baffle?" asked Tony.

The ancient handler of pitchforks scratched his head. " 'E wur a pester, miss," he said. "All boys is pesters."

As they drove off Mr. Milner became aware that a book was lying on the seat between himself and his daughter. He picked it up and looked at the title, *Annals of Breconshire*. Without any comment he put it back, but to himself he said, "This must be looked into."

7

The struggle with the Boers in South Africa, about which no earlier mention has been made because it had in no way touched the lives of the people with whom this story is concerned, was still dragging on. The end was in sight, but visitors in London were constantly reminded that the nation was at war. Recruiting posters occupied all strategic positions, bands paraded the streets, crowds gathered before the bulletin boards of newspapers, and there was much cheering at intervals and constant marching to the beat of drums. The venerable Queen's health was said to have been seriously affected by the protracted nature of the struggle and the heavy loss of life. Opposition to the war still persisted. Tony was present at a lunch one day when her father had the political editor of a London newspaper as his guest. "That fellow Lloyd George is still keeping it up," grumbled the editor. "The mobs have nearly got him twice now. The fellow will come to a bad end." Her father had replied: "I heard him speak once and I must say I was impressed. He's full of fire, that little Welshman."

When they reached their hotel after the long and dusty drive back from Pursley, a Highland regiment was marching past. Tony stood up in the car to watch, thrilled with the fine marching of the Scots to the tune of "Bonnie Dundee," their kilts swinging in unison. Her martial ardor subsided, however, when she reached the lobby of the hotel and heard a voice on a squeaky gramophone intoning the words of the song Rudyard Kipling had written the year before:

"Duke's son—cook's son—
Son of a hundred kings . . ."

Cook's son! The thought flashed into her mind that this would be sung in Balfour with a special application and with malice if the truth about Ludar's mother and the mysterious house with the clock tower ever became known.

She was standing beside her father at the desk while he collected his mail. She plucked at his sleeve. "Daddy," she whispered, "did you hear that song?" She hummed the air in a low voice and then said the words. "Don't you see?" she demanded. "We must never tell anyone what we found out today."

It took several moments for her father to grasp what she meant. Then he nodded in confirmation. "You're right, chickabiddy," he said. "It wouldn't do to let it out. The story of Pursley Castle will have to remain a secret, just between the two of us."

8

This was the story Milner proceeded to tell in the Christian home that evening. He told it in much less detail, of course, and he omitted all mention of Pursley Castle and the visit they had paid there.

Ludar listened with tense interest, but whatever he felt of pride in the descent and position of his family or disappointment at their unwillingness or inability to help him he kept to himself. He made no comment at the finish but continued to stare moodily at the dispatch case the publisher was carrying.

William made no effort to hide his disappointment. "It's not what we hoped for," he said to Mr. Milner. Then he turned to Ludar. "This means, sonny, that you'll have to make your own way in the world."

"But our money!" cried Tilly. "What are they going to do about that?"

"Nothing," said the publisher. "The brother is the only member of the family who's in a position to do anything. He washes his hands of it."

"But," asked Tilly, her face flushing resentfully, "can't we sue them? Can't we set the law on this brother with all the money and make him pay us back?"

"No, Mrs. Christian. There's absolutely no basis for an action at law against him."

"It's lost, Tilly," said William, taking a deep drink from a jug beside him. The visitor noticed that his hand was unsteady. "We might as well get used to that now, once and for all."

"I won't give in!" declared Tilly fiercely. "Does the law say we must sit back with our hands folded and do nothing at all?"

Ludar felt that he could not bear any further discussion. He said to Mr. Milner: "Thank you, sir, for all the trouble you've taken." Then, feeling that more was expected of him, he told them he was not surprised; he had never been very hopeful of the outcome and he had no desire to go back. "I thought they might pay Uncle Billy and Aunt Tilly and all the people who contributed," he said. "That wouldn't have been hard for them. I—I don't think they are behaving very well."

"No," agreed the publisher. "I think they're behaving badly."

"I'm going to keep the name Prentice," said Ludar, coming to a sudden decision. "It's been mine for more than ten years. My—my family don't want anything to do with me and I—I don't want anything to do with them! I don't want their name!"

The publisher got to his feet. "I'm not sure that would be wise," he said. "You're a Redcraft, and it's a good name. Think it over well before you decide one way or the other."

Ludar had been keeping up a brave front, but he was showing signs now of less resolution. Inwardly he was not as philosophic about the blow as he wanted them to believe. Knowing he could not maintain the pretense much longer, he turned suddenly and walked to the door.

"I think I'll go out," he said. "I'll have—I'll have to walk this off."

Chapter VII

I

William was standing at the side door one evening when Ludar returned from work. He smiled and started to descend the three steps to the ground, when suddenly his knees doubled up. He fell full length. Without any delay, however, or seeming effort, he got to his feet again.

"Never had such a thing happen to me before," he said, smiling. "It was just as though I hadn't any knees at all—nothing, in fact, between my hips and my ankles. It was a good thing I wasn't at the top of those steep stairs at the shop. That would have been a fall!"

The shop seemed to stay in his mind, for he began to talk about returning to work. "This fall I just had," he said at the supper table, slowly finishing the dry piece of toast which was all he had taken, "has shown me I can't go on coddling myself. I'm going to be a regular invalid if I do. It's having Street and Walker for bosses that does it. I'm getting soft and I must take myself in hand.

"I've been going over it in my mind," he declared when the others had finished and Tilly was starting work on the dishes. "I'm going back to work Monday, positively and without fail. That will put us on a proper financial basis again. We won't depend on your pay, Ludar, and you'll give up this job of yours and get something better. Then you must start to save what you make, and in no time at all you'll have a tidy little sum laid by." It was clear from the way Tilly began to shove the dishes around that she was in disagreement with this. "When you've got enough saved up you'll buy yourself a new outfit. You'll get a tailor to make you a blue serge suit"—he had acknowledged once that all his life he had wanted a blue serge suit for himself but, needless to state, perhaps, he had never had one—"and you'll get a new tie and a new pair of socks, and a fine christy hat. And then you'll go to England, and you'll find this rich uncle and you'll say to him, 'Here I am.' He won't be glad to see you, being the kind of man he is, but you won't let that bother you. You'll just say, 'I got a right to be here, and here I'm going to stay.' You'll stand right up to him and ask, 'What are you going to do about it?'"

Tilly turned around at the sink. "You'll say to that man, 'Where's the money you and the lawyers stole from me to start this business?'" She gave the towel in her hands a twist, as though she had it around the necks of lawyers who worked on the other side and uncles who took money away from helpless orphans. "You're talking sense, William. You've hit on the right thing to do as sure as anything."

William's thin cheeks flushed with pleasure at this praise. "I guess I've got a little common sense, Tilly," he said. "It's time Ludar learned to look out for Number One. Yes siree. If he doesn't, no one else will do it for him. What I say is this: we've got our plan worked out and now we must stick to it through thick and thin."

Dr. Gibbon came the next day, and he looked startled when William told him about the fall. "Have you noticed this weakness in the knees before?" he asked.

The patient acknowledged that he had. It was hard, because of it, to get out of bed or a chair. He hastened to explain, however, that it could not be serious because the weakness was confined to the knees. The doctor had his own opinion on that point and began to ask other questions with a serious air.

"Do you have any pains in your hands?"

"Well, not pains exactly. Sometimes I feel a tingling in the tips of my fingers. It doesn't bother me much."

"Any pain in your ears?"

"Why, yes. But that's nothing. I feel it only in the lobes. That's funny, isn't it?"

"Yes," agreed the doctor. "Very funny. So funny that you'll forget all about this foolish notion of going back to work. If I hear anything more of that," spoken with an emphasis which had a kind of fierceness about it, "I'll come here and tie you in your chair. Is that perfectly clear?"

On the way out the doctor motioned Tilly to follow him to the door. "Has he had another shock of some kind?" he asked in a whisper.

Tilly nodded. "We had some bad news from England. That *was* a shock, I can tell you."

"Do you mean you've heard something about the boy's family?"

Tilly's feelings always mounted close to the explosion point when discussing this subject. "An uncle, a dirty old thief of an uncle, has gone and gobbled up the whole estate. He won't give the boy a shilling and he won't pay any of us back our money. I know it was a shock to William because I was watching him and I saw him go white while Mr. Milner was telling us."

The doctor shook his head with a kind of sad exasperation. "One shock is enough generally. He's had two. Mrs. Christian, your husband is a sick man. If you don't want something serious to happen you'll have to keep a close watch. Don't let him make an unnecessary move. If he should go back to work—well, I wouldn't answer for the consequences."

Tilly looked white and frightened. "Doctor, you—you don't think he won't get well?"

"There you go, jumping to the worst possible conclusion. All I said was that he was a sick man. Now then, keep him on a milk diet. No meat. One slice of red beef and it might—well, it would be *bad*. Does he like milk?"

"Not very much, Doctor."

"He's got to like it. If it disagrees with him, then we'll have to give it to him sweetened or even fermented. You tell him that, Mrs. Christian, if he gives you any trouble."

2

William gave no trouble. Irritability and stubbornness were symptoms of the disease from which he was suffering, and most particularly an unwillingness to accept assistance or advice. He remained as complaisant as ever and was always cheerfully willing to act on suggestions. He even made fumbling efforts to be useful around the house.

This may have been due in part to his belief that he would soon shake off his bodily ills and become well and strong again. Even when he began to lose weight and became weaker each day he still had an unshakable confidence. He would say to the doctor, "Losing weight means nothing to me. I can put it on as fast as I can lose it. I've always been that way."

The doctor combed his mutton-chop beard with his fingers and gave the patient a half-amused smile. "Well, if that's the case, young fellow," he said, "you had better start to show me how fast you can put it on. It's been too much the other way." His manner became completely serious again. "If we can only stop this loss of weight we'll be on much sounder ground. Don't you think you can get some appetite back? Your system can't do it all, you know. Every engine needs fuel."

William shook his head ruefully. "I'm always sick at my stomach," he said. "It never lets up. It's as though I had a lump of lead down there. If I eat anything I get sick right away."

"How much milk are you able to drink now?"

"Well, Doctor, none. I try it, but it doesn't stay down. A neighbor brought me some ginger ale, and it seemed to suit me fine. I enjoyed it and was able to drink almost a bottle the first day. But now I get sick at the thought of it." His loss of weight had resulted in such a change of physical appearance that a casual acquaintance would have found it hard to recognize him. At this point, however, he smiled and the lines around his eyes drew together in the familiar twinkle, and for the moment he looked himself again. "They say, Doctor, it's always possible to drink champagne when you're in this state. Should I go on a champagne diet?"

The medical man was in no mood for facetiousness. He prepared a dose of medicine with a base of bismuth. "Here, down this. It'll do you more good than all the champagne in the world."

William attempted another smile. "And there's this to be said also, Doctor. It costs a lot less."

Before leaving, Gibbon placed a hand on his patient's throat and located the thyroid gland which was the seat of the trouble. It felt soft to the touch, but it was so active and alive with pulsing motion that it was like a separate organism which had found its way into the neck. He frowned and shook his head. "Do you notice irregularities of the heart still?"

The patient acknowledged that his heart had been "cutting up." Sometimes it would seem to stop and then it would begin to beat very fast, "just like a drummer in a band."

The physician seated himself in front of the old green couch in the living room on which the sick man was lying. He was silent for several moments, his eyes remaining fixed on the floor.

"William," he said finally, "you're in quite a pickle, you know. This isn't the worst disease in the world, but it's bad enough. I should tell you that there's an operation for this particular trouble and that it's being tried quite a lot. The surgeon opens the throat and goes right in with his knife and cuts down the size of the thyroid gland. In theory it's plausible enough. But it results in too many fatalities. Do you feel disposed to try it?"

"No! No siree!" exclaimed William. "I'm not a rabbit or a guinea pig. I'm not going to have my throat cut by any surgeon. I'll abide by what you can do for me."

"I don't believe in it myself," declared the old doctor. "But, William, suppose I can't do enough for you?"

"Is this a warning?"

"Well—in a sense. I feel you should be allowed to face the facts honestly. The operation is a last desperate measure, and the odds are against the patient. You would have to go to Toronto or perhaps across the line to have it done."

"It would cost a fortune," said William scornfully. "And even then I would probably die under the knife. No, Dr. Gibbon, the decision isn't hard to make. Not even with you hinting that perhaps—the odds don't favor me anyway."

"I didn't say that, William. But I would be remiss if I didn't tell you about this gamble you could take on a quick recovery."

The discussion had added to William's sense of nausea. He held a hand over his mouth and breathed heavily, having found that this sometimes relieved the insistent discomfort in his stomach. "I'm not a gambler," he whispered. "I'd rather go on as we are. How long would it take to get better this way?"

"Perhaps two years." The physician began to close his medicine case. He smiled at his patient. "We've still got a lot of fight left in us, haven't we? Tomorrow I'm going to bring a mercurial preparation I haven't tried yet. Perhaps it will do something to stop this pesky habit of yours of losing weight."

He tried the mercurial preparation, and the day after the scales showed the weight of the patient to have remained stationary. Dr. Gibbon refused to accept this as evidence that the drug had asserted itself at once. He reported the fact to William, however, feeling that it was at any rate an indication of some improvement in condition.

William's confidence, which had been wavering, came racing back in full form. "I've turned the corner!" he said. He immediately drew himself up into a sitting position on the couch. "I guess we're going to win this battle, Doctor. And without either champagne or the knife."

The doctor was willing to let the optimistic nature of the sick man take the bit into its teeth, but he injected one word of warning. "We mustn't get too sure. Tomorrow you may lose more weight. We must be prepared for setbacks."

Ludar found the patient that evening sitting up in a canvas folding chair, which had been borrowed from next door, and looking the very picture of confidence. "We're beating old Dr. Graves," William said. "I see you've brought a newspaper. That's fine. I'll have you read it to me tonight, if you don't mind. It's queer how much it bothers my eyes to read."

Perhaps the news of the day was duller than usual or perhaps his strength was not as great as he assumed. Whatever the cause, William soon became restive. In the midst of a story about the efforts of an inventor to fly about the eaves of Notre Dame Cathedral in a contraption which resembled the wings of a bird, he flew suddenly into the first fit of irritability Ludar had ever seen him display. "The fool!" he said, his voice quivering. "It stands to reason a man can't fly like a bird or an angel. He needs a machine equipped with power. This fellow is making the same mistake I did with my roller boat."

Ludar remained silent for several moments, wondering whether to go on or not. Then, as the sick man had lapsed again into a quiet mood, he began to read an editorial about a scheme of the mayor to cut the city into zones so that the business and residential sections could be kept separate. This brought the invalid's eyes wide open, and he proceeded to criticize the measure with even more heat.

"That great jackass!" he said bitterly. "Is he trying to keep the city from growing, that flannel-mouthed tomfool? He's always sticking his tutty nose into things he doesn't understand."

Ludar gave up then. He put the newspaper away and said, "I think you should rest now, Uncle Billy."

When he reached home the following evening there was no sign of confidence on the face of the sick man. William was lying back in the folding chair and seemed to find it an effort to open his eyes. A glass of ginger ale, untouched, had gone stale beside him.

"Come here, Ludar," he said.

The boy drew up a chair and sat down. He noticed with a sinking heart that for the first time William had not shaved.

"I lost weight again today, sonny. I—I must face the facts now. I know that—that I'm a pretty sick man."

"The doctor said you must expect setbacks."

"Yes. Setbacks. I know—that. But it's this weakness which sets me thinking."

"Your condition hasn't changed since yesterday. And he was very pleased with you then."

William ran an apologetic hand over his unshaved cheeks. "I must look terrible. But I was—too busy—to shave. I was making a will." He was speaking in cautious whispers. "I tried it this morning, but my hand was so weak I couldn't hold the pencil. But I always get—oh, very much better in the afternoons, and so I was able to do it." He fumbled in the chair and produced a crumpled sheet of foolscap. "Here. You keep it. I want Mr. McGregor and some—someone else. To witness it."

Ludar took the paper and said, "I'll get them in right after supper, Uncle Billy."

"Your aunt Tilly must have close to—to three thousand dollars left," went on the sick man after a long pause to recruit his strength. "She's always been mighty close about that principal of hers. But she was left two thousand dollars by an uncle—years ago. We've been a very saving pair. The thousand we sent to the lawyer wouldn't eat up all we kept out of my earnings. Then there's insurance. I have two policies. A thousand in the Woodmen and the same in the Foresters." He nodded with an air of dignity and pride. "I guess that isn't bad. The Hulls of Coldwater won't be able to say I—I didn't do well for her.

"I'm leaving her the money, Ludar." He was speaking now in a husky whisper. "But with a—a condition. For two years you're to save everything you make. And you're to go to England just as I planned. I think it will get you what you're entitled to. And I'm leaving you my watch. It's not expensive, but it keeps—good time. And my drawing set, the compass and rulers. It's a very good set." He reached out and touched one of the boy's hands. "I've always said you'll make your mark, with hands like that. It's not much I'm leaving you—but it's all I have."

The next morning he was in better spirits. He called to Ludar when the latter was leaving for work. "I must have been down in the dumps last night to talk the way I did." He smiled in a shamefaced way. "You must have thought I was giving in. Well, I'm not. There's plenty of fight left in the old hulks." He nodded. "But it *was* a good thing to get the will made out and witnessed. A man should never wait until the last minute to make his will."

3

At noon that day Lockie McGregor came to the side door. "Tilly," he said, holding up a large bottle wrapped in tissue paper, "I've brought this for William. It's the only thing he's likely to keep on his stomach, I hear."

It was a pint of champagne. The donor went on to explain, nodding with satisfaction at the common sense he had used in conjunction with his generosity, that he had brought a pint because William would not be able to drink more, and what was left would go flat. Tilly looked at the bottle as though it were a stick of dynamite and apt to explode at any moment.

"Neighbor," she said, "we're strict temperance, Billy and I. You won't be offended, I hope, that I must give this back? I wouldn't hold alcoholic poisoning to the lips of my sick husband."

The Remarkable Man looked surprised. "Do you mean it, Tilly?" he asked in a voice both hurt and incredulous.

"Of course I mean it. Do you suppose I've been Temperance Society and Band of Hope all my life without meaning it? I'm sure it was kindly meant, neighbor, but I must ask you to take it away."

The merchant became angry suddenly. He had been standing on the side stoop, but now he marched into the house, taking the bottle with him. He placed it on the kitchen table.

"Tilly, do you know your husband is in a bad way? You've seen he can't take any nourishment, haven't you? Don't you know he *must* if he's to get enough strength to throw off this disease?"

"I know that," she said in a whimpering voice. "I know my poor husband is very ill."

"This," pointing to the bottle, "may stay on his stomach and give him a little of that strength. Do you mean to tell me you won't try it because of silly scruples of your own? Don't you know there's alcohol in plenty of good medicines? I've been a strong Presbyterian all my life and I'm as good a Christian as you are, but I would break most of the rules I live by to help someone struggling for life. Be shamed to ye, ma'am!"

Tilly was unable to withstand such vigor of argument. She surrendered. "But I won't give it to him myself," she protested. "I won't! I won't be the instrument of his falling from grace."

Ludar came in at this moment, his eyes looking deeply sunk in his black face. He supported the contention of their neighbor. Opening the bottle with difficulty—for none of them had ever seen champagne before—he carried a glass of the effervescent wine in to the patient.

William looked at it with wonder and a touch of amusement. "Champagne? Lockie, you're kind of going it, aren't you?"

"Never mind the expense, William," said the merchant grandly. "It will be good for you, and that's what counts. Down it, man, down it."

William lifted the glass, but his hand trembled so much that some of the wine spilled. He put it down quickly.

"I don't mind telling you, Lockie," he said, "that I've always been curious about champagne. I've always wanted to taste it just once. But aren't you supposed to drink it out of a slipper?"

McGregor picked up the glass and held it to the sick man's lips. "It will taste better this way," he said.

William took several swallows. Then he looked up at them and smiled. "It's nice," he said. "Why haven't I been drinking it all my life? I've a feeling I'm going to be able to hang onto it."

He not only succeeded in hanging onto it, but a little later he took another glass and he maintained a firm hold on that as well. He was alone at the time, but the regular afternoon improvement in his condition was setting in and he managed to hold the glass steadily. "I feel a little stronger," he said to himself with a sense of satisfaction.

Tilly had given in, but the teachings of a lifetime still governed her thinking. When the doctor came later in the afternoon she stopped him in the hall and told him what had happened.

"Has he kept it down?"

She acknowledged that he had. The doctor seemed pleased at this, and then surprised her by saying that it did not matter.

"Mrs. Christian," he said, keeping his eyes averted, for he remained the most humane of men and always shrank from the duty which now devolved on him, "I must tell you the truth. Your husband is in a bad way. I'm afraid it doesn't matter much what we do now. This wine can do him no lasting good, but neither can it harm him. If it brings him some relief—temporary, of course—should we take it away from him?"

"Doctor!" exclaimed Tilly in a shocked tone. "I didn't expect to hear the likes of that from you."

The physician had been through a trying day. He sighed wearily. "I think we'll let the patient decide this for himself," he said. Before going into the sitting room he asked in a whisper, "Did you grasp the meaning of what I said?"

Tilly began to cry. "I think you mean I'm going to lose him."

"His chance is so slight that I wouldn't advise you to count too heavily on it."

Lockie McGregor paid another call during the evening to see how things had gone. He found William in a better state of mind.

"I think," said the patient in a whisper to prevent Tilly from hearing in the kitchen, "that it did me good. Isn't that a terrible confession for a temperance man like me to make?"

The merchant sat down in a state of content over the success of his venture in generosity. He began to talk, and this meant, of course, that he talked about himself.

"I've had no luck about the sign, William," he confessed. "I've thought of a dozen ideas and then chucked them out as worthless. Can it be I'm a one-idea man? Does it mean that my first sign was such a tremendous inspiration, such a masterpiece, you might say, that I was left dry on the subject?" He shook his head despondently and then proceeded to air another grievance. "I'm having no better luck about adopting a substitute for the grandson I might have had if—if things had been different. They see a child, Catherine and her husband, and they tell me it's just what they want. I don't know why it is, but there's always some obstacle they've overlooked and which I see the instant I lay eyes on the little creature. Think of this, William: last week they were certain they'd found the right child at last. His parents had come out from Scotland and he was a fine, healthy little chap. Man, it sounded good. I went back with them and—and what in Tophet do you suppose I found? He was a Colquhoun! The Colquhouns, as you must know, are the hereditary enemies of my family. No McGregor can look at a Colquhoun without shuddering. I said to Catherine, 'Come home at once and read your Scottish history before you look at any more brats.'" He sat in smoldering silence for a moment. "Why is it, man, that there's something glaringly wrong with every child I set eyes on?"

The news of William's condition had been getting around, and the next afternoon Langley Craven called, bringing a quart of champagne with him. Ludar knew that something out of the way had happened as soon as he entered the kitchen that evening. The bottle, nearly empty, stood on the table. Tilly was sitting in her rocking chair with a look on her face which defied all attempts at classification.

"Who brought it?" asked the boy, inspecting the festive-looking bottle.

"Mr. Langley Craven. This afternoon." And then, to his complete confounding, she hiccuped.

"Aunt Tilly!" Ludar was unable at first to believe the evidence of eyes and ears. He looked at her and saw that her cheeks were strangely flushed. "Have you been drinking it too?"

"What if I have?" She gave her head a defiant toss. "What else was there to do? That bottle cost all of two dollars, and Billy wasn't able to drink more than a glass. Wouldn't it be sinful to let two dollars go to waste?"

Ludar had not slept the night before after hearing what the doctor had said to her. He was tired almost to the point of exhaustion. He sank down into a chair and gave way to laughter which had a touch of hysteria in it. He laughed so hard, in fact, that tears came into his eyes and welled down his cheeks, making long gray streaks on the ebony surface.

"I hope you enjoyed it," he said. "But what will the Band of Hope think?"

When she made no reply he realized that her temporarily unbalanced feelings had been hurt. He stopped laughing abruptly and reached into a pocket of his coat for a handkerchief. The pocket was torn and the coat itself, having been splattered often with hot glue, was coated heavily with emery deposits. The handkerchief was no longer clean, but the condition of his face made this a matter of small moment.

"If it gave you a few minutes of forgetfulness," he said, "I'm glad you drank the wine, Aunt Tilly."

Tilly began to weep copiously. "You must never tell," she said between sobs. "You must never tell anyone. I wouldn't be able to hold my head up in this town again if you did."

4

Ludar had fallen into the habit of coming home for the noon meal. The next day he found William in such a weakened condition that the patient refused to leave the folding chair. "I can't lie down," he said. "It makes me dizzy and it hurts my ears. I'll stay just like this, I think."

Dr. Gibbon arrived as the boy started back to work, and they paused outside the door.

"Doctor," asked the boy in a hesitating tone, "will—will it be very hard for him?"

"You mean at the finish? No, my boy, that's one consolation you have. It's quick and it's merciful. Syncope. One moment he may be smiling. The next, eternity. No struggle or pain or even awareness of it."

"Is there—no hope at all?"

"Well, my boy——" The physician paused. Then he shook his head. "No, Ludar. None."

That evening it became clear that William was drawing on his last small reserve of strength and that, moreover, he knew what this meant. He asked Ludar to read the Bible to him.

"The minister was here this afternoon," he said. "He read to me, and I found it good to listen."

Ludar read from the Psalms. William proved very attentive. He nodded his head several times and at the finish he began to sing in a thready voice:

"Oh, I must be a lover of the Lord,
Or I won't go to heaven when I die."

It was one of his small repertoire of songs, and he had always sung it with full reverence. A few moments later, however, the words of a parody which he had heard in a lumber camp where he had worked as a young man came back into his head. Without realizing what he was about, he began to sing them:

"Oh, I must be a lover of the old man's daughter,
Or I won't get a second piece of pie——"

He stopped abruptly and looked at Ludar with startled eyes. "I didn't mean it!" he whispered, panic-stricken. "It was habit. I haven't thought of that for years. It just came out of itself. Do you suppose They were listening? Will They mark it down against me?"

"No, Uncle Billy, of course not. Why should a little thing like that be marked against you?"

The sick man's head vibrated violently on its thin neck. "Everything is written down," he whispered. "I know it is. I had a dream this afternoon. A terrible dream."

In halting tones, and with many pauses for breath, he proceeded to tell about it. He must have died, for he found himself far up in the clouds. So far up, he explained, that it would have taken that crazy inventor many years to get there with his trumpery wings. He found himself in a very large room. It had a row of pillars at least twice as big around as the organ pipes at the Baptist Church. Perhaps they were organ pipes too. They stretched up so far that he did not recall seeing the tops of them at all. And he thought they were made of gold. There was, in fact, a great deal of gold about this room, and diamonds and all other precious stones. He was aware that beyond this room there was a great space and that he would be sent out there for his hearing. It was not closed in because he could see stars, and it was a busy place because he could hear thousands of voices.

There were two angels in the room. They had wings, but, as they were very busy, they seemed to have laid aside their harps. They were talking about *him*. The first angel said to the other, "I hope this one passes." The second angel said, "I'll look at the book." He picked up a scroll and turned over leaves until he found the right place. "It seems," he said, reading in the scroll, "that this one has boasted openly and continuously that he looks out for Number One." The first angel seemed to think this was bad. He shook his head. "What has been your experience," he asked, "with applicants who looked out for Number One?" The second angel seemed less concerned. "I've found that they're often sent in the Other Direction. It's not just the selfishness but the other things that go with it. You know how He feels about selfishness." "Yes," said the first angel,

"I've noticed how hard it is for them to explain it away. I hope this one will know the right thing to say when he goes out there. I *do* want him to pass. So many lately have been sent in the Other Direction."

William was very tired when he finished the story of his dream, and it was clear he believed it had been a warning. "It's true," he said after a long pause. "You've heard me say I always looked out for Number One. But, sonny, what else was there for me to do? How could a man in my position, without education or influence or anything, get along? And the worst of it is, I didn't get along." He twisted his head to one side and then grimaced with the pain it had caused in his ear. "I'm wondering what I'm going to say when I go out there. I don't want to be sent in the Other Direction. What do you think I'd better say, Ludar?"

In a choking voice the boy answered: "There won't be any questions asked, Uncle Billy. What's written on those pages will prove you've been one of the most unselfish men that ever lived."

The patient shook his head. "I wish I could believe that. But I saw those angels and I heard what they said, and I saw the book with my own eyes. Ludar, I'm frightened."

He became more restive as the evening passed, and Ludar begged him to go to bed. This met with a decided shake of the head. He did not dare lie down. There was always a choking sensation in his throat when he tried it. He refused even to have his clothes taken off so that he could lie in the folding chair with more comfort. He was too tired, he said, and did not want to be put to any effort.

Finally he dropped off into a fitful sleep. Ludar, watching him, was astonished at how small he seemed. Curling down into the arc of the canvas, he looked no larger than a boy. Remembering the doctor's words, "One moment he may be smiling. The next, eternity," the boy kept anxious vigil.

Ludar was inexpressibly weary himself. It was hard to keep on the alert. He yawned and rubbed his eyes and stretched his arms. And finally he fell asleep and had a dream himself. It seemed that he and Uncle Billy were on the roller boat. It was not the same one, being very much larger and built of iron or steel. There was a motor in the center and it was humming away, and the boat was rolling so fast that it fairly skimmed over the waves. Uncle Billy was on top of the world. He smiled and almost skipped as he attended to the motor. "This will show them," he said. "Now they won't be able to laugh at me and say my inventions are no good. And, sonny, just think of all the money rolling uphill for us. Perhaps it would be better to have stained glass in those windows instead of colored. It's more tony. And the trim in the library should be curly maple."

Ludar wakened with a start when the clock struck eleven. William also opened his eyes, and it was clear to the boy, although he could not have explained why, that Uncle Billy was happy and quite easy in his mind.

"Ludar," he said in such a low tone that it was hard to hear him, "I've got the answer. I know what I'll tell Him. I think I may pass after all. I'm not afraid any more."

He started to smile. At any rate, the lines at the corners of his eyes drew together. The process stopped there, and Ludar, watching but not daring to breathe, finally realized that he had already gone to present his case.

Chapter VIII

I

When the quitting bell sounded Ludar carried the gloves he had been using, and which had become large and as stiff as metal gauntlets, to a corner of the room where rubbish was deposited, dropping them with a sigh of relief. He looked down from the nearest window and saw that men were already rushing out from the front entrance in a great hurry to get home and dressed for the Saturday night promenade on Holbrook Street. The owl-nosed buffer spoke to him, calling him "Duke" instead of "Glue," and Ludar smiled for the first time in days, because this was a reminder that he was through working at Cauling's.

As he walked down the stairs he realized how completely exhausted he had become with the strain of the past week. All the responsibility for the funeral had fallen on his shoulders. Immediately after William died Ludar had wakened Tilly and told her he was going to use the McGregor telephone to notify the doctor. She had thrown off the covers in a great hurry and had said: "No, you don't. I'll go myself. You needn't think I'm going to stay in this house alone." During the three days which elapsed before the ceremony she had refused to remain unless someone was with her, and she had gone each night to Cousin Celia's to sleep. When Ludar protested there was no such thing as a ghost and that, in any event, there would be nothing to fear from a spirit with a twinkle in its eyes, she brushed this aside. "I can't help it," she said. "That's how I am. When my own mother died I left the house and never came back until the time of the funeral."

In spite of the provision in William's will she had fought the idea of Ludar's keeping everything he earned with such bitterness that he had been compelled finally to give in. It had been agreed between them that she would receive half of whatever he made. He was quitting Cauling's, moreover, against her wishes.

On the way to Diamond Street his step became lighter as he thought of what lay ahead of him. Now he would be able to get his hands really clean and keep them that way as long as he lived. He would be able to go back to the kind of life he had lived before, even to see Tony. Ever since her return he had been desperately afraid of meeting her and had always walked with an anxious eye on the street ahead of him.

In the first few blocks of Diamond Street there was one store only, that of the widow Malone, who was using her house for the sale of yarns and sewing supplies. Ludar was passing here when the door opened and a vision in a jacket of the kind known as a Zouave and a dark blue velvet skirt stepped out briskly. It was Tony, and she was smiling as though she had expected him.

The widow Malone took great pleasure later in reporting what she had seen and what she had surmised. "She come into the store and she gave me a smile and she said, 'How d'ye do, Mrs. Malone?' quite friendly and easy-like. I was sure right away there was something on her mind besides the yarn for the knitting of a hat. She didn't know what color she wanted. She would pick out a skein and carry it to the front window to see better—that's what she *said*—and she'd march it back and say it wasn't the right shade. The last one was a dull brown, but she no sooner got to the window than back she come. 'This'll be the one,' she says. 'How much is it?' I says, 'Nineteen cents,' and she lays down a quarter and she walks out without waiting for change, and stuffing the yarn into a pocket of the cute little monkey coat she was wearing. I says to myself, 'Ye're up to something, Miss Tony Milner. Ye're waiting to meet someone.' And she was. I goes to the window and peeks out, saying to myself, 'Just supposing it's a drummer from out of town or—*a married man!*' For pity's sake, how wrong I was! It must have been something to do with church work or charity. When I gets to the window, there she is, walking up the street with a man whose face and hands are black as tar!"

Tony said, "I've so much to tell you, Ludar. Father couldn't have told you half about your family and about what we saw and heard."

Ludar was keeping his eyes fixed on the plank sidewalk in a state of great embarrassment. "I was hoping," he said, "that you wouldn't see me like this."

"Do you really supppose I care about it?"

"I do." He was disturbed by the appearance of his hands and shoved them into his pockets. "I left Cauling's tonight and I'm going to find a different kind of work. I planned to see you then."

"I've been back weeks and weeks."

They had reached the deserted lumberyard. It became evident now that she had acquainted herself thoroughly with his habits, for she crossed the road unquestioningly and turned with him through the gate.

"Don't you think you've been too sensitive?" she asked. "You're doing a very brave thing and you should be proud of it instead of ashamed. I was proud of you when I got back and was told about it, and I was furious when I heard someone laugh at you because of this way you take of getting home."

Ludar stopped at once and looked at her, the first direct glance he had allowed himself since they met. "I didn't think anyone knew," he said. Then he began passionately to explain himself. "I'm not ashamed because I'm working in a factory. We needed money to keep going while Uncle Billy was sick, and this was the only way to get it. I haven't liked it, but perhaps that's because I'm no good at the work. As I have to work this way and get so black, I wanted to be—to be let alone. When I met people I knew on the streets and they went by without speaking or laughed at me behind my back, I suffered because I knew what they were thinking. They were remembering I used to be called 'Duke' and was supposed to belong to some fine family. I think they were remembering, too, that I was considered smart at school, and it had been expected I would make something of myself when I got out. And here I was, after all, working in a stove foundry and going home at nights with my face black."

"Why are you sure they felt that way?"

He shook his head unhappily. "Some of the things said were repeated to me. Oh yes, there was a lot of laughing going on at my expense. I stood it as long as I could. Then I tried this way." He felt that he had not yet succeeded in making her understand. "Men are doing this kind of work all over the world, and there's no reason why I should be ashamed of it. I don't think it was that. I really believe, Tony, that I was more ashamed of myself than of what I was doing. I have some ability, and yet when the time came to help I'd made so little use of it that the best I could get to do was to glue emery wheels."

It had rained during the day. Water still dripped from the piles of rotting lumber and from the roof of the office, which slanted forward at a more drunken angle. The path was a succession of puddles. Tony paused before jumping over one of them, and instinctively he offered his hand. Then he quickly drew it back.

She stopped at the next puddle also. "I'm sorry if I seemed to be lecturing you. I didn't mean it that way. And I should have understood how you felt."

He was realizing now, however, that she was right. All he had accomplished by nursing his pride in this way was to give a lot of fair-weather friends a chance to laugh at him. He should have had the courage to walk straight home and look everyone in the eye as he passed.

"I've been making a spectacle of myself," he said.

"No, Ludar."

"I'm not the first fellow who's had to forget his ambitions and ideas and take whatever work there was. It's going on all the time. I can see now that I've been pretty sulky. The only excuse I can offer is that I've been so completely unfitted for this kind of work. I've been a joke in the foundry as well as to my one-time friends."

She smiled and held out her hand. "You might help me over."

There were so many more puddles that by the time they reached the other end of the path the only difference between their hands was that hers was smaller than his. She held it up proudly. "We've been friends as long as I can remember," she said. "I'm glad to have this small share in what you've had to do." She seemed suddenly on the point of tears. "Poor Ludar! What a terrible time you've been through!"

They reached the willows, and again she turned in under the long row of trees without either hesitation or explanation. "I waylaid you today. You knew, didn't you, that I was in that store for the sole purpose of seeing you? But I have the best kind of an excuse. I've a message to give you. Father wants to see you tonight. At our house."

Ludar stopped short and looked at her with surprise and even a trace of apprehension. "What's it about, Tony? Something to do with what you found in England?"

"No. Not that."

"Do you know?"

She smiled mysteriously. "Yes, I know. But I'm not allowed to tell you. Father wants to do all the telling himself. And he wants you to come before eight o'clock because he has a meeting later downtown."

"I'll be there on time. Are you sure you can't give me a hint?"

"No. Not a hint of any kind. I promised."

Tony stopped. Looking down at her over his shoulder, for he had kept right on growing through everything and was now perilously close to the six-foot mark, he saw that she was frowning and quite apparently perplexed. "I wish I could——" she began. Then she shook her head with fresh determination. "No, I mustn't. I must give you no hint at all. It's a matter you must settle yourself and without any—any advice."

They had reached the end of the path under the willows. Ludar stopped. "Don't you think you had better go on alone? I'll wait a few minutes here. Then no one will see us together."

She placed her soiled hand on the sleeve of his emery-encrusted coat. "I *want* to be seen with you," she said. "I insist on being escorted to my own front door. And I hope, what's more, that we pass everyone we know."

2

Nothing could have reminded Ludar more forcibly of the loss they had suffered than to find Tilly in the kitchen with the inner door closed to shut off the rest of the house. He had no time, however, to indulge in the saddened reflections which had filled his mind almost continuously. After a hasty supper he spent a full hour in the summer kitchen with a tub of hot water. When he emerged the only trace left of the day's toil was the dark graining of the palms of his hands. He had shampooed his hair, using several changes of water, and had restored it to its natural color.

He donned his new suit, glad now that Tilly had put her foot down on any use of it except for church, because the trousers still had a little crease in them. It was unfortunate that William's plan to get him a new tie had not worked out, but he had a clean shirt which was very little frayed at the cuffs. When he was ready to leave Tilly ran in haste for a bonnet and shawl. "I'll stay with Margaret next door," she said.

Gertie Shinny answered his rather diffident ring of the Milner bell. "Hello, Luddar," she said, having her own system of pronunciation of names. "It's nice to see ya. Are ya here to see Tony?"

Ludar shook his head hastily. "No. Mr. Milner."

"Right ya are. I'll tell Mr. Milner."

He had never been inside before, and he glanced about him with the most intense interest at the interior of the house; this house which had enjoyed the wonderful privilege of sheltering Tony all her days. A giant grandfather clock was striking the third quarter after seven as he came in. It seemed nothing short of eight feet in height, and its weights were as large as the explosive shells being used in the Boer War. At the rear was a winding staircase of yellow oak with a newel post in the form of a cathedral. Ludar paused to look at this with a proper sense of awe and had to be reminded by Gertie that he must not delay. "He's a tartar if ye're late," she whispered.

The room into which he now walked, with shoes which had developed a completely new squeak, was a revelation to the boy. The bookcases ran clear to the ceiling and they were tightly packed. This had not prevented an overflow of books on tables and chairs and on the top of the huge roll-topped desk at which the formidable publisher was sitting. Mr. Milner was smoking a pipe with a long curved stem and had a mass of papers in front of him. He had been wearing a purple dressing gown earlier, for it was draped over the back of one of the chairs.

"Right on the dot," he said approvingly. "A good start. A newspaperman has to be prompt. Did that daughter of mine let the cat out of the bag?"

"No, sir."

"Well, I'm offering you a job as a reporter on the staff of the *Star*." The publisher puffed away at his pipe and watched Ludar sharply through the screen of smoke thus created. "How does the idea impress you?"

"A reporter!" The exultation Ludar felt was evident in his voice. To earn his living by writing, to have a chance to improve and develop the talent he still believed himself to possess! This was a miracle, an answer to prayer. "I'm delighted to have the chance, sir. It's hard to—to say anything." The boy gulped. "I believe, sir, I have some fitness for that kind of work. At any rate, I'll try very hard."

"When could you start?"

"On Monday, sir. It was Uncle Billy's wish that I give up my job at Cauling's and find something else. I left there this afternoon."

"Report tomorrow," said the owner. "A newspaperman has no hours. He must be ready to work all day and all night if necessary. Sundays and holidays are the same as other days to him. Be at the office tomorrow at one o'clock, and Mr. Willing will be there to give you a verbal grooming. He may even have assignments for you."

"Yes, sir." Ludar had already pinched himself to make sure he was not dreaming. His eyes were glowing with enthusiasm and delight.

"The usual salary for a new reporter is six dollars a week. I'm going to pay you seven. You seem to have better qualifications than any others I've taken on, and of course you have heavy responsibilities."

"Might I suggest, sir," said Ludar, "that you pay me six dollars a week and put the other aside to be used to pay off the people who put money into the fund for me?"

Mr. Milner stared at him with an air of disbelief. "Do you really mean that?" He removed the pipe from his mouth and stared at the boy. "Now that's the kind of thing I like. I'll be more than pleased to do as you say, and I'll speak to Willing about it tonight. How will you want it distributed?"

"The smaller contributors first," answered Ludar. "Some of them couldn't very well afford to lose it, and I won't feel comfortable until they're all paid up and satisfied."

"I think that's sensible. You seem to have a level head in spite of what I've been told about a poetic strain in you. Well, then, we seem to have completed our financial arrangements." Mr. Milner indulged in a long pause. "There's another matter. A condition."

Ludar's satisfaction had been so complete that the hint of an obstacle caused him to look up in alarm.

"You seem to have been on a rather special footing with my daughter," said Mr. Milner. "She prattled about you as a child and she was very much interested in tracing out your family connections in England. I want it understood that there must be no thought of romance in either of your heads. In point of years you're mere children still. So, young fellow, if you take this position with me it must be on the understanding that you won't see my daughter, except casually, for a period of time. Say, two years. That will give Antoinette a chance to recover from all this excitement and the interest she took in her detective work in England. I want her to meet plenty of other young men. What happens at the end of the two years will be out of my hands. I'll let my daughter have her own head then."

One thought only was in Ludar's mind: Had Tony known of this condition? If she knew, what would she expect him to do about it? What answer would she want him to make? On second thoughts he realized that a sense of resentment was growing in him. He had never been invited to the house, not even to Tony's parties when she was a small girl. And now he was being told he could have this desirable post on her father's newspaper if he promised not to see her. Only by swallowing his pride could he accept.

"I want you to decide here and now," said Mr. Milner. "My daughter is angry with me. She thinks I'm treating her as a child. You knew that, I suppose?"

"No, sir."

The publisher was keeping a shrewd eye on him as they talked. "Do you have any idea what her wishes are in this matter?"

"No, sir. She said nothing to me about it."

There was a sudden sound of many footsteps on the front porch and a clamor of young voices. The bell rang, and when it had been answered in a flurry by Gertie Shinny the hall seemed to fill with animated youth. From where he sat Ludar recognized some of the boys and girls who had gone to school with him but who had since drifted far away with an amazing abruptness. He felt they were strangers now. The boys wore suits which were slickly tailored, and expensive ties, and they carried bowlers in their hands. He was sure they would scorn to be seen in the high school cap which he still wore. The girls had become sleek and they had their hair up and even had skirts with bustles. The talk, such of it as he heard, was strange to him.

And suddenly it came to him that Mr. Milner's suggestion offered one great advantage. To go on seeing Tony now would mean that he would have to compete for her favor in this kind of company. Inevitably he would be made to look ignorant and tongue-tied and out of his class. On the other hand, two years would give him the chance to learn the many things he did not know now, all the little customs like sending going-away presents to girls. He could find out the things he needed to know about manners and clothes. It should be possible for him to pick up some of the interests about which he now knew nothing.

Two years was not long. At the end of the period of prohibition he would be only twenty. Jacob, on being forced to wait for Rachel, had spoken of seven years as "but a few days." Two years? It would be nothing.

"I accept the condition, sir."

They heard a step on the stairs, and Ludar caught the merest glimpse of Tony's head passing the stately newel post. She called to her guests in a gay tone, "I don't know how you did it, but you're actually two minutes ahead of time."

The voice of Norman Craven answered: "My doing, Tony. I was so anxious to get over here that I kept right after them like a slave overseer with a whip. Did you know that——"

The voices receded as the party trouped to the other side of the house and, presumably, into the living room. A gramophone began to squeak the words of "She Was Bred in Old Kentucky."

"I hoped to get away before anything like this could start," said Tony's father. He tamped out the pipe and got to his feet. "I must run along now. I want to add, my boy, that I'm glad to have you on my staff." Right through the conversation there had been a note of personal interest on his part. At this point, however, he became completely the publisher and employer. "I hope you'll turn into a good newspaperman. If you don't, I'll be quick to let you know. I'm not a believer in coddling and I don't beat around bushes. When a man doesn't fit in I feel it's in his interests as well as mine to let him know."

Ludar crossed the hall practically on tiptoe. His resolve had been to keep his eyes away from the open door of the living room. This proved too much, however, and he could not refrain from looking back. He caught a single glimpse of Tony standing beside Norman Craven and smiling. She must have known he was in the house when her friends arrived and that the footsteps in the hall were his, but she did not turn.

A doubt which had remained throughout at the back of his mind flared now into an active torment. Would she consider him a coward because he had given in? Would she think he had preferred the position on the *Star* to her?

It seemed certain to him that she would have no inkling of his real reason for accepting; that to refuse the offer and to attempt thereafter to compete on even terms with these friends of hers would be the surest way of losing her. No, she could not be expected to understand that.

Two years? It would be an eternity. Perhaps it was less time than Norman Craven would need to press his suit to a favorable conclusion.

Book Three

Chapter I

I

The smell of ink was the first impression Ludar received of the newspaper world. It filled his nostrils as soon as he entered the *Star* building at the appointed time on the following day. He turned in at the side door and walked up a narrow flight of stairs to the floor above. To the left there was a glass partition, and he saw behind it a row of typesetting machines and a couple of printers, shaved and neat and uncomfortable in their Sunday best, jeffing with type on an empty stone table. To the right was another glass partition with a door which had been left open. Recognizing Sloppy Bates, who sat at a desk just inside the door, he turned and went in.

The old reporter was shabbier than ever and he was quite drunk. He was writing news items slowly and painfully on sheets of yellow copy paper. He looked up and grinned and nodded his head.

"Glad you got in before the demon editor," he said. "This gives me a chance to plant some truths in your head. It's important, young man, to start with sound ideas. Willing will tell you a lot of stuff. Some of it will be right enough and some will be so much swill and nonsense. He'll tell you about the importance of the business office and the circulation. My eye and Betty Martin to that!"

Ludar had taken a chair at another desk and had faced around to listen. His eyes were busy also, however, taking in the appearance of the city room. It had three desks, one of them with a typewriter on it (that, he was sure, would belong to Jinks Snider, the senior reporter), the same number of wastebaskets, a table piled up mountainously with exchange newspapers, and at one side an office closed in with glass which no doubt belonged to Allen Willing.

"A reporter," Bates was saying solemnly, "must tell the truth and nothing but the truth. Sometimes, alas, he can't tell the whole truth. That's old stuff. But I want to get this into your callow, impressionable brain: there's always a story behind the story, and that's what you must find and write. Do you understand what I mean, boy?"

Ludar confessed with some reluctance that he did not understand.

"Then listen to me." Bates drew himself up and belched vinously. "A man is hit by a train in a railway yard and all smashed up. The average newspaper report gives the details of what happened, where the victim lived, his age, the size of his family—that's all. But that man may have been saving up to buy a toy horse for his little son's birthday. The horse is in a package in his hand, he is thinking how happy the little fellow will be, and then—— Wham! The engine hits him and cuts him into mincemeat. People never remember stories of accidents, but they'll think of that poor workingman and the toy horse for years. That's what I mean, the story behind the story. It's always there and always to be found if you dig hard enough. Get into the habit of looking for it and you'll become so great that other newspapermen will speak of you with awe, and publishers will take what you write and print it between cloth covers."

Ludar was listening with interest and conviction, but still letting his eyes rove about the room. Most particularly he was noticing a big hook on a table near the door which was filled almost to the top with handwritten sheets of paper. Copy for the printers!

"But listen to this, boy," went on Bates, who had not needed much of such talk to work himself into a maudlin frame of mind. "My advice to you is this: turn around this instant and walk down those stairs and never come back. A thousand old newspapermen have said that to a thousand cub reporters. I'm saying it again. Look at me, John Kinchley Bates, better known as Sloppy. I showed great promise when I started. By rights I should be writing leaders for the Toronto *Globe* today or specials for the New York *Herald*. Yet here I am, the butt of the town, a drunk, getting seven dollars a week, with a wife trying to make both ends meet on that. I stayed in newspaper work." He looked at Ludar and saw arithmetical speculation in his eyes. "You're wondering how a man can get drunk as often as I do on seven dollars a week. I'll tell you, boy. I never pay money in a bar. I cadge my drinks!"

Allen came running up the stairs. "Welcome," he said to Ludar, stopping in the city room. "I'm delighted you're joining us." He looked at Bates, who had swung around to his desk and was putting words down on paper. "I'm sure Sloppy has been giving you some advice. I hope you listened carefully. He knows what he's talking about."

Bates spat noisily into the wastebasket beside him. Then he picked up the sheet on which he had been working and read aloud:

> "Mrs. Alvin Truebody has returned from Hespeler where she visited her daughter, Mrs. Bert Twaddy, and enjoyed the company of her three fine, healthy grandchildren. She reports that Bert is doing well in the barbering business there."

He glared, first at Willing, then at Ludar. "That's the kind of deathless prose I put down on paper these days. Let it be a lesson to you, boy."

Willing said briskly: "Ludar, you're in luck. There's a story for you to cut your teeth on today. Tanner Craven dropped dead half an hour ago at the corner of Grand Avenue and Yarmouth Street."

Both his hearers gasped with surprise. Sloppy Bates got to his feet with some difficulty and fumbled for his hat. "Quicker than I thought. I gave Tanner another year. I'll get right out on the job."

"Not you, Sloppy," said Willing. "You've fixed yourself this time. In the state you're in, Mrs. Craven wouldn't let you put a foot inside the door." He turned to Ludar. "Jinks is visiting his girl in Simcoe. I'm going to collect photographs and take them to Toronto on the three o'clock train. I'm going to stand over the engravers and make sure I have the plates to bring back on the early morning train. That leaves the writing of the story up to you, Ludar. It will have to be a full one. Three columns. We'll go up to the Craven house now, and I'll get you started on the right lines." He paused and grinned. "He was found with an empty berry box in one hand and a rubber without a sole in the other. That's a beautiful bit of color which, unfortunately, we won't be able to use."

Sloppy Bates let out a bibulous and indignant roar. "What did I tell you, boy, what did I tell you?"

When Willing returned the next morning on the early train he went straight to his office and found a towering pile of handwritten sheets on his desk. "What's this?" he asked himself. "The original manuscript of *War and Peace* complete? The setting copy for a new dictionary?"

He picked up the first sheet and glanced at it. Twenty minutes later he carried the copy to the large corner office occupied by the proprietor on the floor below. Mr. Milner had just come in and was scowling at the pile of mail in front of him.

"Boss," said Willing, "we have here the obituary of the late Tanner Craven, written by the latest member of the city staff."

Mr. Milner looked disturbed. "That long? Great Scott, he must have worked all night on it."

"Practically all night. It seems he wasn't content to handle it in the usual way. He went to a lot of unexpected sources: that old school friend of Craven's, Dan Marcy, who's been night watchman at the Craven plant for years; old Rabbit Gains, who taught him arithmetic; his secretary, the headwaiter at the Cameo House, his tailor and barber and, of course,

Langley Craven. This copy is jam-packed with anecdotes. And it's twice as long as any obituary ever published in the *Star.*"

"Tanner Craven was our leading citizen," said the proprietor doubtfully. "We've got to be careful what we print about him. What do you propose to do with this stuff?"

"Use it!" exclaimed Willing. "I wanted to let you know about it before I devoted seven columns of space to a single obituary. I tell you, boss, our subscribers will read this from beginning to end. I don't want to do anything to it except cut out an adjective here and there."

2

Grand Avenue was lined with carriages for two blocks when Ludar made his way to the Craven residence for the funeral, and the employees of the various Craven concerns had filled the yard and were now spilling out on the sidewalk and the road. A suspicious undertaker's assistant stopped him at the door and looked even more unbelieving when Ludar said, "I'm here for the *Star.*" If Norman Craven had not come into the hall at that moment he might not have been admitted to the company of the elect. Norman, giving the impression of being more managing than sorrowful, led the way to the stairs. "My mother wants to see you," he said. "You'll find her in the small sitting room to the left of the upstairs hall."

Mrs. Craven, looking massive in black and wearing a heavy widow's veil, was sitting on a couch, surrounded by relatives and friends. She said, "Come in, please," when Ludar appeared in the door. As a general introduction she added, "This is the young man from the *Star.*"

Ludar was apprehensive. What mistakes had he made that he had been thus summoned? What things had he written to cause offense?

He had been called in for praise, however, and not blame. "I want to thank you, young man, for what you wrote about my husband," said Mrs. Craven. "It was very thorough and sympathetic in tone. I don't know where you got all your facts. Certainly not from me. But I've no fault to find. You wrote it quite well, I thought."

Ludar, very much relieved, said, "I'm glad, Mrs. Craven, that you liked it."

"I must say I was surprised," she went on. "Mr. Willing's explanation of why you were to write the sketch did not satisfy me at the time. But apparently he knew what he was doing. An older man couldn't have done better."

She nodded then, and Ludar took this as his dismissal. He returned to the crowded main floor and made his way to the conservatory which, he had been informed, was being kept clear for necessary clerical work. Here he found one of the undertakers and Lou Bennet engaged in getting up lists of those in attendance. Lou, looking very grave, was wearing a black silk waist. They had been friendly at high school, and she gave him a nod and the half-smile which was all the nature of the occasion allowed.

"Will there be a carbon for me?" he asked.

"It will be ready half an hour after the services," answered Lou. "I could send it down to you at your office, if that would help."

"It would. I could go right back and get my story into shape."

Her busy fingers kept on with their task, but she turned again in his direction. "Mrs. Craven was so pleased with what you wrote in the *Star,* Ludar. Her eyes were red from crying when she got through. I think Norman liked it, too, although he said nothing."

He stationed himself in the conservatory door just in time to witness the arrival of Langley Craven. The latter was looking very solemn and unhappy, dressed in proper black and with the usual wide mourning band on his sleeve. Norman Craven, as it happened, was standing with some friends of his own close at hand, and Ludar heard him say in an angry whisper: "There's nothing I can do about it, but neither Mother nor I want him here. Thank heavens he didn't bring his wife. *That* would have been too much."

"Joe's outside," said one of the group. "I saw him as I came in."

A few moments later Adelicia Craven put in an appearance. It had been supposed that she was on her way home from Mexico and that she would not arrive for the funeral. Fortunately, however, one of the telegrams sent out had reached her at Chicago. She had taken an afternoon train out on the Michigan Central and had left it at Simcoe, the nearest point. All she could obtain there in the way of a conveyance to bring her on to Balfour had been an open buggy, and she was covered with dust from head to foot.

She had been so stunned by the suddenness of the news and so afraid that the carriage would not get her home in time that she was emotionally upset when she came into the house. Spying Langley, she went over at once and put her arms around him. "Poor little Tanner!" she said, dropping her head on his shoulder. "I'm so glad you came, Langley. I was worrying about that all the way up. I was afraid—— Oh well, here you are, old boy. You must keep a chair for me so we can sit together for the service."

Then she saw Norman, and her pent-up emotions could no longer be kept in check. She wept as she embraced him and was not able to stop until there were streaks down the dusty surface of her cheeks. "Is there time to see Effie before they start?" she asked. Norman said there was and escorted her up the stairs. Ludar heard her say as she dabbed ineffectually at her moist countenance, "Just like me to be away on a pleasure trip when a thing like this happens!"

Ludar had continued to stand just inside the conservatory door. One of Norman's friends from out of town said, "That rum old cove must be the wealthy aunt or Norman wouldn't be so attentive to her."

Another voice, a familiar one this time, said it was indeed Miss Adelicia Craven. "Norman's always got his eye on the main chance. Old Lish has quite a pot of it, and he wants his share. I asked him yesterday if he would buy himself an automobile now that he's coming into his money. He's been talking about it continually. All he did was stare at me and say no. Do you think he's going to think twice before spending money now that it's his own?"

A third voice joined in: "He's always thought twice—or three times or four. He's never been one to toss his dough about. It wouldn't surprise me if he didn't buy a car after all. Have any of you noticed a change in him lately, as though he's under his mother's thumb?"

"I don't know about that," said someone else. "But I do know something pretty funny. Will you all promise not to give me away? A girl I trot out a bit typed the will. It has one innocent little clause in it——"

Ludar finished his report of the funeral and carried the copy in to Allen Willing's office. He had been so absorbed in his work that he had not seen Mr. Milner go in before him. This was embarrassing, because there was something he wanted to discuss with the editor. He paused irresolutely.

"Spit it out," said Allen.

"I overheard something at the funeral. It was about Mr. Craven's will. It seemed to me"—his manner suggested that he did not expect them to agree but that he would hold to his own opinion in spite of them—"that if it's true it would be the most interesting piece of news we've published in a long time."

"Well, what is it? Has Tanner Craven cut off his family? Has he left his fortune to the cook, or is he establishing a fund for broken-down newspapermen?"

"He bequeathed the drum table to his brother."

Both men said, "What!" in loud and skeptical tones. After a moment Allen Willing added, "What a dramatic gesture of reconciliation to make at the very end!" Then his eyes began to glitter with excitement. "Boss, I know this is all a pipe dream. It can't be done the way I'm seeing it, but—— Do you recall that Ludar here was one of the participants that day when Tanner whisked the table from under Langley's nose? We could publish his first-person account of what happened and then at the very end quote the clause from the will. Whew, wouldn't that be a story?"

"One," said Mr. Milner, "which would so antagonize the widow and her son, and all of Tanner Craven's friends, that they'd never forgive us."

"I know. I know. I was just indulging in a happy vision."

"We have to consider the people who support us," went on the proprietor. "We live so closely together in a place of this size that an intimate story like that can't be printed. But I'll go this far. I'm in a position to get the details of the will. I'll get right after it, and if possible we'll publish the news today, including this very good bit about the drum table. With no comments, of course. . . ."

"Good work, Ludar," said the editor in dismissal.

When the two older men were alone Mr. Milner leaned across the table. "Say, Allen," he whispered, "that boy was pretty sharp to snap up the news about the table."

3

It was clear to Ludar that there had been a change in personnel at the Waifery as soon as he arrived the following evening. Miss Craven had prepared him for it when inviting him to dinner. "Velva is gone," she had said. Velva was the girl from the orphanage. "She married the milkman. I more or less, as you might say, insisted on it. Under what might be termed the Circumstances. Albertina has taken her place. Some people like Albertina when they get sort of hardened to her."

The roaring sound he heard as soon as he stepped off the hoist was Albertina. She came over to greet him, a voice surrounded by a sweater-enclosed bust, a wild fringe of frizzled hair, and eyes like a lion tamer's.

"You're the little reporter, are you?" she boomed at him. "Did you see me in *Nid, Nid, Noddin'?*"

"No," said Ludar. "I'm afraid I didn't."

"Not surprised that you didn't," went on the Voice. "Couldn't hardly expect you'd be one of the seven brave souls who did. It was a rotten show and it was no surprise to little Albertina when we went bust here."

She shook her head at him understandingly, as though assuming that he understood all the ins and outs of show business. "I don't need to tell *you* that we were getting in a bad way when we didn't take in enough to pay for cartage. We'd beaten two hotel bills. *You* know we couldn't hope to get away with it a third time."

"Of course not," said Ludar.

"I knew we were dished as soon as I peeked through the curtain and saw that our audience was Miss Craven, sitting there in her own box, four men in the pit, and two boys eating peanuts in the Gods. The manager vanished before the first act was over. All the rest of the company had enough hidden away in the sock to get back to home base on their own, but, as usual, I was as broke as a poached egg on the floor. Miss Craven offered to take me in until I got an answer from my agent in New York. That was before Miss Craven went on her trip." She tossed back her aboriginal mop of hair and laughed delightedly. "The answer hasn't come yet. Do you know what I think? I think that small-souled, penny-pinching, rump-sprung agent of mine is *never* going to send me any money."

"What part did you have in the play?" asked Ludar, feeling that politeness demanded the display of some interest.

"I played the noisy part." Miss Dare exploded into a loud roar of amusement over her own definition. "It's an odd thing, boy, but I always found myself cast for the noisy bits. I'll tell you, it was just one long hullabaloo when I was on the stage. I'd go into my dance while the others were working on the dialogue. They used to say I was stealing their scenes. I didn't steal their scenes. I mangled them, I pulverized them, I murdered them. When I got through, their scenes could be swept up in a dustpan and carried out to the garbage.

"Perhaps you've noticed it already, but I'm a noisy person," she went on. "I was the squallingest infant ever born. When I started school the music teacher resigned. She said she couldn't cope with me. When I was eight my mother—she was in the profession, too, a club swinger—took me to the manager of a Tom show and said I could play Little Eva. He said no, but he was willing to give me a tryout as one of the bloodhounds."

Miss Craven rescued him at this point and took him to the head of the table. "Albertina is one of my mistakes," she whispered. "I should have bought her a ticket to New York when the show bust up, but I thought it would make things lively to have an actress with us for a while. The trouble is that it's too lively."

There was a rib roast of beef for dinner. Miss Craven went right on eating and cutting off more slices, most of them for herself, and she had no time for talk. It was not until the dessert arrived that she resumed the conversation.

"I want to warn you," she said in a whisper. "My nephew Norman's getting terribly down on you, Ludar. It used to be Joe, but now he gets into a rage whenever you're mentioned."

"I don't know why he should single me out."

She winked at him. "Yes, you do. Two young fellows of your age can't be interested in the same girl without the air getting full of feathers. I shouldn't tell you this, but yesterday, after everything was over, he said he and Antoinette were both young, but after all he was now head of the family and a rich man in his own right and he didn't see any reason at all why they shouldn't get things settled about the future. I tried to tell him it wasn't proper for him to speak so soon, but he paid no attention. Out he went, and I guess he saw her. He was very glum when I went over there this afternoon. I think he'd said his piece and she told him no."

"I hope so," said Ludar fervently.

"He was in a rage about you. You see, Norman's always had things his own way and he doesn't like opposition. And he was mad as hops about that piece in the paper on his father's will. He blamed it on you and he said it was damned impudence. Did you write the piece?"

"No."

"Did you have anything to do with it at all?"

"I heard about the drum table being willed to Mr. Langley Craven and I told them at the office."

"Well, he's going down tomorrow and raise a fuss with Mr. Milner."

The dessert was one of her favorites, a variety of apple dumpling called a cock robin which was served with cinnamon sauce. Miss Craven took a bite of the extra-large one in front of her and then desisted long enough to slosh the contents of the cream pitcher over it.

"You two are putting me in a fix," she said, frowning uncomfortably. "I like you both so much. I've got one suggestion for you, Ludar. If there's anything more to be written about Effie Craven and her son, don't you do it. Keep away from them. The less you see of Norman, the better it will be. I don't like to say this, but he—well, he won't hesitate to do a little hitting in the clinches. Try not to meet him at parties and things like that."

"I won't be going to parties."

Miss Craven reached again for the cream pitcher. "It's all very aggravating. Sometimes I feel like batting some sense into the head of that Norman. But I—I must own up, I'm getting fonder of him too. He's going to have some of the drive and the ability of my father. Not as much, of course, but some. Will you do me a favor, Ludar? Kind of—well, walk on eggs for a while."

Miss Dare had been getting noisier as the meal progressed. The owner of the Waifery regarded her with a critical eye. "Albertina has had more than a nip," she said. "She seems to have been getting into the habit since I went away. She's damn well got herself a snootful this time. Before the evening's over she'll be singing that 'Boom-de-ay' song and showing us her clog dance. There's just nothing else to it. I'll have to saw her off right away."

Perhaps this mention of her need to do some sawing off aroused memories of Cousin Hubert. She turned to Ludar with a broad smile. "Here's an item for you. Cousin Hubert's going to get married."

Ludar was too astonished to make any immediate comment. His hostess threw back her head and laughed unrestrainedly.

"It's hard to believe of that dried-up, spindly-legged runt. But it's a fact. Next week he leads Cissie Anderson to the altar. We used to call her Silly Cissie when she was a girl. She played tennis and always screamed when the ball came her way. *That* will give you an idea of her. Her father left her quite a wad when he passed on." She finished the last crumb of the cock robin and decided reluctantly not to have another. "It seems she's kind of musical. She and Cousin Hubert, that wolf on the fold, were playing a duet, and they got started on some kind of nonsense which didn't have a thing to do with the score. Cissie tipped over backward off the piano bench. Because he helped her up, or because he didn't—I don't know which—they decided they were made for each other. And there you are."

Tea had been served instead of coffee. It was steaming hot, but this did not prevent the hostess from imbibing hers in quick sips. "There's one more thing," she said. "About your name. I hear you're angry with your relatives—I don't blame you—and going to stick to Prentice. You're wrong about it, my boy. Prentice isn't your name. It's just something they made up when your father came out to this country. But you *are* a Redcraft. There've been Redcrafts, I hear, since the Flood or nearly that far back. Your relatives in England may be a sour lot, but that doesn't mean you should forget your heritage. Think of all the fine ancestors you've had down the centuries, thousands of them, and all Redcrafts.

"From now on," she declared emphatically, "you should use your real name. People will get onto it gradually, and in no time at all Prentice will be forgotten as completely as you forgot the real name when you were a poor little put-upon coot back there in England."

After a few moments of consideration Ludar agreed. "I think you're right," he said. "I've been foolish about it. Well, from now on I'm Ludar Redcraft."

4

Publication of the terms of the late Tanner Craven's will had stirred up a great deal of discussion in town. He had left two thirds of everything to his son, ten thousand dollars to be paid in cash at once, the rest unreservedly when he came of age. The balance went to Mrs. Craven after a few small bequests had been taken out. Two hundred and fifty dollars were left to each of the two servants, the same amount to Dan Marcy, a silver inkstand to his secretary. To charity and the church, nothing.

It was the giving of the drum table to Langley Craven which set tongues wagging excitedly. This, people said, was the end of the feud, the hand of friendship extended from the grave after years of bitterness.

But, in this, public opinion was at fault. It was not the end of the feud. The morning after Ludar's appearance at the Waifery, Norman paid a call at the offices of Dengate, Jones and Dengate, who were his mother's lawyers.

"I have no more intention of letting that table out of my hands," he said to Mr. Harris Dengate, the senior partner, "than I have of making over all my inheritance to a fund for immigrants who come out to this country with signs on their backs."

"And how," asked the lawyer, "do you propose going about it?"

"We must have that silly clause thrown out. Father wasn't in his right mind when he did it. Why, it stands to reason he wasn't, because the will was made after he took the stroke."

"You mean, then," said the lawyer, tapping one hand with a paper cutter, "that you want the whole will thrown out?"

"No!" cried Norman. "What do you think I am? I'm satisfied with the will as a whole. It's just this one clause I want chucked out."

"Do you want me to go into court," asked Mr. Dengate, "and say your father was mentally responsible when he made this will except for just one short moment of aberration, and that he conceived and executed the clause about the table in that single flash of irresponsibility? It can't be done, my boy. Either you accept the will as it is or you upset the whole applecart."

Norman sat for several moments in sulky silence. "You're a lawyer," he said finally. "You tell me what other ways there are. I know there always *are* ways. I'm ready to try anything to keep that table in my hands, where it belongs."

"There's one thing you can do right away." Harris Dengate did not like Langley Craven, and there was a barely perceptible glitter of satisfaction in his eyes as he spoke. "The settlement has never been made for the furnishings of the Homestead. Take that matter into court, and I guarantee you'll win."

"Now you're talking! I'd like nothing better, Mr. Dengate, than a lawsuit with that uncle of mine. That is, if you're sure we'll win."

"It's an open-and-shut case. Your father knew that all he ever had to do was to start legal proceedings."

"Then why didn't he?" Norman was puzzled. "We often talked about it, but the pater would never say very much."

"Your father had a mental reservation in the matter," declared Dengate. "He didn't want your grandfather's wishes carried out. He didn't want money. He wanted the stuff. He kept holding off and holding off and hoping that some way would be found to change things. If your uncle had actually gone to the wall—he was mighty close to it a couple of times—your father would have stepped in and settled with the court to take over the collection intact."

Norman grinned delightedly. "Let's get this started, Mr. Dengate."

It became apparent at once that the lawyer had been right. Langley Craven did not contest the claim. He said, in fact, that he wanted to make the settlement, that he had always wanted to make it. Accordingly a board of three experts was appointed, consisting of George Satterlee, who owned a furniture store, and two antique dealers from out of town. After an appraisal of everything which had been moved out of the Homestead the board decided that Langley Craven should pay the sum of seven thousand dollars, to be divided evenly among his sister and the heirs of the late Tanner Craven. The sum, perhaps, was a little high (Adelicia Craven said it was much too high), but Langley Craven did not protest. He gave a check to his sister at once (and she promptly found some new charities to absorb it) and deposited the other half with his lawyers, to be handed over as soon as all conditions had been fulfilled; which meant, of course, when the drum table had been returned.

History then began to repeat itself. Norman found all manner of objections. The figure settled upon was not a fair one. It should be much higher. There should be compensation for all the years that his uncle had enjoyed possession of the furnishings. Interest should be added to the money for the time that it had been held up. Finally he agreed to accept the settlement with one stipulation: that the drum table remain in his hands. Langley Craven, through his lawyers, rejected this curtly and irrevocably. He had no intention of disregarding the wishes of his father, who had stated that the possessions he prized should be kept intact.

There the matter remained for quite a time. Norman moved the drum table into the front parlor and put a handsome piece of statuary on it and some valuable books in limp leather bindings. He always called it to the attention of visitors. "There it is," he would say. "The famous drum table. It has my father's initials scratched underneath it, and those of whole generations of English boys whose parents owned it, some of them going back two hundred years. I'll never part with this piece. I'd like to see anyone get it away from me."

When it became certain, however, that his uncle did not intend to let him have the money until he relinquished the table, and that there was no way to force him to do so by law, he became less assertive on that score. Finding that the value placed on the table had been seventy-five dollars, he wrote one day on a sheet of paper:

$3,500 $75

and he studied it thoughtfully for a long time. He fell into the habit of carrying this slip with him and taking it out every now and then. Each time he looked at it he would frown and shake his head angrily. Finally he could not stand it any longer. He went to his mother and laid the slip down in front of her.

"We must throw in the sponge," he said. "There's just no sense to it. I can't chuck thirty-five hundred dollars out the window for the sake of a measly table which wouldn't fetch more than fifty dollars under the hammer."

Mrs. Craven was very much disappointed, as her manner and voice showed. "Are you sure, my son, that you need the money more than you do the table?"

"I want them both!" cried Norman. "If there was any way of getting both I would take it. But that lawyer of ours can't seem to do anything about it."

So a few days later the table was delivered at the Langley Craven residence, carefully packed up in canvas sacking, and the check was duly transferred. Mrs. Craven was ill at the time of a curious kind of fever which was baffling the family doctor, and Langley was so worried about her that he paid no attention to the returned heirloom. It was nearly a week later that it was unpacked, and then it was found that a blow from some sharp implement had opened a wide crack across the handsome top from side to side.

He did not discuss the matter with his wife until he had allowed himself time to cool off. Going to her bedroom then, he found her sitting up but looking far from well. She had never been really sick before in her life, and it was hard for him to believe that the pale and listless figure in the bed was his usually lovely and fresh and sparkling wife.

"Well, Langley, what's wrong?" she asked, knowing from his manner that he was upset.

"That insolent pup has played me a pretty trick!" he said. He told her briefly and curtly what had happened. "He has the money," he added, "and there's nothing I can do. If I sued him he would have witnesses, I'm sure, to swear the table was in good condition when it left his hands. I've decided instead to put an expert to work on it. These fellows can do remarkable things at restoring antiques. The miserable cub will be expecting an explosion; I'll disappoint him by saying nothing."

Mrs. Craven smiled wanly and drew about her shoulders a handsome blue shawl, against which, alas, she looked faded and sick. "That's very sensible, Langley. Don't give him the satisfaction of fighting over it. I'm sure he'll be very much surprised when he doesn't hear anything at all."

Norman *was* disappointed. A week or so later he met Meg at a small party and went over at once to talk with her.

"Hello, Cousin Meggy," he said. "Are those beads around your neck or the hearts you've been collecting? And by the way, I hope that much-fought-over table reached your father safe and sound."

"Oh yes," said Meg casually. "I think I heard it was damaged a little, but nothing to worry about."

"Oh!" Norman rubbed a thumb over his chin and frowned down at her from his full six feet of height. Was she making fun of him? Then for the first time he became fully conscious of the power of her charms. She was wearing a polonaise dress of biscuit-colored foulé, with a high silk-lined collar against which her small head nestled like a lovely exotic bloom. "Meggy, our families have been fighting like fury for as long as I can remember, haven't they? Is that any reason why we should be enemies?"

"Well," said Meg, letting her eyes linger on his for a moment before turning them away, "I think that depends."

"What does it depend on? You and me? See here, Meggy, there's no law that says a fellow can't give a cousin a kind of a rush."

"Isn't there? Perhaps there should be, Norman."

"Well, Miss Cleopatra, we'll just give it a try, anyway."

In the meantime Felix Rosmann, an old German cabinetmaker, was working on the table. With loving care and magic touch he drew the wood back together and fitted in the splintered parts and then slowly but surely applied the glue. In ten days he returned it, and it was hard to tell that any serious damage had been done. The top carried a partly obliterated line through the leather, and that was all.

Thus came to an end the episode of the drum table.

Chapter II

I

On a bright warm Sunday in early May of the following year, when Strawberry Hill was dotted with the white heads of trillium and grape hyacinths stood in stiff rows along each side of front walks, Ludar was summoned from a deep immersion in *Henry Esmond* by an automobile honking in front of the house. The horn did not belong to any of the cars in town. He jumped up excitedly. "I'll bet the Cravens have their new automobile," he said to himself, dropping the book on the floor and reaching for his hat.

He had been right in his guess. A dark green monster was drawn up at the curb. It had a snout like a battleship and the wheels of a juggernaut. Joe was seated at the wheel, looking handsome in a new brown suit, and very much excited. Beside him sat a resplendent creature in gray serge with gigot sleeves (which took up almost as much space as the wearer) and a saucer hat with rosebuds. "That Meg is certainly growing up," Ludar said to himself as he climbed into the tonneau, where Mr. and Mrs. Craven sat facing each other.

Joe leaned back over his shoulder to explain: "A Daimler! Holy man, it's got enough power to climb right up Sand Hill in high. I tell you, Ludar, you've got to watch yourself when you're driving this old rarin'-to-go car!"

"The house is finished and we're going up to inspect it," said Langley Craven. "We thought you'd like to go along and give us your opinion on certain points. Some of the furniture was moved in yesterday. The rest will go in tomorrow. With, I hope," smiling, "less excitement than we stirred up the last time we moved."

The walls of the new house on upper Fife Avenue (two blocks beyond the last house and therefore still terra incognita) had risen during the late fall, and the roof had been on before the first snow fell. All through the winter there had been work going on inside and, as soon as the weather permitted, the contractor's men had swarmed out all over the place. Ludar had followed progress with the most intense interest. It was different from anything in town; a wide and gracious house of white brick with severe wooden pillars over the entrance and wings on each side. It was so different, in fact, that he had been puzzled at first and not sure that he liked it. It had been by gradual steps that his feeling had changed to one of admiration for the simplicity of its lines and the promise it gave of spaciousness and comfort within.

"We haven't seen much of you lately, Ludar," said Mrs. Craven in a plaintive voice.

He gave her all his attention at once, becoming aware that the fever from which she had suffered some months before had left its mark. She was still rather lovely, but in a subdued way which he found hard to reconcile with the sparkling vision of her he always carried in his mind. Even her eyes, he saw with a sense of sadness, had lost some of their liveliness. Her hair was lifeless and most of the color gone from her cheeks. Any other woman would have said (and they all did) that she was beginning at last to look her age.

"They keep me on the jump at the office," he explained. "I have meetings to cover every night."

"You might have said hello to me," said the resplendent vision on the front seat.

He switched his gaze to Meg and found her well worth his full attention. Her eyes, which had seemed to hover uncertainly between blue and green, were now unmistakably of the latter color. Let bards sing their monotonous songs to eyes of blue and brown and black, but permit the remark that they may have been overlooking something, that a green eye, when it is of the bright emerald of the sea on a sunlit day, can be absolutely devastating. This may be accepted as an accurate description of the orb which looked back at him over a gray serge shoulder.

He said, "I can see it's true, Meg."

"What is true?"

"The things I've been hearing about you. That you're beautiful enough to inspire the efforts of poets."

"I'm more interested in inspiring telephone calls," said Meg.

"Our daughter is in a quandary at the moment," said Langley Craven.

"She can't make up her mind whether to stay at home and captivate all the men in town or go in for a career as a dancer."

Ludar said without hesitation, "That should be an easy decision."

Mrs. Craven did not care for the turn the conversation had taken. "A life on the stage, even in ballet, isn't one for a young lady. They say it's all hard work and dieting and practicing over and over again. There isn't any glamor in it. Certainly I wouldn't consent to anything of the kind for Meg. She's to stay home and come out in the usual way. I hope she'll have plenty of beaux and enjoy her youth as she should. That's what you meant, of course, Ludar."

He answered in an apologetic tone, "I'm afraid, Mrs. Craven, I was going to cast my vote for a career. Meg would make a wonderful dancer."

The girl turned her saucy hat and smiled brilliantly at him. "Thanks," she said. "The compliment was slow in coming, but it was nice when it finally arrived."

Now that it was finished, the house proved to be a thing of unmistakable beauty. Ludar looked at it with an unqualified admiration which carried at the same time a sense of guilt. This unadorned dwelling was the exact opposite of the one Uncle Billy had dreamed of building. It had no stained glass, no verandas with high carved roofs, no balconies. It even lacked a tower, that capstone of all architectural beauty in the Gay Nineties. Here before their eyes was the tangible proof that taste in the world was changing and that it was getting far, far away from the era of curlicues and tasseled chenille and tear-bespangled art.

"I can see you're taken with it," said Craven, quite pleased. "It's my own plan—but worked out in detail by the architect, and improved, of course. Joe rather resents the lack of a tower, and Margaret would have liked something in the nature of a ballroom. Otherwise we've all been in agreement."

The house had a central hall with glass doors opening into the gardens at the rear. The rooms to right and left were large and beautifully paneled. There was a painting over the fireplace in the living room, but it was the portrait of a Georgian general with a porcine nose and a powdered wig and not the once much-loved Barbudo-Sanchez. Looking at the red-coated figure, Ludar asked, "Where is it, Mr. Craven?"

The owner sensed what was in his mind. "The Barbudo-Sanchez? It's hanging in the library. I'm still fond of it, but I've been picking up some fine things—masterpieces, really—and they have to be displayed."

"Where's the drum table?"

Mr. Craven had to give some thought to this. "I don't believe it's been brought over yet. I suppose I'll find a quiet spot for it somewhere. Under the circumstances, I don't want to place it too conspicuously. But the truth of the matter is, I've been caught in the Queen Anne net. What wonderful furniture they made in those days! Wait until you see the secretary-desk I'm buying in Montreal. Made in 1710, of seaweed mahogany. It has a lot of secret drawers. And the card table I'm getting at the same time. Irish, 1720. It comes out of a house where two old men, each over ninety, have been living alone."

"I'm as sentimental as a girl in her teens," complained Mrs. Craven. "I love this house and all the new things, but I've still such a warm spot in my heart for the Lester place. We've been so happy there."

"I'm sentimental about it myself," declared her husband. "We lived in the Homestead during the years of the Decline and Fall. We lived in the Lester place through the period of the Rise. I love every brick in the old rookery, every windowpane that rattled during the night, every crack that came in the plaster as soon as we painted and papered." He smiled and ran his fingers through his beard, a sign that he was going to say something daring or improper. "But, my love, I must tell you that I'm glad to leave. It was an eccentric house, to put it mildly. For instance: having the main bathroom open off the landing of the front stairs. As we're on the subject, I confess to having been tempted often, when the front doorbell rang, to open the door a mere fraction of an inch to see who it was."

"Heck!" said Joe. "I did that a hundred times."

Having made this confession, which caused his mother to exclaim in a shocked tone, "Joe, you didn't!" the son of the house motioned Ludar to follow him on a tour of inspection. They went from top to bottom, Ludar in a state of delight at the tangible evidences he saw on every hand of the splendid new status of the family. He thought of himself as belonging, almost, to this charmed circle. Joe had been the first friend he had ever had, and he would always be the best. When he, Ludar, had been no better than the son of an immigrant who had made a failure of life in Canada and had killed himself, Joe had taken him in and the Craven family had been cordial and had always let him see that he was welcome. The gratitude he had felt because of this had now grown into a much deeper feeling. He loved them all, Mr. and Mrs. Craven, Meg, and, above all, Joe. To see them assuming again the position they had once held was more gratifying than any good fortune which might come to him; even if he were to receive the much-longed-for letter from his people in England claiming him as one of them.

Joe was to have his own bedroom and bath in a bright corner of one of the wings. None of the furniture had been moved in yet, but he had brought over many of his special belongings: his skates and hockey stick, his lacrosse stick and cricket bat, a couple of old sweaters, a pair of running shoes, some books (a very few!), and many pictures of athletes and prize fighters.

The two friends sat on the window sill here and talked.

"Is your father going to let you keep the red car for your own use?" asked Ludar.

Joe nodded his head happily. "Yes, sir!" he said. "The little bus belongs to me now. I hope it keeps running forever." He paused, and a self-conscious grin spread over his face. "Lude, I've a confession to make. I—I kind of think I'm in love."

"You've been in love a dozen times, my man. Is this something extra-special?"

"Ludar, this is *it*. This little babe—Ludar, she's lovely, she's sweet, I'm mad about her! I'm never going to look at another girl as long as I live."

"Do I know the paragon?"

Joe shook his head. "And that's not surprising. I don't know her myself! But that doesn't matter. I'll get a knockdown soon. Just seeing her has made a reformed character of me."

"Well, Joseph, let me have a look at her before you go proposing or anything like that. What's her name?"

"Jessie Slaney. She's the younger sister of the girl Pete Work wants to marry. Well, Pete and I are thinking along the same lines. I want to marry Jessie."

"I've seen her," said Ludar. "She's as cute as a bug's ear. But you listen to me. You mustn't go rushing in with a ring in one hand and a marriage license in the other. Take your time. Let yourself cool off a bit."

The voice of Mrs. Craven reached them from below. "Boys, where are you?"

As they walked to the stairs Joe said in a whisper: "Mum's the word. I don't want the parents to know anything about it yet. Nor Meg. She'd make life miserable for me if she knew."

Meg was dancing in the hall as they came down the stairs, and Ludar stopped to watch in a state of awe and wonder. Joe answered another summons from his mother at the rear of the house, saying, "The kid can dance, at that."

The daughter of the Cravens had become more than a beauty. She was grace itself as she wove slowly across the drawing room and then back into the hall, her skirts swaying rhythmically as she moved, her arms beating time lightly. She was humming in a high, sweet voice, the rare kind which never seems thin or strained. Seeing Ludar at the foot of the stairs, she opened her eyes wide (a new mannerism which she would use often thereafter and to deadly effect) and gave him a smile which lit up everything like a flash of lightning.

"It's only because there's no one else here to try her tricks on," said Ludar to himself. But he had to acknowledge that it was exciting to watch her and pleasant to have himself noticed in this way.

She stopped and made a gesture of invitation. "It's no fun dancing alone," she said.

Ludar, ashamed of his limitations, answered, "I'm sorry, Meg, but I can't dance."

"You dummy!" she exclaimed scornfully. "You've got to learn sometime. You can't go through life without dancing like a—a gillygawker."

"But," he pleaded, "I'd make such a rotten dancer! I'm slow on my feet and as clumsy as a bear. If I tried to waltz I'd step all over my partner's feet."

"You should have seen Joe when he first went to dancing class." She held out her arms. "Come and have your first lesson," she commanded.

He moved toward her reluctantly. "Now," she said, slipping into his arms, "all you do is move your feet in time to the music. I'll do the leading. See, like this. That's easy, isn't it?"

She led the way, humming "The Blue Danube." Ludar realized that this was different from the kind of thing which was usually called dancing. It was like teaming with a wood sprite, a spirit skipping around a fairy ring, a wisp of moonlight. He struck out with an experimental foot. Three times he went wrong and lost step, making it necessary to pause. Once his toe collided sharply with her instep, and the wisp of moonlight said, "Ouch! You great ox!" Then, by some magic agency, no doubt, he found that he was waltzing; painfully, it was true, stiffly and self-consciously, but waltzing, nevertheless. After several moments more the stiffness began to leave his legs and he found himself turning and reversing and swinging about with some ease and a mounting sense of pleasure.

"Meg, you're a wonder!" he exclaimed. "You're actually teaching me to dance."

"Well," said Meg guardedly, "we're making a start."

"And by the way, you don't seem to mind being called Meg any more. What's brought that about?"

She was succumbing to the rhythm and dancing with her eyes shut now.

"I was just a silly child when I said I didn't like it. Meg is a sweet name. It has charm, and I love it. A dozen times a day I say to myself, 'I am Meg Craven,' and I always love the sound of it. I'm so glad you like it too, Ludar."

In a moment or so her mood changed. "Mother's right about this matter of being a dancer," she said. "It's a pleasant thing to dream about, but I've heard you have to practice hours at the bar every day of your life, and at first you have to go out with road companies and sleep on trains and have chorus men make love to you. No, it wouldn't do, I'm afraid. Besides, it makes your legs heavy and muscular."

"It would be sacrilege," he declared, "to let anything change their present slim perfection."

"Well!" She opened her eyes at that. "So you *have* noticed, after all."

They were waltzing under the Georgian portrait in the drawing room, and it almost seemed that the gouty subject was watching her with protuberant eyes and puffing out his mustache in gruff approval.

"Ludar," asked Meg, "do you see a great deal of Tony nowadays?"

"No."

"Is that so! Why not?"

"I'm busy all the time."

"A lame excuse. Do you know that a certain cousin of mine is *most* attentive to her?"

"Yes." Ludar felt depressed immediately, the invariable reaction when this matter came up. "I hear all about it regularly."

"And isn't Sir Galahad doing anything to keep his fair Guinevere for himself?" Meg's Arthurian references were badly scrambled, but her meaning was clear enough.

Ludar did not answer.

She opened her eyes again and studied his face. "Good old dog Tray," she said. "Faithful unto death. And what a silly goose I am, taking you in hand, Mr. Redcraft, and making a dancer of you, and all for the benefit of another woman!"

Mr. and Mrs. Craven, who had been settling some problem of furnishing in another part of the house, returned to the hall in time to observe this stage of Ludar's first dancing lesson.

"I catch my breath every time I look at the child," said Mrs. Craven. "Langley, she's becoming unbelievably lovely."

He nodded, letting his eyes follow Meg with delight and pride. "She's her mother all over. And yet she doesn't really look like you, Pauline."

She was watching the dancers with a line of worry between her eyes.

"I must see that nothing comes of *this,"* she said. "It wouldn't do, Langley. It wouldn't do at all."

"Pauline! She's just a child still."

"I've great plans for this beautiful daughter of ours. My own looks are fading, and I don't mind telling you that it's a blow——"

Langley slipped an arm around her shoulders. "You look as beautiful as ever, my own."

"I like to have you say so, but I know it isn't true. Mirrors are always to be believed and husbands never. I—I'm getting old, Langley dear. I'm becoming faded and plain. It's terrible and I suffer—ah, how I suffer!—but there's one consolation. I can put all my pride into the beauty of our little Meg." She turned to her husband and said in a voice which verged on fierceness: "Langley, nothing but the best will do for Meg! She must have a chance to shine. When she marries, it must be to someone who can provide the setting she deserves. I even dream of a title for her."

"My Lady Meg!" Langley smiled. "I sympathize with your point of view, my dear, but I'm afraid it may lead to trouble."

"At any rate," said his wife, "I'm going to put a stop to this dancing lesson. It's been going on long enough."

2

Catherine had become the one link left between Ludar and Tony. The latter was out of town a great deal, but when she was home she paid regular visits to the invalid. It was understood that he was never to be there at the same time, and as a signal that Tony had arrived the blind in the sitting-room window would be raised three quarters of the way up. He would have to wait until it had been drawn down to the usual halfway before leaping the fence and hurrying in for a report.

"How was she looking?" he asked one evening in the following fall when this procedure had been observed.

"I think she gets more attractive all the time."

"That's a pretty mild way of putting it. What was she wearing?"

"As usual, she was beautifully dressed." Catherine found this an interesting question to answer. "She had a dress of chestnut serge. It had a bodice and gigot sleeves, if you know what they are——"

Ludar nodded. "Leg-of-mutton."

"It was a most becoming dress. Tony certainly has style. *Such* style! When she gets a little older she's going to be the best-dressed woman in Balfour."

This suggestion removed Tony so definitely into the sphere to which he lacked access that he hastened to a more congenial topic. "What did you talk about?"

"Well, let me see." Catherine patted the head of her companionable little Peke, which was called Lily. "We discussed the best way of making fudge. She told me she was beginning to play bridge and had been praised by Mr. Langley Craven. And, oh yes, she heard you reply to the toast to the press at the banquet of the Anglican Brotherhood at St. Paul's Church. She worked in the kitchen and didn't do any of the serving, but she happened to be looking through the door when you were called on. She waited and heard what you said."

"I didn't speak until near the end. I thought all the ladies had gone home by that time."

"Tony had volunteered to stay and put things in order."

He was disappointed that it was duty which had kept her in the Parish Hall so late (in this he was wrong) and not a desire to hear him respond to the toast. Catherine went on, however, to tell him something which more than made up.

"I did something I shouldn't. I've always wanted to ask her if she thought you did the right thing in taking the position on the *Star* when her father made the conditions. Tonight I got up my courage and came right out with it. She said she had wanted you to take it, that it would have been wrong not to. Of course she was angry about it at the time—at her father, that is, for thinking it necessary. Once we got started, we went right on talking about you. She was delighted it was working out so well. Her father is pleased with the work you're doing and thinks you're going to turn into a fine newspaperman."

Ludar gave a deep sigh of relief. "I couldn't be sure, and there was no way I could find out. I couldn't ask Tony herself. You've taken a great load off my mind."

He scrambled over the fence immediately after returning from work one evening in early October. Catherine's husband and father were due to arrive at any minute (having a partner made it possible for Lockie McGregor to get home earlier), and she had dressed herself with special care to welcome them. She was wearing dark green trimmed with narrow braids of the scarlet and blue tartan of their clan, and her hair had been brushed until it had a golden sheen.

He held out a square envelope addressed to Mr. Ludar Redcraft. "It came this afternoon," he said. "An invitation to Tony's dance."

Catherine smiled and said, "I knew you were getting it." She went on to talk about this important occasion. "It's going to be the event of the season. The coming-out tea will be for the older people, and then Tony's friends come in to dance in the evening. The orchestra is going to be made up of all the musicians in Balfour. A juggler's coming from Toronto and a special violinist as well, and the Four Boys' Quartet will sing at the piano during supper. There's to be a policeman on duty outside to direct the carriages."

"Carriages!" said Ludar in a startled voice. "Will—will some of them be going in carriages?"

"All of the girls, of course." Catherine then asked, "Ludar, have you bought your tuxedo?"

He shook his head. "It would cost twenty-five dollars with all the fixings. Aunt Tilly put her foot down hard. She figured it would take six months for me to pay it off. I wanted to get one, but she raised such a fuss I didn't." An uneasy feeling took possession of him. "Will many of the fellows be wearing them?"

Catherine was beginning to look worried also. "They all will. Oh, there may be a few without. Two or three. Not more." She frowned as she thought the situation over. Then she sighed. "Ludar, if you can't go in evening clothes I think it would be wise not to go at all."

This advice was so unexpected that Ludar was startled. "Not go!" he exclaimed in a horrified tone. "I've wanted to be invited to Tony's parties for more than ten years. I'm invited to one at last and now you say I shouldn't go!"

"Is there any way you can get evening clothes?"

He shook his head slowly. "Even if Aunt Tilly gave in, there isn't time to have a suit made now. I'm so tall I'm hard to fit, and I wouldn't go in borrowed stuff. Boys who wear their fathers' evening clothes always look like scarecrows, and I would be the worst of the lot. Anyway, who could I go to?" After some further uneasy thought he asked, "Don't you think I'll look all right in my blue serge suit?"

"Ludar," said Catherine, "the two or three unfortunate boys who appear at this dance in business clothes won't enjoy themselves at all. They'll feel terribly out of place. The girls won't want to dance with them, not even the wallflowers. They'll stand around in corners." She shook her head with decision. "That isn't the reason why I'm against your going. You could survive that part of it. But Norman Craven will be there, of course, and he'll be looking very handsome in his dinner clothes. If you're in a business suit Tony will feel sorry for you." She sat up as straight in her chair as the condition of her back permitted. "Ludar, that must never happen! It would be better for Tony to be angry at you for staying away than to feel sorry for you. You're handicapped enough as it is."

Ludar was half convinced but still rebellious at this unexpected blow from an unkind fate. "But what can I tell her?" he asked in a desperate voice. "I can't say to her, 'I'm not going to your dance because I love you and don't want to do anything to lose you!' That would be breaking my promise. Catherine, there's nothing I *can* say! All I can do is send my regrets and leave it at that. What will Tony think? She'll think I'm angry at her and very rude into the bargain."

Catherine was getting tired. She sighed. "Tony is coming to see me tomorrow afternoon. I'll tell her, Ludar. I promise to make her understand. You must send her flowers, of course. Something simple. Perhaps a bouquet of sweet peas."

Ludar dropped the square envelope into his pocket. "Life certainly is complicated," he said bitterly. "I was so excited about going. I've become a pretty fair dancer and I wanted to show them. I wanted to dance with Tony."

3

Ludar waited until he was sure the tardiest guest would be safely inside before he allowed himself to walk past the Milner house, and even then he took the other side of the street. Light streamed from every window, and he could hear the thump of piano and drum and the sweet whine of violins. Joe was there, he knew, but Arthur had decided sensibly (or his parents had decided for him) that he could not afford a trip back for a matter of such small importance as a dance.

Ludar was now besieged with doubts. Had Catherine been wrong in the advice she gave him? Perhaps he should have gone and faced it out. The right kind of moral courage would have made it possible to accept the handicap of improper garb without giving in and standing in a corner all evening. But the difficulty was that he did not possess that kind of moral courage. Perhaps he had lost the chance for reinstatement with his friends of high school days by not appearing at the most important event of the year.

What did it matter? He raised his head with a sudden fierce determination. Soon he would have a book published, and then they would come seeking him. *The Bad Samaritan,* by Ludar Redcraft. He pictured in his mind a pyramid of them in Haverstraw's window and the sensation it would create in town. Yes, they would be glad to recall old friendships then. In the meantime Catherine had been right. No matter what it might cost him, he would not stand in a sack suit in a corner and see all the fathers' sons and incipient counter jumpers go whirling by him in evening clothes. That he would not, could not, endure.

A girl carrying a sheaf of letters in one hand was walking up Grand Avenue in front of him. As he drew abreast she turned her face away, as though anxious not to be recognized. Something about her carriage and walk, however, had already made him certain that it was Lou Bennet.

"Hello, Lou."

She seemed a little startled. "Oh, hello."

"It's quite a big party down the street."

The girl gave a reluctant nod. "Yes, it's a big party," she said. After a pause she added: "I've been working late for Mrs. Craven, but there's no use pretending I'm on my way home when I live in the opposite direction, is there? I walked up three blocks to have a look. Just as you're doing, Ludar, I think. Silly of us, isn't it?"

"Yes. There's no sense to it, but I couldn't stop myself."

"Neither could I. I wasn't invited, of course. Were you?"

They had reached the end of the block, and he could see by the street light that she was in an agitated state of mind. At the same time he realized that she was looking very well. She seemed to have acquired a dash of style since leaving school. Her hat was almost frivolous, and he suspected she had helped nature a little, as most girls were now doing, in the color of her cheeks.

"Yes, I was invited, but I couldn't go. No tux." After a moment he went on in bitter tones: "I haven't the courage, or the brass, to go like this and stand along a wall and feel out of place. Do you think I should have gone anyway?"

"No!" She spoke vehemently, fiercely. "You were right. It's better to go nowhere, to stay at home, to suffer in silence, than to give them a chance to draw comparisons. You're as good as anyone if you have enough sense to stay out of things." She had stopped while delivering this outburst of opinion, but now she resumed her slow walk. "You and I, Ludar, are in somewhat the same fix, aren't we?"

He nodded. "Yes, I guess we are."

"I can't tell you how much I suffered at high school because I wasn't invited to the parties which counted." She was speaking in a low but tense voice. "There wasn't any reason why I shouldn't have been asked. My father was a minister, even if his church was small and unfashionable.

My mother came from a good family in Huron County, the Carpenters. I suppose I should have put myself out to be nice to the right girls. But I hadn't the clothes and I couldn't hold up my end, so what was the use?" She laughed, not at all happily. "There! I've never said it before, not even to my mother. But we *are* in the same position, and I know you'll understand me. I'm worse off than you are. The fence between your yard and the Milners' is high——"

"So high," he said, "that when I was a boy it seemed to reach right up to the stars."

"But still you can climb it if you try hard enough. I'm sure you will sooner or later. But a girl isn't allowed to climb."

They walked along in silence for several moments. Then he asked, "Did you see Norman before he started?"

"Yes."

"He's inclined, you know, to give Tony a rush."

Lou Bennet answered after a pause, "That's what I've heard."

"I suppose that everyone in town knows how I feel about Tony."

"Yes, Ludar. There's no secret about it."

A spark of suspicion had entered his mind, and he wanted to say, "I think you feel that way about Norman," but he did not do so. They reached the next corner and came to a stop under the street lamp. "He may be dancing with her this very minute," he said.

All his life he had been subject to sudden surges of anger. They came seldom—and only when the provocation had been great—and they were of short duration. While they lasted, however, they were quite violent. His hands would rise in the air slowly, as though forced up against his will. One of these brief seizures came over him now. He wanted to say, "I'd like to go back and cut off the lights and put an end to it!" Instead he allowed the spell to pass and then said instead, "I hope for Tony's sake that the dance is proving a great success."

A streetcar was coming down Grand Avenue, clanging and clumping along on worn wheels. Lou stepped out and signaled it to stop. She smiled at Ludar and said, "Misery loves company, but I'm afraid I must go home now." He smiled and answered, "I'm feeling better. I'll go past the house once more and then I'll go home too. Good night, Lou."

He went past the house more than once, however. It must have been a dozen times that he had paced up and down before he saw Gertie Shinny come out from the side yard and start homeward. He crossed the road at once and caught up with her.

"Hello, Gertie," he said. "It's a very fine party, isn't it?"

"Hello, Luddar. Why didn't ya come? I saw the list, and you was on it."

"I'm always busy in the evenings with meetings."

"Oh, Luddar, ya missed it! The grand music! The food! The chicken and the charlottie russ! I had some of everything, and they said I was to take some charlottie russ home for my little brother."

"Gertie, how does she look?"

"Just lovely." The faithful domestic's voice rose to a rapt pitch. "She has a white princess dress, with lace flounces and a—a something-or-other collar, a furring word. And she looks wonderful in it. When she first came down her eyes was shining like the angels. You would just fall down and worship her if you saw her."

"Of course I would." He felt capable of falling down and worshiping the image which had been conjured up in his mind. "Gertie, did she get many flowers?"

"Heaps and heaps. We thought the bow-kays would never stop coming."

He was tempted to ask what flowers she was wearing, but Gertie made the question unnecessary. "She picked out a small one to wear. Miss Milner was *furyus*. It was sweet peas."

Chapter III

I

The beauty of the daughter of the Langley Cravens burst abruptly on the consciousness of Balfour. She had always been considered a pretty girl, but suddenly people were asking each other in awed tones, "Have you seen Meg Craven?" The widow of Tanner Craven studied her intently one morning at church and brought the subject up at dinner, which was served at one on Sundays.

"You noticed the daughter of the Langley Cravens, I suppose," she said to her son.

"I couldn't help noticing her. The beauteous Meg was getting more attention than the minister. And with good reason."

Mrs. Craven looked at him with a hint of sternness. "You consider her good-looking?"

"No," answered Norman. "Not good-looking. Beautiful! She's turned into a regular stunner."

His mother transferred her attention back to the well-filled plate of stewed chicken and dumplings in front of her. "It seems to me there's a cheap note about her."

Norman grinned. "Couldn't be that you're prejudiced, could it?"

"No." Mrs. Craven rejected the suggestion flatly. "I've nothing against the child—in spite of the things I've heard. I'm sure you won't be foolish enough to get interested in her." She helped herself to yellow pickles. "Her mother seems to have stopped coming to church." After a moment, during which she waited for some comment from her son, Mrs. Craven went on: "I hear her looks have suffered. Could it be that vanity keeps her so much at home these days?"

Norman reached for the plate of hot biscuits. "Very likely. I've seen her once or twice. She didn't look well at all, I'm glad to say."

"She had a long turn," said Mrs. Craven. "I needn't tell you how tired I got of hearing her beauty praised. Well, it won't be mentioned any more, from what I hear. Life has a way of evening things up." She sighed dismally. "Nothing much has happened to even things up for *me*."

Joe was proud of his sister at first, but he soon began to tire of the role of brother to a demanding beauty. "Cleopatra's mad at me," he would report to Ludar. Or, "This beauteous creature we have in our house is running the whole show. What do you suppose Mother's doing? She's bringing out a Frenchwoman to give Meg lessons. And she doesn't trust the clothes you can buy in this country. Pretty soon they're going to New York to buy a whole season's outfit for Mistress Meg."

One morning Joe was at breakfast and doing justice to the sausages and pancakes, with the morning edition of the Toronto *Mail and Empire* turned back to the sporting pages in front of him. Meg came into the room in a discarded negligee of her mother's which trailed along the floor after her.

"Well, Circe!" Joe was very much surprised, because Meg was not an early riser by habit. "What's on your mind?"

"You."

"Me?" Joe, still more surprised, desisted from reading and looked up at her. "Have you got a bone to pick with me, kid?"

"Joe," she said, sitting down at the table and popping a small link of sausage into her mouth, "you've got to spruce up. You're going to be a drag on me if you don't. The company you keep is simply *ghastly*."

A chip appeared immediately on the shoulder of the son of the house. "And what's wrong with the company I keep, Miss Snooty?"

"The girls! For heaven's sakes, Joe, why can't you get interested in someone in our set? These cheap little creatures you run around with all the time! Everybody talks about it. I know to my certain knowledge that quite a few of the best girls in town would be only too glad if you'd take an interest in them."

Joe was taking the lesson with considerable self-restraint. "You're all wrong, Queen of the May. I haven't been doing any running around for weeks and weeks. I've been a regular hermit."

"There!" triumphantly. "You've acknowledged it. That's what I've been driving at. It's true, then, that you've given up playing around with all the riffraff in town because you're stuck on some little squib of a girl in the East Ward."

"Go away back and sit down, Meg Craven! I used to warm your behind with a shingle whenever you needed it, and you're not so grown up that I can't do it again."

"You've been giving it out that you're in love with her," charged Meg. "Joe, you've just got to stop this sort of thing. It isn't fair to me when you get interested in the wrong kind of girls. You'd better come to your senses, Master Joseph."

Joe got up from the table and glared down at her. "You keep your dainty, designing fingers out of my affairs. You may seem like Snow White or Helen of Troy or somebody to other people. To me you're just a kid sister, and I know how to deal with you."

That evening the little red car stopped in front of 138 Wilson Street, and a prolonged pressure on the horn brought Ludar out promptly. Pete Work, who was sitting beside Joe on the front seat, grinned nervously and didn't say anything.

"Hop in," said Joe. "We're going to make a call, the three of us together. On the Slaney family. Pete's going to give the knockdowns—cheer up, Pete, it won't be as hard as you think it will—and you're going along as ballast."

"I've got goose-flesh all over me," said Pete, who looked the picture of misery. "I'll make a mess of things sure."

When Ludar had ensconced himself in the rear seat Joe proceeded with an explanation. "The great beauty sailed into me this morning," he said. "It got my dander up. I said to myself, 'What are you waiting around for, you frozen stump? You'll be left, if you don't look out, like the last piece of gristle on the platter.' Besides, I wanted to show that sister of mine."

"I've nothing else to do, as it happens," said Ludar. "Let 'er rip. The curtain rises on the comedy romance of the season."

The Slaneys lived in a dark little cottage behind a high lattice fence, with a mass of shrubbery, mostly lilacs, filling all the space between fence and house. The door, from which the paint was peeling, had no bell. Pete rapped with his knuckles and looked as guilty as a thief caught in the act when a nice-looking woman with a young face and white hair opened it.

"Good evening, Peter."

"Mrs. Slaney," stammered Pete, "I've brought my friends. This is Joe Craven. And this is the Duke."

"How do you do, Mr. Craven? And you, Mr. Redcraft. You *are* Mr. Redcraft of the *Star,* aren't you?"

Ludar was growing so tall that he bumped his head on a lamp of yellow glass suspended from the ceiling in the hall. Mrs. Slaney led the way into a large room to the right which would have seemed very plainly furnished had it not contained two very pretty girls who were introduced as her daughters, Mary and Jessie. Mary seemed a little flustered by so much unexpected company. She was a sweet-faced girl, wearing her hair in a high pompadour above an intelligent forehead. Jessie, the younger, put aside a geometry book which she had been using and did not seem concerned at all. She had black hair and dark blue eyes and freckles on the bridge of her short nose, and was small and trim and lively.

"I've a piece of news for Mr. Redcraft," said Mrs. Slaney. "I was hearing it over the telephone when you arrived." There was in her voice a slight suggestion of pride in the possession of a telephone. "This evening, when Alf Braden was sweeping out his store, he found what he thought was a dollar bill sticking to the end of his broom. It wasn't a dollar though. It was a hundred-dollar bill!"

A gasp of surprised interest greeted the announcement. A one-hundred-dollar bill was symbolic of all the wealth in the world. The listeners felt as though they were peering over the shoulder of Ali Baba at the treasures in the cave of the Forty Thieves.

"It wasn't his, of course," went on Mrs. Slaney. "He soon found out who it belonged to. Henry Smedley, the bookkeeper at Hence's Bakery."

Ludar had become entirely the reporter. "I wonder how it happened that this Henry Smedley was carrying a one-hundred-dollar bill? Does anyone know where he lives?"

"Up on Sand Hill," said Pete. "I know where his house is."

"Can't we all pile into the car and go up there now?" Ludar was anxious to get the facts while the people involved would still be in a state of mind to talk. "Would you mind, Mrs. Slaney, if we stole your daughters for half an hour and ran this story down?"

"Oh, let's!" cried Jessie. "I've never ridden in an automobile. Will it be enough if I put a veil over my hair?"

"Go along, by all means," said Mrs. Slaney. "But please, young man, no speeding. I'm frightened to death of the things. I don't want to lose my daughters in an accident."

Pete said he would drive, so he and Mary and Ludar occupied the front seat, while Joe and Jessie were most pleasantly crowded in the back one. As Pete had a wholesome respect for Mrs. Slaney's wishes, they went at a careful fifteen miles an hour, but even that was enough to threaten the veil on Jessie's hair. Clutching it with both hands, she looked at Joe and said in a delighted voice, "I knew it would be wonderful!"

Ludar, excluded from a low-voiced conversation between Pete and Mary, heard some scraps of the talk in the rear seat.

"Jessie," said Joe, "I've been in a funny state of mind about you. I saw you first six weeks ago, and I said to myself right away, 'There she is. That's the girl for me.' Always before when I got interested in a girl—and I might as well tell you that I've been interested in a heap of girls in my time—I'd take after them like a man chasing his hat in a windstorm. But I kept hesitating. I'd say to myself, 'Perhaps I'd better wait until I get my new serge suit,' or 'I haven't a tie to my name fit to wear when I meet *this* girl.' It came down to this: I was scared. It wasn't until to-night, when I got both Pete and Ludar to help me out, that I dared come."

"I didn't know," said Jessie, "that I was such a terrifying person."

"You," cried Joe, "are the loveliest girl in the world and, now that I know you, I'm going to be on your doorstep all the time." He paused. "Jessie, do you know how many you have of those cute freckles on your nose?"

"I haven't the least idea."

"Six!" said Joe triumphantly. "I've been counting them. And I wouldn't be content with one less. Which do you prefer, red or white brick for houses?"

"Why, red, I guess. I really don't care very much."

"Do you like high ceilings or low?"

"Oh, high, of course."

"How many fireplaces do you think a house should have? A medium-sized house, I mean."

"As many as you can afford," answered Jessie.

Joe nodded his head. "Fine. We think along the same lines. Now it will be an easy matter to plan the house I'm going to build for us."

Jessie was both startled and puzzled by this time. "Some girls," she said, "would be foolish enough to think you were—well, proposing to them."

"Not exactly," explained Joe. "I'm just making a statement of future intentions."

The Smedley house proved to be a small frame cottage surrounded on three sides by a wide veranda. Ludar went in alone and remained for ten minutes. He was aware, when he came out, that he had not been missed.

"It's a good story," he said. "The Smedleys, if any of you are interested, are a nice couple, rather like a pair of lovebirds. They've been saving up for two years to get the hundred dollars needed as down payment on a house of their own. Mrs. Smedley gave up sugar in her tea and her yearly jaunt to Port Dover over the First of July. Henry gave up smoking. It was a terrible pull, because he was quite an addict and smoked his ten cigarettes a day as regular as clockwork. Finally they had enough saved up and Henry went to the bank yesterday to get the money. He could have issued a check, of course, but he has a dramatic streak in him, and nothing would serve but to take a one-hundred-dollar bill home to his wife and flash it in her face. He got it, and tonight as he passed Alf Braden's it occurred to him that he could start to smoke again. He went in, bought a pack of Sweet Caps, and paid for them out of a dollar bill. When he got home, smoking like a furnace and striding like a conqueror, he had in his pockets only what was left of the cigarettes and the change from the dollar bill."

"But they have the money now, haven't they?" asked Jessie in an anxious tone.

"Mrs. Smedley has it. Put away safely in *her* purse. Henry is very humble and agrees that he shouldn't be trusted with money in large amounts. Tomorrow they buy the house."

"Will Mr. Smedley go on smoking now?"

Ludar shook his head. "No. Not for twenty-two years. By that time the mortgage will be paid off. He's going to finish his pack one a week."

They returned to the Slaney cottage, consumed the better part of a very fine banana cake, and started for home at nine-forty sharp. Pete climbed into the rear seat. "I certainly appreciate what you fellows did tonight," he said. "You certainly broke a lot of ice for me. Duke, what's your honest opinion now? Do you think Mary will ever like me well enough?"

"I think, Pete, she likes you well enough right now."

"There's four cars in town," exclaimed Pete excitedly. "As soon as there's six I'll ask her. When there's eight I'll be able to afford to get married."

"'What care I how sound cars be, eight of them will marry me,'" quoted Ludar. "If eight won't, a dozen shall. . . ."

Joe had been driving and paying no attention to them. "She's a little bit of heaven!" he said at this point. "Someday, fellows, I'll come along in the car and Jessie will be with me, and I'll give four toots on the horn. Four, mind you. That will mean we've gone and been married."

2

The fascination that Meg had for men was never more manifest than in the case of Arthur. He arrived home for the holidays the Saturday before Christmas, looking very sober and important and full of talk about life at varsity. Joe had driven Ludar to the station to meet him, and the trio rode away together, all sitting on the front seat, the horn pleeping at every imaginary need.

"You're both coming home with me for lunch," announced Joe. "I phoned your mother, Arthur, and she said it was all right." He gave a grunt of disgust. "We'll have to put up with the m'amselle who arrived last night. She's Meg's French teacher. Imagine, a teacher brought all the way from Paris for that flighty sister of mine! Holy man, it's terrible what's going on in our family right now. Everything is Meg, Meg, Meg. You'd think she was a prima donna."

"Meg's a nice little girl," remarked Arthur. His voice was indifferent, even a shade patronizing. Why should a varsity man be interested in girls he remembered as nice and little?

"Yes, she's a nice little girl, isn't she?" said Joe, winking at Ludar.

The latter winked back delightedly. "But won't you feel sorry for her when she encounters the cold, superior man from varsity? He'll freeze her with one look."

"I guess," agreed Joe, "that I'd better speak to Meg and suggest she stay up in the nursery so she won't get her feelings hurt."

"The very thing to do, Joe." The two friends laughed uproariously and slapped each other on the back.

Arthur looked soberly first at one, then at the other. "What's the meaning of all this high school hilarity?" he demanded.

Meg came to the lunch table a few minutes late. "Making an entrance!" muttered Joe disgustedly. If it was a planned entrance, it was, at any rate, a successful one. She had been outside and she was wearing a white sweater, which fitted her snugly, and a white skirt. Her green toboggan cap, which when removed revealed that her red-gold hair was quite artfully disheveled under it, made her look like the spirit of ice and snow. Her eyes were glowing.

"Mums," she said, pausing in the doorway, "it's such a wonderful day! I was skating on the ponds." Then her eyes lighted on the guest from college. "Why, Arthur! What a delightful surprise!"

"Surprise, in a pig's ear!" whispered Joe to Ludar. "I told her he was coming."

Arthur stumbled to his feet. All he could say was "Meg!" staring at her and swallowing hard in amazement and a mixture of other deep emotions. In describing the scene later Joe said that he "gawped" at her, and the term seems the only adequate one.

On the opposite side of the table from Arthur, Meg seated herself beside M'amselle, a middle-aged woman who talked a great deal in what she assumed was English but which missed its mark by the full width of the Channel. "I'm *so* glad you're here today," Meg said with a confidential suggestion which seemed to exclude everyone else at table. "You see, I'm starting to take my education much more seriously. M'amselle is not only going to teach me French, but—well, general culture also. You'll be able to help me, Arthur. We'll compare notes."

Arthur said, "Yes!" eagerly. He had intended, clearly enough, to say more, but his mind was not capable of finding and assembling words.

Meg talked a great deal during lunch. Arthur sat and worshiped in a mental state which could only be described as drugged. His appetite, usually quite good, had deserted him. His soup went back to the kitchen, cold and untouched. The chops had no interest for him, and it was doubtful if he knew there were magnificent muffins on the table. Even the homemade ice cream, for which he had a passionate liking, was allowed to be whisked away from under his nose after no more than a perfunctory spoonful or two.

"This is going to be a severe case," whispered Joe to Ludar.

Joe himself was in his mother's bad books. She disregarded him pointedly. Once, however, when something had been said about the need for having the best kind of friends, she looked meaningly in his direction and remarked, "Some people seem to prefer the wrong kind." He told Ludar later that there had been in the family for some time a concerted campaign against his "rushing" of Jessie Slaney. He was quite cheerful about it. It was natural for them to feel that way, he said, but it would make no difference. His mind was made up.

After lunch Mr. Craven took Ludar into the library to look over some special books which had just arrived. Most of them had been ordered by catalogue from England. There was a first edition in blue leather of Strickland's *Queens of England,* some first editions of George Meredith, a George Gissing novel which had been out of print for years, and a copy ("One of the first off the press!" exulted the happy owner) of the recently published *Kim* by Rudyard Kipling.

Ludar had no idea how much time had been passed in looking these acquisitions over reverently until Arthur put his head into the library and said: "I'm going home now. How about walking over with me?"

They left the house together, and Ludar noticed that his friend's face was actually pale and his manner unaccountably solemn. "Aren't you feeling well?" he asked.

"No," answered Arthur. "This has been a shock."

"Are you referring to Meg?"

"Of course I'm referring to Meg. Only things which refer to Meg will be of importance to me from now on."

"Come off! What about this career you were discussing on the way from the station? Isn't that important any more?"

"No." Arthur reached into a pocket for his pipe. "I'm a fatalist in a way, and so I shouldn't feel badly over what has happened. It has its bright side. But I realize that sooner or later I may have to do away with myself."

"I see we're going to have to do something about you."

Arthur proceeded to fill his pipe from a leather tobacco pouch. In spite of the state of mind he was in, he was quite methodical about it. "I know that I'll never be able to win Meg," he said. "She's intended for a more brilliant life than I can give her. But I can't live without her. That's the situation." He was speaking slowly and with complete composure. "I couldn't go on living a life which had lost all meaning. That was what I meant when I said things might come to—well, to a sudden ending."

Ludar studied him seriously for a moment. What was the best attitude to adopt in a case like this? He decided to treat the matter lightly. "Well, Arthur," he said, "if it comes to that, I think you should consider your parents as much as possible. Don't make it a messy business. How about taking a wad of blotting paper and stuffing your throat full of it? You would smother quietly and nicely that way, and you could be fixed up to look fine for the funeral."

It was snowing lightly and the street was filled with cutters. One came down the avenue behind them with a loud jingle of sleigh bells, and for one exciting moment they thought the driver had lost control of his horse. It had galloped up so close to them that the runners slid over the edge of the sidewalk. Ludar was on the outside and he felt something rip through his trouser leg (it was the metal step of the cutter) while a voice shouted at them, "Get out of the way!"

It was Norman Craven. He grinned at them as he hauled at the reins to get the horse back on the road again. His eyes conveyed the message, "That's what I think of the pair of you."

One of those sudden surges of rage took possession of Ludar, and he started to run after the cutter. The necessity of getting the vehicle back on its proper course had slowed things, and he was able to catch up. Springing on the rear runner, he launched out with his fist and caught the driver a glancing blow on the side of the head.

"Hey, what's this!" cried the Craven heir. He dropped the reins and leaned back over the seat, getting a grip with one hand on Ludar's coat collar. They wrestled angrily, striking at each other with their free hands, while the horse, out of control, began to gather speed down the frozen surface of the road.

Ludar, who knew nothing whatever about boxing, received two sharp blows on the face in as many seconds. This, to his own surprise, only heightened his desire to go on with the struggle. His inhibitions had vanished and he felt an exultant thrill when he succeeded in planting his fist squarely between his opponent's eyes. He received one in return which found exactly the same mark and which nearly jarred him from his hold on the back of the cutter.

"Watch out, you young fools!" someone shouted from the sidewalk. Another cried, "Runaway!" A man on the nearest crossing reached out to stop the horse, but without success. Arthur was racing down the street after them, trying hard to keep up, his sorrows and troubles forgotten in the excitement.

Norman turned halfway around in an effort to get the whip from its socket. The cutter skidded sideways as he did so and ended up with a crash against the trunk of a tree. It came to a stop, Norman shooting out one side to the road with the whip in his hands, Ludar on the other into a pile of snow. The latter was the first to get to his feet. He could not see out of one eye.

The Craven heir rose up wrathfully. "This is a livery outfit!" he said. "You'll have a pretty bill to pay for the damage you've done."

Quite a crowd had gathered around them by this time. "You started it." Ludar stuck out his leg so everyone could see the rent in the trouser. "My suit's ruined. You drove up on the sidewalk deliberately and tried to run us down."

Arthur had caught up with them, carrying his pipe in his hand. "I'll swear to that!" he shouted.

"I saw the whole thing," said someone in the crowd. "He drove right up on the sidewalk at them."

Norman seemed to realize that he did not have a good case. He glanced about him, scowled, and then tossed the whip on the seat of the damaged cutter. "Someone hold the horse," he said, "while I go and telephone the livery."

"You're going to have a beaut of a black eye," said Arthur as the two friends turned onto St. Stephen's Avenue, where he lived. "We'll put some raw beef on it as soon as we reach the house. Does it hurt much?"

"A little." Ludar was breathless but triumphant. He was proud of what he had done. With an exploring finger he went over his face, feeling the bridge of his nose, which was very painful, and his other eye, which was aching also. "I took quite a beating," he said. "But I hung on just the same."

When they had told their story to the members of the Borden family and a piece of steak had been obtained and put on the bad eye, the two friends ensconced themselves in a corner where they could talk. Arthur's mind had gone back to his own tribulations.

"There's one ray of hope," he said. "It's small, I guess, but I can't help thinking about it and clinging to it. I had a talk with her while you were in the library. It was a wonderful talk. She called me *'Arturo mio.'* Don't you think that means something?"

"I hate to tell you this," said Ludar, shaking his head, "but you should know the truth. There was a time when she called me *'Ludar mio.'* "

"Were you in love with her? I thought you still——"

"No, I wasn't in love with her at all. That makes no difference to Meg."

After a moment's silence Arthur asked in a dismal voice, "What does she call you now?"

"It all depends on the mood she's in. When she's angry at me she has a long list—Stupid, Donkey, Dummy. When she really wants to take me down a peg she calls me 'the self-elected genius' or 'the great author.' When she's feeling friendly, which is most of the time, she calls me 'Old Ludar' and 'The Man with the Sign.' " Feeling that he should do something to save his friend while there was still time, he proceeded to give some advice which he knew to be sound. "She's a great beauty and she's certainly bright. I'm very fond of her. But, Arthur, she's an out-and-out flirt. The best thing you can do is put her right out of your mind."

"That is just what I can't do." Arthur spoke calmly. "As a matter of fact, I don't want to put her out of my mind. This is a wonderful experience, even if the situation is so hopeless." He smiled for the first time since they had started the discussion. "I've just thought of a good excuse for going over there after dinner."

3

One evening well along into the winter Ludar paid a visit at the Craven house to get some advice about a lawyer's letter he had received. At first Norman had seemed disposed to pass over the cutter incident without taking any action, but the receipt of a bill of twenty dollars for repairs had proven more than he could stand, and his lawyers were now trying to collect from the other participant. Ludar was worried. He did not like any form of litigation.

All members of the family were out with the exception of Meg, and she was sitting rather mysteriously by herself in a white dress. In a high-backed chair, with her bronze head reclining on the needle-point panel, she looked like a queen without a court or a Renaissance beauty waiting, not too patiently, for the noblemen in silks and satins, the cardinals and great painters and hired bravoes to start dropping in.

"I'm delighted to see you, Ludar," she said in a dreamy voice. "I've been alone for an hour and I've been doing some serious thinking. Give me your opinion. Is it better to have the love of one good man than the admiration and perhaps love of many?"

"Why don't you cast your vote for both? That's what you're after, isn't it?"

"But can you have both?"

"Of course. The only question is, for how long? I suppose that would depend on the patience of the one good man. Some might be willing to let you go on flirting with all the eligible men in town. Others might get nasty about it right at the start. Now if it's Arthur you have in mind——"

"It isn't." Meg's voice lost its dreamy note. "Why do you and Joe keep throwing Arthur in my face? Are you trying to be matchmakers?"

Ludar drew up a chair in front of her. "It's time you did some serious thinking," he said. "It seems to me you are branching out too far. Why, you even seem willing to welcome me to the ranks of the suitors of Meg Craven. You make me think of those powerful countries which can't bear to see some poor little state maintaining its independence; like Austria-Hungary saying, 'Now this Sanjak is altogether too free and perky, I guess I'll just annex it.' You take a look at me and you say, 'Here's that friend of my brother's always coming around here. He doesn't amount to much—in fact, he's a pretty insignificant guy. But he's around and I might as well reach out and scoop him into the net.' So you give me the wide eyes and the thoughtful glance and the slight catch in the voice, and all the time you don't want me any more than a small girl wants last year's doll from which she has knocked all the sawdust."

"Is that so?" said Meg. "You really are quite a talker, Ludar. Are you finding the wide eyes and the thoughtful glance disturbing, by any chance?"

He responded by quoting the old poem which began, "Blue Eye, beauty, do your mother's duty," and when he came to "Green Eye, greedy gut, eat all the world up," he stopped with a gesture.

"That's you, Meg. You want to eat the whole world up. You want every man you see to fall in love with you."

"That's an insulting thing to say," declared Meg, sitting up indignantly in her chair. Then her mood changed suddenly. She became dreamy of eye again and she spread both of her arms out wide. "Ludar, it's true. I *am* greedy. I *do* want to eat the whole world up. I want everyone to admire me; I want all men, even you, Ludar, to fall in love with me. You, perhaps, more than all the rest because you're so aloof about it." She glanced at him appealingly. "I'm telling you all my faults and all my secrets. Is that wise, do you think?"

"No," said Ludar hastily, "I don't think it's wise at all."

"Oh, Ludar, why can't you take me tonight to an embassy dinner in London where I'll sit beside a duke and hear secrets of state discussed? Why can't we start out for a wonderful ball where I might meet the ruler of some small kingdom in disguise and have him make love to me? I want to know kings and earls and counts and great millionaires and oriental potentates. Why can't you and I see the kind of life we read about instead of the dreary things which go on in this poky little hole? Why can't we have great adventures?"

"I'm afraid Balfour doesn't offer much in that line, Meg," said Ludar. "The only suggestion I have is that we drop in at the opera house, where they're having the full-dress rehearsal of *The Little Minister*. I don't think much of the cast the Society selected, and it will probably be a pretty dismal business."

"We have tickets for tomorrow night." Meg got up from her chair abruptly and briskly. "But I think it will be fun to watch. Let's go right away."

She went upstairs to get ready and came back quickly in a black velvet coat with ermine trimming and a small black hat with ermine tails hanging over one ear. Ludar looked at her intently and then gave his head a shake. "I'm sure all this was an accident," he said. "The good Lord wouldn't have done it on purpose."

"Why, what do you mean?" asked Meg, making her eyes seem wider than ever before.

"You know damn well what I mean, Miss Meg Craven. Come on, let's start."

The opera house was an unpretentious building on Holbrook Street with an entrance between two stores. The stage was lighted when they walked down the aisle, but not as well, it developed immediately, as the amateur stage manager considered necessary. In an important voice he was demanding to know: "What in Sam Hill is the matter with the lights? You can't see the expressions on the faces of the cast." Mrs. Lundy Weaver, who was in charge of the lighting and who, moreover, was very much inclined to be arty, hurried to defend her conception of dim illumination by remarking from where she sat in one of the boxes, "There are no expressions on the faces of the cast."

Ludar escorted his companion to the front row of the orchestra, where there was already a huddle of silent spectators. As she took a seat Meg said in an audible voice, "I've always had a feeling, Ludar, that I was intended for a life on the stage."

The cast was stumbling rather woodenly through a situation in the first act and doing very little to offset the charge brought against them. It was apparent, in fact, that the production would have other faults besides the predilection of Mrs. Lundy Weaver for uncertain lighting. The part of the little minister was, of course, being played by Lance Cropley, who had always been given the male lead in amateur efforts. Unfortunately he inclined to the tall side. The Lady Babbie of the cast, moreover, was Mrs. Jackson, another veteran. It could not be denied that in the abbreviated skirt which had been deemed necessary for the part of the gypsy she looked quite heavy of knee and thigh, and some fault might reasonably have been found with her conception of the part which caused her to walk with the suggestion of a hippety-hop.

"This casting committee!" said Meg indignantly. Then she added in a whisper: "Why didn't they think of me to play the gypsy? If I had the part I'd singe the hair off the audience."

"I'm not at all sure," said Ludar, "that they want any hair singeing done."

"I hear the part *was* offered to Tony Milner. She refused it because she didn't think she could act well enough."

Ludar had been uncomfortably aware that Meg was making the most of the fact that he was her escort. She leaned toward him confidentially while they talked and once had touched his sleeve to emphasize her point. He now became aware of the awkward fact that Tony was also sitting in the front row, although on the other side of the aisle. Her eyes met his, and she looked away quickly without giving any sign of recognition. A few moments later she got up and left the theater in the middle of a speech by the Lord Rintoul of the cast, who was having difficulty both with his accent and his Dundreary beard.

Ludar's heart sank. For a moment he was too upset for coherent thought, but when he found it possible to take stock of the situation he found little which offered consolation. "Surely," he said to himself, "she'll know this was all an act of Meg's." He looked sideways at Meg and thought, "She knew Tony was there, of course."

Meg, he realized, had made her byplay most convincing. Why should he assume, therefore, that Tony would be understanding enough to see through it? Was it not more likely that she would think rather of the effect which had been produced on the other spectators?

He said, "Excuse me," hurriedly, and went stumbling down the dark aisle in the hope of overtaking Tony and making an explanation. The lobby, however, was empty when he got there.

He stood beside the ticket taker's box and reflected bitterly that this might prove the last straw; that Tony, who had remained loyal in the face of so many difficulties, might now decide to put him out of her favor. This possibility was so distressing that he said to himself, "Now you know how it feels to see your chance for happiness stolen from you." It crossed his mind that he had been standing smugly at one side and finding amusement in the plights of Pete and Joe and Arthur. To make matters completely hopeless, he realized that the two years of his promise were not yet up, and he was barred from finding Tony and explaining what had happened.

"I've lost her!" he said in a half-audible whisper. "Now what am I to do?"

4

A month later saw the expiration of the two-year period of probation, and Ludar decided he should be declared free of his promise. Accordingly he found himself one afternoon knocking on the door of the publisher's private office and wondering which made the most noise, his knuckles or the beating of his heart.

"Come in!" called Mr. Milner.

The owner of the *Star* had a copy of the day's paper, which had just come off the press, spread out on his desk and was reading it with an intent frown. He apparently discovered the identity of his visitor without looking up, because he asked in a grumbling tone, "You writing the police-court news now?"

"Yes," answered Ludar anxiously.

"It's not bad. I like the facetious note you're getting into it. Men have mentioned it at the club. Keep it up."

"Mr. Milner," began Ludar, "when you made me a reporter you imposed a condition——"

William Pitt Milner lost all interest in the day's paper. He turned around and looked intently at Ludar over the top of his glasses.

"That's so."

"You said two years, and they're up now." The reporter hesitated, not being sure how he should word his request. Finally he managed to get out, "May I assume, sir, that the—the restriction no longer applies?"

The owner took off his glasses and rubbed the bridge of his nose with a handkerchief. "Why is it necessary to bring this up?" he asked. "You're not prohibited from seeing my daughter."

"But I am. At least I can't consider myself free to see her on—on the basis I would like."

One of the girls on the business office staff deposited a sheaf of invoices on the desk. The publisher picked one up and began to examine it with absorbed attention. After a few moments, however, he continued the conversation.

"Look here. You've been doing good work during the two years we've had you on the *Star,* and you've shown a real interest in the work. I've been keeping an eye on you, and my impression has been that you've been quite content. Why not let things continue as they are? My daughter's still too young to be thinking seriously of courtship, let alone"—he cleared his throat—"marriage."

"But I haven't been quite content, sir."

The Milner eye paused over one of the invoices, and he dashed down some furious comment in red pencil on the border.

"Are you trying to tell me you're in love with my daughter?"

"Yes, Mr. Milner. I'm deeply in love with her."

"Nonsense! Couldn't be anything but puppy love. You're not quite twenty. I didn't fall in love until I was twenty-six. I was married when I was twenty-nine. And that was the right age for such a serious step."

He pushed the pile of papers impatiently to one side, but he still did not turn to look at his employee.

"I can't build a wall around my daughter. I'm not going to make a fool of myself by acting like the hardhearted father in one of these silly plays. Antoinette has a sensible head on her shoulders. If you're in love with her, I suppose you must have a chance to tell her so. Is that what you wanted me to say?"

"Yes, sir. And—thank you very much."

Ludar was on the point of leaving when the proprietor stopped him by raising his head. "Now for business. Here's something I want you to do. Mayor Sifton has a headful of ideas for the future development of the city. Some of them are sound and progressive. I want you to interview him and get the whole program down on paper. There's no time like the present, and I'll see if he can talk to you this afternoon. If you can get it in shape I'd like to see it on the front page tomorrow."

The interview was arranged, and Ludar spent the evening putting on paper nearly three columns of description and exposition of the mayor's plans. It was good copy and he was quite proud of it, particularly when Allen Willing licked his chops over it and said, "Under a three-column heading this will immortalize our worthy chief magistrate."

Ludar's content was badly jolted, however, when he got on the telephone as soon as the paper was off the press and called the Milner home. Gertie Shinny, who answered, informed him that Mr. Milner and his daughter had left at noon to make connections for New York. How long would they be away? Oh, a long time, said Gertie, because they were taking a boat and going to a place called Cuba.

5

More time passed. Ludar was being given most of the assignments which offered opportunities for a display of writing ability and was throwing himself into them with an enthusiasm which made nothing of long hours and physical effort. He received another raise and, as Tilly continued to receive her full half of what was paid, he was certain she was now making weekly additions to her principal. There was no time left for work on his novel. The manuscript lay in a drawer of his bureau. Sometimes he would take it out, read a page or two, say, "Not half bad," and decide that someday soon he would resume work on it.

Tony seemed to be away most of the time. She went again to Europe with her father after returning from Cuba and spent some months in France and Italy, coming home direct from Naples and not touching England at all. In addition, they went twice to New York and on several occasions to Ottawa. Ludar had few glimpses of her, and he was overwhelmed by the fact that a beautifully dressed and obviously sophisticated young lady had taken the place of the Tony he had known.

Meg, in the meantime, had become even lovelier, if such were possible, and the circle of admiration in which she existed grew wider by the day. She accepted devotion from any quarter, no matter how humble. She even flirted with married men; quite discreetly, however, because she was sensible enough to realize that to be caught at it would bring her into conflict with what she called the A.W.B.—the Associated Wives of Balfour—a powerful body which would never relent if once roused to hostility. Ludar was in and out of the house a great deal, and their relationship had developed into something offhand and casual. He had a new name for her every time he called—Lucrezia Borgia, Lola Montez, Mademoiselle de Coverley, Fanny Elssler. This pleased her secretly, although she pretended to be annoyed. She had fallen into the habit of calling him Charles; an allusion, he supposed, to his desire to emulate in some small degree the success of Dickens in his youth.

She relied on him in many ways, however, and sometimes asked his advice. Once she came to him and said, "I'm very curious about a certain man in this town."

"How unusual!"

"This *is* unusual. I've seen him twice and I haven't the faintest idea who he is. He's tall and thin and dark, and he has a strange way of smiling. It's as though he's enjoying a joke he won't share with anyone."

"That sounds very much like Rupert Corliss. Keep away from *him*, my young friend."

"Why?" Meg's interest, it was clear, had increased.

Ludar gave his head a disgusted shake. "I see I've made a mistake. I should have told you this man with the mysterious smile, this cheap male Gioconda has eight children and bunions on both feet. *That* would have ended your interest in the fellow."

"But who is he?"

"No one seems to know much about him. How old he is, where he comes from, whether he's married—no vital statistics at all. He's a special kind of accountant, and some of the manufacturers hire him to work out cost systems. He boards with an old couple on Dunkirk Street. There's one thing about him I should mention. When a girl's inclined to be flighty, he always knows it. When a young wife—or a middle-aged one, for that matter—is dissatisfied with her lot, the eager Mr. Corliss knows about *that*. Do you remember the scandal about Mrs. Gadsby?"

"Oh! *That* man." Meg smiled audaciously at her informant. "He sounds a little vulgar. But fascinating, Charles."

"You have nothing to do with him!" said Ludar shortly.

Strangely enough, Meg did not meet George Francis Cannon for sometime after his much-heralded arrival in town, although she heard of him early. Her father mentioned him at dinner one night. "A scion of the English aristocracy is being sent here by the Confederated Bank," he announced casually.

"Young?" asked Meg.

"I suppose so. He's to be the junior clerk."

"What's his name, and when does he arrive?"

Joe suspended operations on the steak and mushrooms. "What are you going to do, meet him at the station?"

"Joseph!" admonished Mrs. Craven sharply.

"His family name is Cannon," said Mr. Craven. "His father is Viscount Alderstone and *he* is the only son. He's out in this country to get a taste of colonial life before returning and plunging into politics. At least that's what I'm told."

Meg and her mother looked at each other. "He sounds," said Mrs. Craven, trying not too successfully to sound casual, "like a very eligible young man."

"I'll tell you an eligible young man!" exclaimed Joe in a sudden fury of partisanship. "His name is Arthur Borden, and he's going to be a very fine engineer. That's the kind of eligible man you ought to be interested in."

"Mind your own business, little boy," said Meg loftily.

"You must find out when the young man arrives in town, Langley dear." Mrs. Craven was very much interested, it was clear. "It's time Meg met someone besides these home-town boys. I want her to have a much wider circle of acquaintances before she comes out in the fall."

"If she does," grumbled Joe, "you'll have to hire the fairgrounds."

George Francis Cannon, however, showed no tendency to make acquaintances other than business during the first months of his stay in Balfour. He was a rather pleasant-looking and very blond young fellow who read a great deal and went for long solitary walks. He was accustomed, no doubt, to being angled for and had developed considerable skill in evasion. Fall came, and then the early winter months, and the future Viscount Alderstone had not yet been induced to accept any form of hospitality. The introductions which he had not been able to evade had led to nothing. After one cautious overture made by Langley Craven to the manager of the bank, and as promptly rebuffed as all similar ones had been, Mrs. Craven wisely gave over any idea of further efforts. "Someday he'll see Meg," she said to herself. "Then nothing more will be needed."

As the time drew closer for Meg's coming out, which was to be at least the equal of Tony's, the eyes of the unsociable Mr. Cannon had not lighted on her, however, and Mrs. Craven began to feel some uneasiness. She questioned Ludar about the austere bank clerk, but he was not able to supply anything of value. Young Mr. Cannon, he stated, was interested in birds and was supposed to be doing a paper on Canadian varieties for an English publication. He confined his churchgoing to the eight o'clock Holy Communion service and he never went out in the evenings. He had not brought his evening clothes with him because he had been sure there would be no need for them in a country as wild and unsettled as Canada.

The young Englishman's first view of Meg came about through the happy fact that a need for additional hospital funds became felt. A committee was formed, with Mrs. Langley Craven as president, to put on a kermess. It was to be managed by a promoter from out of town who made a business of organizing such events on a percentage basis. It would consist of songs and recitations by local talent, assisted in all cases by choruses, and a variety of marches, dances, tableaux, and short comedy sketches. Meg was selected for a major role. She was to dance while a group of assistants sang "That's What the Daisies Say," picking the petals from flowers as they did so.

On the opening night she arrived at the armory a little late. She was already dressed in her wide flounced skirt for the daisy number, and it was no wonder that the people who saw her come in at the side entrance reserved for performers nudged each other and said, "There she is!" There was a space used for the storing of props at one side of the stage, and here she encountered Mr. Rupert Corliss. He was idling about in a costume which resembled that of the Mad Hatter; in a few minutes he would be going on for a pantomime turn at which he was reputed to be very good. Meg turned her back on him and began to examine some baby carriages which were to be used by a later chorus.

"Hello, Rupe," she whispered.

Corliss answered without looking in her direction. "Hello, Meg, little one. You look like a throne toppler in that wonderful yellow dress. Am I to see you after it's over?"

"No, Rupe. Some of the crowd are coming back to the house. I'm sorry."

He whistled a bar of the number being played on stage while two of the hands busied themselves near him. When they were gone he said in the same guarded manner: "I'll be at the rink early tomorrow afternoon. Couldn't manage to drift in, could you? We could take a whirl or two together. Ah, Meg, to hold you once! Just once. And for just one moment."

"Perhaps. I'll try."

"My sweet, my lovely, my fascinating Meg! I'll live in hopes. Well, better get in there. Knock 'em on the Old Kent Road tonight."

"I will," said Meg, who was not lacking in self-confidence.

A voice hailed her from the dressing rooms, so she placed the carriage she had been examining back against the wall and hurried away.

Ludar arrived half an hour later and was just in time for the daisy number and to see Meg drift across the stage like thistledown, making her aides seem no better than awkward little beginners at a dancing class.

"Good girl!" he said, and was one of the loudest clappers for an encore. While she obliged, smiling and flushed, he became aware that a pleasant-looking young man was standing beside him and watching her with amazement. At the end of the number the young man turned to him and said: "Aren't you Redcraft of the *Star?* You've been pointed out to me as you go by the bank where I work."

"Yes. And you're George Francis Cannon, aren't you?"

"Right," said the stranger. "Good Gad, who is that beautiful and radiant creature?"

"That's Meg Craven. You've heard of her, perhaps."

The young Englishman was watching the side of the stage, in the hope, no doubt, that Meg would respond with another encore to the spatter of applause which still came from all parts of the high-domed assembly hall. "Yes, I've been told about her. But I—I had no idea she was anything like this."

"She's quite unusual," conceded Ludar.

"Unusual? Now that *is* an understatement. She—well, fancy coming out here, to the ends of the earth nearly, and finding *her*. I can't quite believe my eyes yet."

Lance Cropley, who was acting as master of ceremonies, came out on stage and began a long announcement. The future Viscount Alderstone shifted his position to face Ludar. "I say, we must have dinner together some night and get acquainted," he said. "I've been a little standoffish. Haven't met many people. Time I began to, I think. In fact, I should like very much indeed to be introduced to Miss Craven. Do you happen to know her?"

Ludar nodded, suppressing a desire to smile. "Her brother will be putting in an appearance soon," he said. "We'll be going home together. Why don't you join us? It wouldn't be safe for you to go backstage with me now, you know. A lot of girls in this town want to meet you, and every last one of them is playing some part in this show. We'd start a riot."

The young Englishman smiled self-consciously. "Awkward, that sort of thing. Your plan will be much better, I'm sure. Look me up when you're ready to start, will you? I'll be most grateful to you."

It was during the glowworm number, which closed the show, that Joe arrived. He looked decidedly glum.

"Got a dressing-down from Mrs. Slaney," he explained. "I said to her when I went in, 'Good-evening, Mother-in-law,' and she started in. Said I was taking things too much for granted and that Jessie might decide she didn't want to marry me. I knew what was wrong. Mother and Meg have just refused to recognize that I'm serious about Jessie. Yesterday Meg met her on Holbrook Street and sailed right by with her snooty nose in the air. She had the brass to tell me what she had done, and so I was prepared for squalls." He shook his head and frowned unhappily. "I thought Mother would come around by this time. It certainly beats hell how stubborn women can be over matters like this." He gave a glance about him which suggested distaste for such ephemeral things. "How was the show? Pretty rotten?"

"Pretty rotten," agreed Ludar. He meant it, for Tony was away on one of her long trips with her father and so was not taking part. "Your sister made the one real hit of the evening."

"I'll fix her," said Joe bitterly. "I'll bring Jessie to it tomorrow night and I'll take her backstage and introduce her to everyone. It's time I did that anyway. That is," with a doubtful frown, "if she'll come."

Meg and a rather large group of the cast had already arrived at the house when Joe and Ludar, with the future viscount in tow, put in an appearance. Joe disappeared immediately, so the duty of introducing the young Englishman fell on Ludar. He drew Meg to one side. "This is an admirer of yours," he said. "He thought you danced wonderfully tonight. Miss Meg Craven, Mr. George Francis Cannon."

Meg had been taken by surprise. She looked at the viscount's son and said in a somewhat breathless voice: "I'm glad, Mr. Cannon, that you liked my act. I don't think Ludar cared for it much. At any rate, he hasn't given me his opinion."

"Didn't think you would be interested." Ludar grinned at them both. "However, I may say it was favorable. I may even put a few grudging words of praise in the report I'm going to write of the affair."

Soon afterward he made his way to the hall, observing that Meg and George Francis Cannon were still talking, the latter clearly in a bemused condition of mind. The inevitable was happening. Before he reached the door, however, Meg had followed him and had slipped a hand under his arm.

"Oh, Charles, Charles!" she whispered. "I love you, I love you."

"But only because I brought you the greatest reward for a fine dance. A head on a charger."

Meg said in a confidential whisper, "He's very nice."

"And think what he has to offer! A gold circlet to be worn at court. A big fortune. A castle or two. A house in London. A shooting box in Scotland. A chance, perhaps, to become the great political hostess of the new century, a glorified Georgina of Devonshire. Meg, Meg, it's colossal, isn't it?"

Meg's green eyes indulged in one of her enigmatic smiles. "All material considerations, my dear Charles. Perhaps I'm looking for something quite different out of life. There are things to be said in favor of love in a cottage, I hear. And why are you running away so soon?"

"Because," said Ludar, wriggling into his overcoat and opening the door, "I'm wiser than any Greek sailor and I know when to get away from Circe's palace." He closed the door after him.

Chapter IV

I

There were perhaps a dozen men in the city room when Sloppy Bates came stumbling up the stairs. He knew them all, of course. Allen Willing seemed to be presiding over the gathering, which included Ludar, Jinks Snider, some men from the business and mechanical departments, three or four hotelkeepers, several assorted citizens, Alderman Eph Sikes, who owned a grocery store, and Sergeant Doyle of the police force. The old reporter stood in the doorway and swayed slightly, having completed his afternoon tour of the bars and being, as a result, well saturated.

"What's this prayer meeting been called for?" he asked.

"For you," answered Willing. "For the John Kinchley Bates mentioned in the agreement of sale when the *Star* became the property of Mr. Milner."

"Pah!" said the old reporter. "That means I'm going to be talked at, and lectured, and made to feel my worthlessness. I'm pretty well tanked, so I guess I can stand it."

"John Kinchley Bates will please sit down," said the editor, winking at the company. "This meeting has been called to consider the dilemma in which he's placed by the agreement insisted on by the late Daniel Cape when he sold the paper. He hasn't received a raise of salary for ten years and won't ever receive one under present conditions." He turned to the inebriated Mr. Bates, who was pretending not to hear by busying himself with some papers on his desk. "Sloppy, there's nothing else to it. You've got to go on the wagon for three months."

"Can't be done," said Sloppy. "I've been drunk continuously for thirty years. If I tried to quit now I'd blow a boiler. I'd go crazy. You'd have to pack me off to the pogie."

"That's what you think," said Willing. "The rest of us have a different idea of it. We've just formed an association to help you. We'll probably call ourselves the S.F.H.J.K.B.G.U.D.R. That means—Society for Helping John Kinchley Bates Give Up the Demon Rum."

"You can all go to hell!" said Bates.

"Now, Sloppy, that's no way to talk," expostulated Alderman Sikes. "We're all good friends of yours. We want to help you. You shouldn't turn sour on us like this at the very go in."

"Aw, shut up, Eph!" said the old reporter.

Allen Willing tapped for order. "Sloppy," he said, "you're in no position to be obstructive. You've got to fall in with our idea and help us in every way you can. We already have a plan worked out. Knowing what a stubborn and prickly old customer you are, we're not going to take any chances. We have the assurance of every hotelkeeper and bartender in town—in writing, mind you—not to let you have as much as a swallow of beer for three months. Wherever you go, one of us will trot along with you, or some volunteer worker in the cause. You won't be able to leave your house without a bodyguard, and no one will be able to slip anything in to you. Alderman Sikes will send two pounds of candy every week free gratis, so you'll always have something sweet to allay the craving for liquor. The Fernley Grocery and Tea Store will keep Mrs. Bates supplied with coffee, and the pot will be on the fire at all hours of the day and night. Haverstraw's will supply free reading matter, so your mind can be kept occupied. Burnley and Son are going to loan you a gramophone, so you'll have music to soothe all the savage instincts in you."

"All your friends," he went on, "are rallying around and trying to make things easy for you."

"When does it start?" asked Bates in a sulky voice.

"Right now. The S.F.H.J.K.B.G.U.D.R. is officially launched as of this minute."

"No, no!" cried Bates in a panic. "Didn't you hear me say I've been drunk for thirty years? I've got to start at this slowly and cut down by degrees." Seeing no signs of agreement on the faces about him, Sloppy burst into an alcoholic rage. "Why, you smug, righteous, psalm-singing Job's comforters! You weeping Elishas! What do you know about the gnawing hunger for liquor that gets into a man's flesh? What will it matter to you if I go out of my mind? None of you will be suffering or going crazy."

"You've had your last drink," said the editor firmly. "Mr. Milner's car is outside now, and you're to be driven to Hudson for a night in a Turkish bath there. As a first step we think you should have as much as possible of the alcohol in your system sweated out."

John Kinchley Bates—it had been agreed by members of the association that the name Sloppy must be abandoned—was brought back from Hudson the following afternoon, looking white and thin and drawn. He slumped down in the front seat beside Philip Beal and did not move a muscle. "Bub," he said, "they told me this Turkish-bath business would sweat the alcohol out of me. Bub, it did more than that. I figure they kept me in it too long. That heat got right to work on my insides, and it dried up the handles and hinges and giblet pins which hold me together. If I as much as raised a finger you'd hear a rattling inside me like a bunch of dancers with castanets."

The car passed Sergeant Doyle on Holbrook Street, and that officer raised a hand as a signal for them to stop. He walked over to the car and inspected the haggard face of the old reporter. "What would you give right now for a drink, John?" he asked.

"What would I give for a drink?" Bates roused himself from his reclining position. "I would give what's left of my life except the time it would take me to finish, slowly and comfortably, a scotch and soda. A double scotch, mind you! And the best money can buy!"

"A handsome offer, John!"

"I might do better. I might throw in a shriveled-up, mangy, channering thing, for which I gave up hope long ago, called my immortal soul. I might even offer you something of real value. My gold watch and chain!"

2

The next morning was spent by the probationer at the office, although he did not accomplish much. Ludar escorted him home after the paper went to press and stayed for the balance of the afternoon while Bates grumbled and groaned and crunched the better part of a pound of candy. The second afternoon his sufferings were so intense that Jinks Snider, who was on duty, thought he would have to send for help in restraining him.

Bates, his yellow, freckled face sunk between clenched fists, said to the senior reporter: "Jinks, the scientists claim there are eight billion particles of original matter in the human frame. At this moment, son, all eight billion particles in me are standing up in rows and yelling in chorus: 'Isn't it bad enough to belong to a body like this stringy maggot bait without being treated this way? If this son of a gun doesn't feed us our daily dose of rat poison soon, we'll go out on strike.'"

It did not help at all when a deputation of ladies, nearly a dozen of them, put in an appearance, headed by Miss Magna Readlong. They came in majestically, their long skirts sweeping up the dust from the carpet (Mrs. Bates was not a very good housekeeper), their eyes fixed with purpose on the sufferer.

Miss Readlong, who had gone to school with him, said: "John, we represent the combined temperance workers of Balfour. We've come to impress on you the importance of what you're doing."

Bates raised his head high enough to see through the thick thatch of his sandy eyebrows. "What's so important about it?" he demanded.

Miss Readlong seated herself in front of him and dipped a hand into the box of chocolates open beside him. She proceeded to explain, punctuating her remarks with what was left of the candy held between a finger and thumb. "All our lives we've been working against the drink evil. We've prayed and sung and preached and fought for a temperance law. We've tried to redeem tipplers and backsliders. And now here you are, the very worst of the lot, a man who has steeped himself in rum all his life and has sunk almost as low as the beasts of the field——"

"Hey, wait a minute! Why don't you say I'm as low as a cankerworm and as sour as snake bile and be done with it!"

"You're all of that. I'm sure your stomach must be almost eaten away——"

"My stomach is a poor thing, Magna," said Bates wearily. "I can't rally myself in its defense. I'm sure it's as rusty and full of holes as an old colander."

"*You,* John Bates, have elected to make the fight to redeem yourself. If you succeed, we'll have the most wonderful weapon ever forged for temperance hands. We'll be able to say to the world, 'If this man, this sot, this rum-soaked wreck, can do it, there's no broken-down specimen anywhere who can't do it!' "

"A lovely thought, Magna, and phrased in the most friendly and considerate way." Bates raised his head and grinned around the circle of earnest faces. "Still, I see what you mean. My sodden brain is still capable of grasping a point so delicately conveyed."

"So now you see, John, that you *must* succeed. You *can't* fail us. You *must* fight the good fight to the bitter end. You *must* keep in mind our motto, 'Choose! The Comforts of Home or the Torments of Hell. The Baby or the Bottle.' "

Stimulated by these words, perhaps, he lasted through the night. The next day he felt so bad that he remained in bed until after dinner, of which he did not eat a bite, and then dressed and crawled out wearily to the side veranda. At three o'clock his wife said in an urgent whisper, "Here's that pest of a woman again!" in the same tone she would have employed had the message been, "There's a boa constrictor crawling up the path." Bates got hurriedly to his feet, but he was too late to get away. The Readlong face was already peering over the side gate.

"John," she called briskly, "how do you feel?"

"Go away. I feel sick."

"Splendid!" Miss Readlong beamed with satisfaction. "Before you can get well you must be very sick. Before you can be cured you must feel as though stretched on the rack."

She came back later in the afternoon to demand over the top of the gate, "Are you still sick, John?" On receiving no more than a grunt by way of reply, she added, "I feel that the least I can do to help is to come in and see you at all times of the day." She was as good as her word, but fortunately she made her visits short. Sometimes she would do no more than wave a hand at him and deliver a message such as "Dare to be a martyr, John," or "Don't drift away, brother."

His temper grew short as he realized that all supplies delivered to the house were stopped at the gate by unknown females and inspected to make sure that the forbidden rum was not being smuggled in to him. He ran out in his nightshirt one morning and screamed profanity at a stout lady who was siphoning milk out of the daily Bates quart for the purpose of determining whether or not it contained any percentage of alcoholic content.

The phase of surveillance which finally broke his spirit was the encirclement of the house by youthful members of the Band of Hope. He discovered it the second day, when he went out to the back yard to visit a small wooden structure in one corner with a lattice screen in front of it over which morning-glories were growing profusely. The tow head of a boy of seven or eight showed above the rear fence, and a pair of bespectacled eyes watched him soberly. When he emerged the towhead was still there. He stopped.

"Sonny," he said, "I'm going to tell you a story. Once there were seven noblemen who attended the court of a great king, and their chief duty was this: whenever the great King roused in the dark hours of the night—because even kings are subject to certain weaknesses of the flesh—these seven noblemen would be summoned. They would come yawning in bare feet and nightgowns and they would follow the King in the dark and cold to a place on the back stairs, and they would witness what transpired. Well, in the fullness of time there was a revolution in this country and the King was killed, and the seven noblemen were seized by mobs who cut off their heads. It so happened that there were exactly seven institutions in the city of the kind known as a public jakes. The mobs set up one of the heads on a long pike in front of each jakes." He resumed his walk back to the house. "Think that story over carefully, sonny. I'm not sure what the moral is, but there's one somewhere if you look for it."

One afternoon the craving of the flesh for alcoholic stimulant became almost unbearable, and he decided to give up the struggle. He was alone in the house at the moment. Concluding that his friend Pinkie Harris, who kept bar at the Occidental Hotel, might take pity on him, he walked to the back yard, furtively climbed the fence, and emerged with a casual air through a neighbor's gate farther down the street. His nerves were so jumpy by this time that he started back when a small girl materialized at his elbow.

"Where are you going, Mr. Bates?" she asked in a piping voice.

He muttered, "I'm going for a walk, little girl."

"Then I'll have to go with you, Mr. Bates."

The sufferer stopped and stared down at her. "Who are you, anyway?"

"I'm Lucy Marie Cudding, Mr. Bates. I belong to the Special Squad. We've been organized and trained."

"What for?"

"To watch you, Mr. Bates. We have signals. If I needed help I could get plenty right away."

Bates decided to try guile. He managed to smile cheerfully at the youthful sleuth. "There's a lot of candy back in my house, Lucy. Why don't you go and get some? It's in an open box on the kitchen table."

The girl shook her head resolutely. "We can't be bribed, Mr. Bates. That was the first lesson Miss Readlong taught us."

Bates rubbed a hand over his unshaved jaw. He needed what he called a good long hooker of whisky so badly that his stomach was turning somersaults. His nerves were completely out of control. He could not keep his arms still, and his muscles were jumping and jerking.

"Little girl," he asked, "what was that you said about getting help if you needed it?"

"Well, Mr. Bates," said the child, "if you should decide to go to a bar, for instance, there would be twelve of us following you by the time you got there. If you should get a drink, for instance, we would walk right in and sing."

"Sing? What would you sing?"

Lucy Marie Cudding struck an attitude and raised her voice in a familiar air.

"Mr. Bates, dear Mr. Bates, come home with us now,
The clock in the steeple strikes one."

Then she smiled, shook her head confidently, and said, "I don't think you would like it very much, Mr. Bates."

After summoning up a mental picture of himself entering the bar of the Occidental Hotel with twelve small girls singing in his wake, he decided he would not like it at all. He turned and went back to his veranda, muttering to himself.

3

The second month of the old reporter's sojourn on the top of the water wagon was nearly gone, and he was still fighting the good fight. The members of the S.F.H.J.K.B.G.U.D.R. had fallen into the habit of winking at each other and saying, "Well, Sloppy's over the worst now." It had already been settled between the proprietor of the *Star* and the editor that the rejuvenated Mr. Bates would start on the expiration of the three months at the quite handsome salary of ten dollars a week.

Ludar had finished his supper one evening and was feeling quite happy that there were no meetings to cover when the telephone rang. It was Alderman Sikes.

"Young fellow," he shouted at the top of his voice, "I just saw that Sloppy go driving past my house. He was in a farmer's wagon, and they were headed down the Hudson Road."

"He's giving us the slip!" exclaimed Ludar. "I suppose he's going to Hudson for a real buster."

"That's what he is. You better get into one of these horseless carriages and stir up a juice of a dust after him."

Ludar called the Langley Craven house and learned that both cars were out. He then got Pete Work on the line and was told that none of the other cars in town, with the possible exception of the Milner Mercedes, was available. With some hesitation he lifted the receiver off the hook a third time and said to Central, "Number 89, please." It was Tony's voice which answered.

"This is Ludar speaking."

Tony said, "Hello, Ludar." Did he fancy it, or was there a trace of coolness in the voice? "Is it Father you want?"

"No," he answered. "There's a crisis. Sloppy Bates has run away. He was just seen driving out of town with a farmer. He has too good a start to be caught unless we can follow in a car, and yours seems to be the only one not out."

Her voice became brisk at once. "The driver's off today and Father's at a meeting. Ludar, I can drive. Would you trust yourself with me?"

He broke into a relieved laugh. "Of course! I know you drive; you've whizzed by me a dozen times. Pete says you're one of the best in town. Shall I come over?"

"I'll have the car out in front of the house."

The car was out and the engine was purring when he rounded the corner of Grand Avenue on the run. Tony had fitted a veil over a small motoring hat and was wearing leather gloves with fringe. She said, "Hop in," and they were off.

"Which way did he go?" she asked as they rolled down the avenue at an unusual rate of speed.

"The Hudson Road."

She gave him a sidelong glance, and he saw that she was excited. "The bloodhounds will soon be on his heels then," she said. "Isn't this quite unexpected? From what I've heard Father say, Mr. Bates seemed to be pretty well over it."

"We were all sure he had finally lost the appetite."

Driving through town monopolized her attention for the next few minutes. The veil she wore was transparent, and he could see the fine line of her brow under the brim of her hat. It was surprising that anyone as feminine could handle the ponderous car with such sureness and ease.

Suddenly he burst out, "I was twenty-one last month."

Tony nodded but did not turn in his direction. "Yes, I remembered. Did *you* know I was twenty last month?"

"Yes, of course. I telephoned your house. You were out of town. You're always out of town when I try to speak to you. Or not at home."

She gave him a quick glance. "I'm never told when you call. I believed you had given up calling. . . . I suppose we might as well recognize that there's a purpose back of this."

"I know there is. You see, Tony, I spoke to your father when the two years were up. Did you know?"

She shook her head.

"I was given a grudging permission to see you, but—the next day you were hurried away to Cuba. It seemed to me a broad and unmistakable hint."

Tony indulged in a moment of thought. "Would it make you feel better," she asked, "if I told you he's the same with everyone else? He growls every time Norman Craven comes to the house, and when we're away together he's like a policeman where men are concerned." She paused. "Ludar, are you fond of Meg Craven?"

"Yes," he answered. "I'm very fond of her. She's the sister of my best friend. But I'm not fond of her in the way you mean."

"I'm glad," said Tony after a moment. "I didn't think you would be happy as one of her hundreds of admirers. Or is it in the thousands now? Poor Arthur Borden isn't happy."

"I'm not sure. He's very quiet and philosophic about it. He smokes his pipe and says he isn't foolish enough or vain enough to hope he'll be the lucky man but that he'll just stay around. He told me the last time he was home that life was too serious and important to hinge entirely on a man's romantic interests. He says a man can turn his life to good purpose even if he doesn't get the girl he wants."

They had passed through the outskirts of the town and were bowling along the broad main highway. It had a gravel surface, and they had already acquired a tail of dust like a comet. Tony rested one arm on the wheel to keep it steady and turned her head to look at him. With the other she raised the veil and threw it back over her hat.

The comment has been made that there was a grave note in her appearance, but it was only when her face was in complete repose that the gravity was apparent. Most of the time she was full of animation and, when she smiled, her eyes became gay. But as she turned now and looked at her companion she was completely grave.

"I guess Father knew what he was doing," she said. "You and I have become strangers, Ludar. We've spoken to each other less than half a dozen times in the last three years. I've changed in that time. I've changed a very great deal. You don't know anything about me any more, what I think and what I like. And I don't know anything about you. You've changed just as much. You look so very much older and different. I'm afraid, when I look at you, that you've become someone I don't know at all."

"What you are telling me, Tony," he said after a long pause, "is that your father was right. I think you mean that, since we've become strangers, we better continue that way."

"No, no! Please! That isn't what I mean at all." She slowed the car down as they passed a buggy hitched to horses which were rearing and pawing in the air. "What I mean is that we can't expect to start back as we were before. We'll have to become acquainted again. These years that we've been strangers have been such important ones. Perhaps the most important."

Ludar knew that she was right. Loving her as much as ever, he nevertheless recognized the truth in what she said. He was a different being from the very young man who had accepted a post on the *Star* and the restriction which went with it; and, on the other hand, this beautifully dressed young woman so different from the grave-eyed Tony was a stranger to whose mental reactions he had no sure key.

"You've stated only part of the case," he said. "The three years have brought other friendships which may be very important."

"I've made friendships," she said after a moment. "But they haven't changed anything that went before."

That was what he needed to hear. Norman Craven had continued to pay her steady attention during the three years. He had given up college at some stage of his first term, saying that, as he was not going in for a profession, he did not need more education. Instead he had devoted himself to the Craven interests, working closely with his mother. He had been abroad twice and he made frequent trips to New York. He was considered cultured and even sophisticated.

A paean of thankfulness was on the tip of Ludar's tongue. He wanted to say to her: "Then I still have a chance! It's still possible to make up for all the time that's been lost." There were so many things crowding into his mind which he wanted to tell her: that he had been given another raise, that the debt was nearly cleared off, that the week before he had been measured for a tuxedo and would have it in a few more days, that he had been watching and listening and reading and learning, that he had been improving himself in every way. Above everything else, he wanted to tell her that he loved her, that the years had not made any difference in his feelings toward her, that nothing else counted in life but the fact that he loved her with all his heart.

Even as these thoughts took form in his mind, however, he knew this was not the time to say them. Tony had told him they had become strangers and must find each other again. He contented himself, therefore, with, "I hope you mean to let me come and see you." Then his feelings got a little out of control and he added, "At once, tomorrow, please!"

She nodded and smiled. "Of course I want you to come and see me."

"Tomorrow evening?"

"I'm going to Toronto for the week end with Father. But Miss Antoinette Milner will be free on Monday evening and glad to receive Mr. Ludar Redcraft."

A farmer's wagon, drawn by a team of steady bays, had come into view ahead of them. There were two occupants on the front seat, one of them a hunched-over figure with sandy hair. "There's the runaway," said Ludar.

John Kinchley Bates was not pleased when they drew up abreast of the wagon and stopped. He said to the farmer: "Whoa, friend. I said there might be a benzene buggy out after me, and here it is. That young fellow is a colleague of mine, and the beautiful young lady with him is the boss's daughter. Don't you think I ought to be proud because they've come chasing after me?" He turned to his pursuers. "Might as well turn that devil's contraption around and go right back. I have a very special errand in Hudson."

Tony, leaning over the wheel, said, "Tell us what the errand is first, Mr. Bates."

Bates answered sharply, "It's not what you think, young lady. I'm not running away to Hudson to go on a spree. I'm through with that. I haven't had a drop in two months. The way I feel now, I'll never touch another as long as I live." His voice rose suddenly. "This, I tell you, is a gesture of independence. You don't know how I've been harried and pushed around and insulted. 'Poor John' one minute and 'This sot' the next. I got tired of being poor-johned and this-sotted. I got tired of the whole thing. This is what I'm going to do. I'm going to Hudson and buy the smallest bottle of whisky I can find. I'm going to take it back to Balfour and I'm going to the Occidental Hotel and I'm going to stand out in the middle of the floor and take a swallow of the stuff. That will wipe out everything we've done. Then I'll smash the bottle and say, 'Down with the demon rum!' Then I'll go on the wagon again for the full three months, and I won't take any help from anyone and I won't have temperance workers bothering me or Band of Hopers keeping watch on my movements."

"But, Mr. Bates!" cried Tony. "If you do this you'll have everyone laughing at the friends who've tried to help you. I'm sure you don't want that to happen."

He thought this over. "Well, I don't know," he said. "They've made people laugh at me."

"But what you don't see," she persisted, "is that they've done this because they like you so much. You've been one of the best-liked men in the place. But if you do what you say you're going to, people won't like you as much. You might even get to be unpopular."

The old reporter's blotched and freckled face broke into a reluctant grin. He turned to Ludar. "The boss's daughter can nail you down, can't she? Is she going to be a lawyer?" His glance wandered to the car. "I've never ridden in one of these things."

"Get in and we'll go for a ride," said Tony promptly.

"Now, young lady, don't go jumping to conclusions. What *you* don't see is that I can't turn around and go back with you now without becoming the laughingstock of the town."

Tony shook her head. "I've just thought of a way out of *that*. The Reverend Dr. Torrington lives in Offaxel, which is right around here. He's head of the temperance movement for the whole of Canada. He stayed with us once. We'll all go and see him now and then we'll say we drove out for the purpose."

"You'll introduce me as Exhibit A, I suppose," grumbled Bates. He gave in, however. "Your father ought to have you in the business. You're a mighty smart girl. How fast did you say this benzene buggy would go?"

Chapter V

I

"Norman," said Mrs. Craven that same evening when her son came downstairs dressed for social activities in the baggy trousers which had become all the go and a stiff linen collar four inches high, "I want to talk with you."

She was standing in her office door. He recrossed the hall with a sulky air and followed her in.

"Norman, you're making a great mistake." Mrs. Craven took her own chair and regarded him with a serious and severe eye. "You took special pains not to consult me, but I've heard all about what you're doing. I mean in connection with the Langley Craven business."

Norman looked crestfallen as well as surprised. "How did you find out?" he asked.

"Never mind. The important thing is that I know—and that I most thoroughly disapprove. You can never get control of the stock in your uncle's company. Things aren't ripe for a move like this. Take my advice and give it up while there's still time."

Norman flushed. "I'm going to win," he declared. "If I get proxies from Mrs. Gatchell and that man Payne in Mount Harris, I'll have control. In fact, I may be able to swing it with just one of them. Anyway, it's too late to back out now. I'm committed to go ahead."

The maid brought in a large cup of black coffee and placed it on the desk. Mrs. Craven began to sip it, watching her son intently over the rim. "Why did you decide to do this?" she asked.

"Because it's time I did something on my own. My grandfather had a business which he had started himself and was doing well with it at my age. Father was a success from the very start and was worth a million in his thirties. So far all you've let me do is sit around and help you in a few matters of detail. I'm not much more than an office boy."

"You must learn first." Mrs. Craven studied the rebellious face of her son carefully. "Perhaps I've kept too much control in my own hands. We'll see what can be done about that one of these days."

"Look at Joe Craven!" declared Norman explosively. "Since his father has become so interested in the making of automobiles Joe has been running the Specialty plant. Everybody's saying he's going to be a big man. What do they say about me? Because I'm interested in other things besides business, they think I'm just like my uncle was. I heard a rumor a week ago that Joe was to be made president and his father moved upstairs to the presidency of the board, and that was when I decided to show what I could do."

"I thought," said his mother, "that you didn't want to be too closely connected with business. You remember, Son, the course we mapped out for you? It seemed to me we had been following it."

"What I don't want is Joe getting ahead of me!" declared Norman.

"But is this the way to stop him? I'm very much afraid not. When your father took the Craven interests away from his brother it was done with the greatest care. Every move was thought out in advance and then carried through in absolute secrecy. Langley Craven had no idea of what was under way until he came to the meeting of the directors that morning. Instead of that you rush in without even telling me about it——"

"I've planned each move carefully."

Mrs. Craven waved his defense aside. "There's something still more important to be considered. Your father made no move against his brother until Langley had alienated the shareholders by his lack of attention to business. They were in a state of mind to welcome a change when your father stepped in. If you had consulted me first, Norman, I would have pointed out that things are different today. The Balfour Specialty Company is doing extremely well and paying dividends. I know that some of the shareholders don't like Langley very well——"

"Plenty of them don't like him at all. He's too high and mighty. They've resented the time he gives to this proposed automobile company and they're beginning to say that Joe, after all, is a boy. There's been lots of talk."

Mrs. Craven shook her head at him. "It's too big a chance to take," she declared. "You're so young that only success will justify you. If you fail, people will say you're a smart aleck trying to imitate your father. Failure in a move of this kind will be a hard thing for a young man to live down."

Norman got to his feet. "I don't care what you say, Mater, I'm going on with it."

Mrs. Craven remained silent for several moments. "Very well, Norman, it may prove a good lesson." She looked at the straw sailor hat in his hand. "Where are you going now?"

He threw both arms above his head so violently that the hat soared across the room. "Why do you want to know? You seem to think I'm a schoolboy, the way you keep tabs on me. If you must know, I'm going over to see Tony Milner. I'm not rushing into things either. I planned this call carefully in advance and telephoned to her this afternoon. I think I'll propose to her again. Anything else you want to know?"

"No, Son. I'm sorry you have to be rude to me when I do no more than show a proper interest in you."

2

The next day was a Saturday, and the paper went to press at twelve o'clock. Ludar secured one of the first to come rolling off, smelling deliciously of ink, and brought it back to the city room. Here he spread it out on his desk and gloated over the introduction he had written to the report of a railway accident.

> The first collision in four years occurred in the Grand Trunk yards this morning when the local from Hudson ran into the rear of a freight. No one was killed, but several passengers were badly shaken up. What made the accident noteworthy was that Jerry Fowley, the local authority on all matters pertaining to wrecks, was not on hand. For the first time in four years he had overslept . . .

Joe Craven came up the steps two at a time and charged into the city room. He leaned over Ludar's desk.

"Are you through for a while?"

"I'm a free man." Ludar began to toss newspapers off the desk in search of his cap. "What's up?"

"I need some help."

Ludar noticed an unusual air of gravity about his friend. Joe, who was quite a dandy, looked as though he had dressed in a great hurry. His hair needed the attention of a comb. Had a crisis arisen in his romance? Ludar waited until they were on their way downstairs before asking, "Something wrong, Joe?"

"There certainly is." The roar of the presses made it necessary for them to speak in loud voices but at the same time saved them from any danger of being overheard. "Have you any idea what that know-it-all, that—that redheaded sneak of a cousin of mine is up to? He's trying to repeat the trick his father played on mine. He's been quietly buying up stock in the Balfour Specialty Company and now he's collecting proxies for the annual meeting on Monday. He's going to chuck Dad out of the presidency if he can."

"He's been so sly," went on Joe, "that we didn't get wind of it until this morning. Dad's sick in bed with the grippe. He couldn't get up, but we went over the list of stockholders together. You know he's buying a large share in this company they're forming in Toronto to make automobiles, and he's had to part with a lot of his other holdings, including Balfour Specialty. Big-Ears Junior got wind of it."

"Are you in any danger?"

"It all depends. After we went over the lists Dad and I felt sure we could count on the support of most of the small people and that we could keep control if we lined up two or three of the largest. Ludar, you know Mrs. Gatchell pretty well, don't you? Seems to me she was one of the people who put up the money for you and that she's said nice things since you started paying back."

"I've been to tea with her a couple of times. She's a fine old lady."

"Then do this for me. Go to her right away and explain how things are. I'll drop you there on my way to see old Mr. Payne."

"I'll do what I can. I suppose you want me to get her to sign a proxy."

"That's it." Joe produced some forms from a pocket and handed them to Ludar. "I think it will be safe to tell her the whole story if necessary."

They jumped into the car and drove briskly down Holbrook Street. "The way I feel right now," said Joe, "I would like to have a pistol in my hand and his damned red head as a target. Or a sword to carve my initials on that covetous gut of his."

"Dangerous talk, Joe," said Ludar, laughing. "Better keep that kind of thing to yourself."

Joe swore still more explosively. "Look what he's trying to do to my dad!"

3

Mrs. Edwin Gatchell was a widow and lived in a large red brick house surrounded by the highest fence in Balfour, having a healthy dislike for the neighbors on each side. She was sitting in a bay window, which commanded a view of Albemarle Park, and was sipping a dark brown liquid from a small glass when Ludar was shown in.

"My Sure Cure Bitters," she said. "Better have one with me."

"No, I think not, thanks." The liquid in the glass, he knew, was straight rye whisky.

"Can't get along without it," declared the widow, winking at him slyly. "Have one before each meal, even breakfast, and sometimes I feel the need during the afternoon and the evening. I tell you, my boy, it picks me up something wonderful."

When Ludar had explained his errand she set the empty glass down on an occasional table at her elbow and snorted furiously. "Why, that young pup!" she exclaimed. "It would be a real pleasure to go to the meeting and tell him right there what I think of him. But I never go out any more. I'll sign a proxy instead. Have you got it for me? Good! I'll take it right back to my desk and sign it for you."

She returned in a few minutes and handed him the slip. "My father," she said, taking the same chair and sipping at a refilled glass, "was John Breck West, the manager here for the United Provinces Bank. People called him 'Hardshell' West, and he was proud of it. But hard as he was, when Joseph Norman Craven wanted to start making carriages instead of farm wagons, my father let him have a loan. Mr. Craven was only twenty-four at the time, and it looked like a big risk. I remember hearing Father say to my mother, 'If this loan goes sour, those stuffed dummies at headquarters will lay me over a barrel.' But it didn't go sour. In three years Mr. Craven had paid it off and was employing a hundred men.

"I was just a small girl with pigtails down my back when this happened, but I don't mind telling you, Ludar, that I had an eye on Joseph Craven. It broke my heart when he up and got married to someone nearer his own age. When Edwin Gatchell came along—this was quite a few years later—I held him off a long time, hoping Mrs. Craven would be obliging enough to pass to her reward and give me a second chance. Well, she didn't, and Edwin and I were quite happy together. I guess it was all for the best."

She ruminated over her glass for several moments. "I have an argument once a month with our pastor about my Sure Cure Bitters," she said. "I always win."

Norman drove up in a hired buggy as Ludar emerged from the Gatchell gate. Since coming into his share of his father's estate he had lost interest in automobiles and, in fact, was showing a rather marked tendency to watch pennies as well as dollars. He brought the horse to a stop at the hitching post and stared at Ludar.

"What do you think you're up to?" he asked.

"If you must know," answered Ludar, feeling the twitch in his arms which always accompanied one of his infrequent outbursts of temper, "I'm trying to buy a drum table."

"That," cried Norman, "is a completely unnecessary remark!"

"Every remark you've ever made to me has been unnecessary." Ludar closed the gate behind him. "It's no business of yours, of course, what I'm doing here."

"I'm not sure of that." Norman threw the reins around the post and sprang out. "I suspect you're mixing into something which *is* my business."

"If I am," said Ludar, "you're too late to do anything about it."

Norman's eyes were beginning to match the red of his hair, and it was apparent that he was close to an explosion. "If you mean," he began, "that you've talked Mrs. Gatchell into——"

Ludar turned and started to walk away. "I refuse to say anything more. On advice of counsel."

"Look here, you!" cried Norman. "I've got something to say to you that I've been storing up for a long time. You've been getting in my way altogether too much. Everywhere I go I find you. I don't like it. I don't like you." Two men on the other side of the street, both grinning with delighted interest, had stopped to listen. Norman paid no attention but went right on in a voice which grew louder with each word: "In fact, Mr. Ludar Redskin, or whatever your name is, I hate you! I'm beginning to think this town isn't large enough to hold both of us."

Ludar turned his head to say, "Better take good care of that livery outfit or you'll be having another painful bill to settle."

4

Norman felt in a much better mood when he found his mother was going out to lunch. "Swell!" he said to himself. He waited until Mrs. Craven had left and then hurried back into the office, where the busy clicking of keys announced that the indefatigable Lou Bennet was at work.

"We're going to have lunch together," he announced. "It will be most pleasantly *en famille*. I'll pretend you're my wife and I'll look across the table at you and say to myself, 'The missus is looking very pretty today.' "

"I'm busy," answered Lou, going on with her typing. "I've already told Delia that I'll have a sandwich and a cup of tea in here."

"No, you don't!" cried Norman. "I won't be fobbed off like that. I'm entitled to a warm reception today, not just a repetition of your well-known Miss Porcupine act. I'm putting over a big deal. I've every reason to anticipate that on Monday I'll bounce Mr. Langley Craven out of the presidency of the Balfour Specialty Company, his own offspring and special pride. I haven't gone over all the votes yet, but I rather think the rabbit is safely caught in the snare."

Lou got up, walked over to him, gave him a pat on the arm, and returned to her work. "There," she said. "When I'm an old woman I'll be able to boast that I touched the sleeve of Norman, the greatest of the Cravens, in the moment of his most spectacular triumph."

Norman smiled at her, being pleased as usual with her somewhat acid method of dealing with him. "This won't be my most spectacular triumph," he declared. "Not by a jugful! Lou, I'm going to tell you a big secret. I'm working on a merger of all the implement manufacturers in western Ontario." Noticing the interest with which she looked up from her work, he began to exaggerate and invent in the manner of a boastful boy. "Andrew Gale of the Gale Implement Company has agreed to go in. John Philip Soames of the Soames Plow Company in London is teetering on the verge. Three of the smaller people are ready to sign the agreement any time. Lou, my sweet shrew, it will be the biggest deal in the history of Canadian business, and I'll make a hundred thousand dollars out of it for myself. It was my idea and I've done all the work. What do you think of that?"

"I think," said Lou, "that Norman Craven is a very smart young man. When it's all settled I'll ask him to let me touch his sleeve again."

"If the deal goes through," cried Norman, "I'll be entitled to something a lot better than that."

"You'll be married by that time." Her fingers were hitting the keys with more emphasis than usual. "Your wife will give you a chaste salute on the brow as a reward for your victory."

Norman glanced at her shrewdly. "I think I detect in that remark a shade more feeling than you've ever displayed before, my sweet Kate."

"Heavens!" said Lou. "He knows his Shakespeare."

Norman seated himself in his mother's chair, swinging it around so that he could watch her. He apparently found this a pleasant diversion because he continued to observe her in silence for several minutes.

"I'm sitting here and looking at you," he said finally.

"Yes, I know."

"It's interesting and even—well, exciting. I could go on watching you for a very long time."

"You'll be tired of it before I'm through with all these letters."

"The line of your throat is nice," he went on. "Your back suggests a ramrod at the moment, but even at that it's well worth a look. And the mere glimpse I have of your left ankle, with the slightest hint of a white frill under the skirt, is positively enticing." He tilted back in his chair until the springs squeaked loudly. "But pleasant as this view may be, Miss Still Waters, I would prefer it very much if you would turn around and honor me with a little conversation."

She swung about in her chair and faced him, her hands dropping into her lap, palms up. "Norman!" she said in a tone of appeal. "You must leave me alone. You really must."

"Must I?" He got to his feet abruptly. Reaching down, he captured both her hands and dragged her up ceremoniously beside him. Then he drew her into his arms and kissed her. "That," he said, "is the way I'm going to leave you alone."

Chapter VI

I

When Ludar reached home that day for his noon dinner, after reporting by telephone to Mr. Craven his success in the matter of the proxy, there was a letter for him postmarked Little Flanders, Kent. It was addressed in an almost minute feminine hand. "It will be from Aunt Poppy," he thought with an intensity of excitement which surprised him, because he thought he had succeeded finally in forgetting his relations in England and the hurt of their silence.

It was, as he had supposed, a letter from Mrs. Tatterton, a long and rambling and affectionate letter. It apologized rather abjectly for not writing him as soon as the visit of the Milners had established a connection for her to follow up, but plaintively advanced as a reason that there were only twenty-four hours in a day given over to the care of a helpless husband, two demanding sons, and an ever-increasing pack of long-haired wards. She conveyed some scraps of news of the rest of the Redcraft family. Cyril was going to be married after all, to a widow with quite a fortune of her own ("She's bossy and she would make two of Cyril, but she has *such* a lot of the stuff"), and Nancy was becoming interested in the teachings of an Indian swami.

Then she came to the important part of the letter.

And now, my dear little nephew—I think of you as a small boy even though you may be over six foot now—I have good news for you. Did that very pretty and most beautifully dressed Tony tell you we were sure there was a document in existence which would prove (if we could get our hands on it) that all Mother's property did *not* go into the annuities? It has been found! Two months ago, and where do you suppose? It was in the safe, the bottom of our grandfather's clock where we keep all of our valuable papers.

This was how it came about. Some money was needed to send the boys back for another term, and I went to the safe to get our one-hundred-pound bond. I didn't find the bond, perhaps because I didn't really look. The first paper I drew out of the hole was the statement of Mother's property!

Nancy's husband, who is a lawyer, took matters in hand. He finally succeeded in getting three thousand pounds out of dear Cyril. I got one, Nancy one, and the third, as Vivien's heir, is yours. A draft for one thousand pounds is going to you at the same time as this and should reach you by the same mail. I think it was the distressing cost of this transaction which persuaded your dear uncle Cyril to pop the question to the widow . . .

"One thousand pounds!" exclaimed Ludar, looking up from the letter. "Nearly five thousand dollars!"

He was so filled with emotion that for some time he could not force himself to go on reading. What he had needed to put himself at rights with the world had come about. It had been established that his family was one of real substance. He had been acknowledged legally. One of his aunts, the one on whom his hopes had always centered, had written him this kind and warm letter, welcoming him as a member of the family. He was no longer alone in the world.

Then real tears came into his eyes. "If only Uncle Billy had lived to see this!" he whispered, skimming through the balance of the letter. "He could take as much of the money as he needed to make another roller boat, with an engine, and show everyone how good he was!"

Tilly could not contain herself any longer. "Well, what does it say?" she demanded, coming in from the kitchen.

"Aunt Tilly," he cried, waving the sheets above his head, "what would you like above everything else in the world?"

"First," answered Tilly without any hesitation, "to get my thousand dollars back. Second, to spend the rest of my life on the Hull farm at Coldwater."

"Read this letter! Read it! It's from my aunt Poppy in England. *It has news!*"

Tilly read the letter through slowly, her lips moving as she struggled with the very English handwriting. When she came to the information about the draft she looked up and calculated what it meant in terms of dollars and cents.

"Four thousand eight hundred and sixty dollars!" she said in a tone which combined joy, awe, and amazement.

"The first thing I'll do will be to make you out a check for a thousand dollars," said Ludar. "To go back into your principal."

Aunt Tilly's mind had leaped ahead to other considerations. "Ludar," she said, shaking her head at him, "there's no two ways about it. I must have the handling of the money."

He had expected this and was ready for it. "No, I'll open an account of my own."

"You haven't any sense about money," she declared. "Why, you don't even carry a wallet, you're that careless."

"I've never had enough money to make a wallet worth while. Anyway, I've a perfectly good system of my own. I put one-dollar bills in my right-hand pants pocket and any bigger ones in the left. I never lose any money."

"How do you know?" She was becoming openly belligerent. "I tell you, this money can't be trusted in your hands. It would all be dribbled away in no time. You'd rush out and buy all kinds of extravagant things. You'd begin *giving presents to girls!* Oh, I know you, young man."

"There's one extravagance I'm going to indulge in right away," declared Ludar. "I'm going to Pete Work's shop and buy a bicycle. I need it the worst way. I sometimes use Jinks Snider's, and it's wonderful how easy it makes the job of news gathering. I think, Aunt Tilly, I'll get a Gendron. It's the silver one and the nobbiest machine on the market."

"You'll do no such thing!" cried Tilly. "Buy a bicycle! I never heard of such foolishness."

"The bicycle, Aunt Tilly, is as good as bought. I'm going right down and pick it out."

She looked ready to burst with the intense emotion which had been created in her by the thought of so much money coming into the family and only a part of it remaining in her hands ("Just my own back!" she said indignantly to herself), and she repeated half a dozen times during the meal, "You'll just hand the money over to me when it arrives, young man!" When he returned later in the afternoon, proudly astride his glittering Gendron, she relented sufficiently, however, to say, "I hope all the neighbors saw you riding it home."

"I'm going to talk to Miss Craven," declared Ludar, taking off a new bicycling cap of gray tweed and tossing it exuberantly, but unsuccessfully, at the hook on the wall. "She'll tell me what investments to make. This money isn't going to lie idle in a savings account, making profits for the bank and not for me." He added fervently, "Praise God, I can now clean off the last of the contributors, and what a relief that will be!"

The bicycle was shining and spotless, but, to escape further discussion, he found an old piece of chamois and went out to give it a thorough polishing. Feeling a drop of moisture on his cheek, he looked up and saw that a shower was blowing in from the east. He hastily wheeled the bike in through the kitchen door and made a place for it behind the chairs on one wall. Tilly left the stove and confronted him with her arms folded across her bosom.

"There's another reason," she said. "Borrowing. All your friends—and a lot who aren't friends—will be at you for help. There's nothing like falling into money to bring them flocking around. And you'll be weak enough to give in to them. You've got no gumption at all about money matters, none whatever. If I'm doing the handling you'll be protected from them."

"I don't need any protection, Aunt Tilly," he answered cheerfully. "Besides, I may decide to get married now that I have this fine nest egg. Who knows?"

Her voice rose to a screech. "I knew you had that in your mind! I knew it! You'll forget all I've done for you and go and marry your Tony Milner, or someone else if she won't have you. Oh, it's always the way. *That* is why you're being so stubborn about the money." She was waving her arms in the air as though shadow boxing. "Well, I'm going to tell you something, young man. I've been thinking of moving back to Coldwater Farm. My uncle Tower Hull is old and he needs more care than poor old Aunt Julia can give him. They've been at me to go and live with them, but I've held off, thinking you needed me. Well, now I know what to expect from *you*. I think I'll write to Uncle Tower tonight and tell him I'll go. He'll put me in his will if I do. It would mean a lot to me to be put in Uncle Tower's will."

Ludar's reaction to this was twofold. The charge of ingratitude disturbed him. He had always realized how much had been done for him and he was prepared to make any return short of entrusting his patrimony to her and to what William had once called her bottomless pit of a pocket. The threat to return to Coldwater Farm, on the other hand, made him realize, as he had not done previously, that his feelings toward her had changed. Her inflexibility in the matter of his employment at Cauling's had weakened whatever bond there had been between them. On top of that, her refusal to remain in the house after William's death had given him a feeling of antagonism.

It would be a relief, he thought, if she should decide to return to the home from which all the Hulls of Coldwater had sprung. Immediately afterward, however, he felt guilty that he had allowed such a thought to lodge in his mind. This, he said to himself, *was* ingratitude, and he must cure himself of it.

Chapter VII

I

It rained during supper but stopped about eight o'clock. Ludar was wheeling out his bicycle for a spin when the telephone rang. Mrs. Langley Craven was on the line.

"Ludar, could you come over?" she asked. "I want to talk to you. I need you."

It was hardly more than a minute's ride from Wilson Street to the upper end of Fife Avenue. Mrs. Craven answered his ring herself and led him to her own special room which was located off the hall at the rear, a small apartment which served the double function of sewing room and household office. He was distressed to notice how thin she looked and how nervously abrupt her movements had become.

"Ludar, I've been a very foolish woman," she said. "I've quarreled with my son. He came in for dinner in the best of spirits because he had talked that farmer man in Mount Harris into giving him a proxy. He had counted everything up and was sure that, since you had secured old Mrs. Gatchell's, there would be no trouble on Monday. I had the bad sense to choose this time to talk to him about the girl he's running around with. He became very angry and went away without eating his dinner. He didn't even wait to talk with his father. Mr. Craven's in bed with the grippe." She picked up an embroidery frame but found it impossible to resume work on it. She gave it an impatient toss into the work-basket beside her. "Ludar, what's this girl like?"

"She's very pretty."

"That's not what I mean. I know she's pretty. She was pointed out to me some time ago. I thought she dressed poorly."

"They're a poor family, Mrs. Craven. Her mother was left a widow when the two girls were quite young."

"What I want to know is about the girl herself. Is she nice? Is she intelligent? What is there in her to attract my son besides the fact that she's pretty? Joe, heavens knows, has run around with dozens of pretty girls."

"He took me there once. It was his first call, in fact. I haven't seen any of them since, but I liked them all very much. Mrs. Slaney has made a living for them by helping on the books at McCordle's Planing Mill. She's a very capable woman."

Mrs. Craven shook her head despairingly. "That's what everyone says. They're so nice. They're so clever. The girl is so pretty. But, Ludar, this Jessie Slaney wouldn't do for Joe. She's not eligible. I think that must be clear to everyone—except Joe himself."

"I've talked to him several times because I could see he was getting in pretty deep. He paid no attention. Mrs. Craven, I'm sure he's really in love with her. He hasn't bothered about any other girls for a year at least."

"I suppose he's with her now." Joe's mother spoke with a degree of irritation which surprised Ludar, who had always thought of her as nothing short of angelic in her disposition. "What I said to him was that his father and I would never consent to any such match. I told him we had talked things over and thought he should go away for a long trip. To Europe. But he wasn't interested in the least. He said he couldn't get away while things were so unsettled in business. He seemed very much disturbed over the suddenness of what Norman was doing and said he would see Norman and tell him he was beaten and that, for the sake of appearances, he had better drop the matter before Monday. But he refused the trip because of the girl. I'm sure of that."

She was so disturbed that she got to her feet and began to walk about the room. Ludar noticed that she glanced in the small mirror over the fireplace as she passed and made a wry face at the reflection. Her loss of looks was still a matter of the deepest concern to her.

"I wish," she began, dabbing a handkerchief to her eyes, "that we could have stayed as we were, when Joe was such a manly little fellow and Meg was just like a pretty little butterfly. Since they've grown up and we moved here there's been nothing but trouble, it seems to me. I sometimes think I dislike this house." She made a determined effort to control her nerves and came back to her chair. She smiled forlornly at Ludar. "And now I must talk about Meg. There's something you can do for me."

"I'll do anything I can."

"She's—she's gone too. It is *most* distressing. Mr. Cannon has been very attentive, as you probably know. In fact, Ludar, he wants to marry her. I tell you this because I know you are discreet and will keep it to yourself. He telephoned a short time ago to say he was coming up. Meg was away, so I answered. I told him to come. And now Meg must be found and brought back. We can't have anything happen to spoil her great chance. He's such a fine young man, apart from the fact that he'll come in for the title."

"Where is Meg?"

"I don't know. She put on a raincoat half an hour ago and slipped out before I could stop her. Ludar, she's with that man. I'm sure of it."

"What man do you mean?"

"That terrible man. That Rupert Something-or-other. You know him, don't you?"

"I've run into him once or twice. But I haven't spoken more than half a dozen words to him."

"Do you know that Meg's been foolish enough to be seen with him? Why, why must one's children do such things, such insane things? If she's seen alone with him on a dark evening like this her reputation will suffer. I shudder to think what effect it might have on Francis Cannon."

"Most girls in Balfour seem to be fascinated by this Corliss fellow. I wish he could be kicked out of town."

"Well!" Mrs. Craven sat up with sudden determination. "She must be found and brought back. My husband is too ill to go out in this weather. Will you look for her?"

"Yes. It's lucky I have a bicycle now. I bought it this afternoon." This necessitated an explanation, and he told of receiving the letter from England.

Mrs. Craven's manner lost its tensity immediately, and she became completely her old self. She leaned over and kissed him with delight in his good fortune. "How wonderful!" she said. "Ludar, I'm so happy about it. I know what this means to you. It's not just the money, is it? It's the recognition. But the money's not to be overlooked, my boy. I'm sure you'll become a rich and important man with such a good start."

She accompanied him to the front of the house and expressed the greatest admiration for the bicycle. She even smiled and said: "That was a most interesting piece you had in the paper tonight, Ludar. About poor Jerry Fowley running around year after year to imaginary wrecks and then oversleeping on the one day when there is one."

As he was preparing to ride away on his quest, however, she laid a restraining hand on his arm and began to speak in a low tone. "I didn't tell you everything about my talk with Joe. I'm afraid I made a scene. I told him if he married this girl he might as well pack up and go because I would never have anything to do with either of them. I didn't mean it, of course, but he didn't know that. He looked at me in such a stunned way. He didn't seem able to believe I had said such a thing. My sweet son! Oh, Ludar, I'm afraid I've caused a lot of trouble. Perhaps he'll never forgive me!"

2

Ludar rode off on his silver bicycle in a depressed mood. The exultant frame of mind in which he had existed since the arrival of the letter from England had left him. Life seemed to have become disturbing, even frightening, and this was not entirely due, he realized, to the troubles of his well-loved friends, the Cravens. It lay with himself, a sense of foreboding; and to this there was little reason, for the future held nothing he need fear.

Depression rode with him during the whole of his search for the indiscreet Meg. As he pedaled up a wooden sidewalk which climbed a rather steep hill to an institution maintained for unfortunate children, the wind, carrying rain with it and so having a substance of its own about as thick and dark as smoke, took hold of the tall trees lining the walk and turned them into unnatural shapes. Usually the grounds were filled with couples, but the weather had kept them in. At any rate, he saw no one and he was glad to swing about and glide swiftly down the hill, watching the fierce commotion in the branches above him.

He made a very thorough search, pedaling along all the side streets and skirting the open lots and spying into quiet corners and shaded nooks. He looked in at the Spickledee, a restaurant run by a stout German named Adolph and made out of two old circus wagons. He found none of the usual youthful habitués there, only a rather villainous-looking stranger eating raisin pie with knife and fork. He saw nothing of Meg and her escort.

Once, when the winds veered and brought the rain back, he took refuge on the veranda of a small house which lay close under the railway embankment on North Street. It was undoubtedly prosaic enough, but he began to feel uncomfortable almost at once. There was a dim light at the rear only, and he fancied he could hear stealthy footsteps and whispers in the dark rooms immediately behind him. This affected him so much that he decided to push on in the rain but paused to watch a freight train which came roaring out from the yards, belching smoke and sparks and dividing everything in halves like a giant pair of tailor's shears ripping through shoddy. The house trembled as it went by, and cinders rattled on the roof.

The streets were almost deserted, and the only direct contact he had was when he emerged from the path through the deserted lumberyard. It is not easy to be skeptical in the dark, and he had gone quickly by the tumble-down office, seeing in his mind's eye the sodden gray face and decomposing body of the dead proprietor, and so he came out quickly on the street, nearly running into an elderly woman.

"Now, then!" she exclaimed in an indignant voice. "Why don't you look where you're going?"

"I couldn't see you in the dark, ma'am," said Ludar. "I'm very sorry."

"Seeing as you've apologized, Mr. Reporter," said the woman, her good nature restored, "I take it back. It's a fine night for a murder, isn't it?"

He gave up the quest at this point and rode downtown to the office. Here he called Mrs. Craven on the telephone, and the cheerful note in her voice banished all his doubts and uncertainties at once.

"She's back," said Meg's mother. "There was even time for her to do a little primping. And, Ludar, *they're engaged!* You were so kind that I feel you ought to know, although it must be a deep secret for a while."

"I'm delighted, Mrs. Craven. Well, not entirely. I can't help feeling sorry about poor Arthur."

"He'll get over it." Mrs. Craven's voice rose to an ecstatic note. "It's all just too wonderful! Frank brought up pictures of Brazen Tor, the country seat of the Cannons. Just think, Ludar, our little girl will be the Viscountess of Alderstone! It's like a dream come true."

"Is Meg happy?"

"Of course! He's such a fine young fellow. They're in the library now, making plans for the future." Mrs. Craven paused. "Ludar, I didn't tell Meg about you going out to look for her. Under the circumstances it would be better not to tell her. That foolishness is all over with now and must be forgotten."

"I'll say nothing. It will remain a deep secret just between you and me."

"You can't tell how such a thing might affect a properly brought-up young man like Mr. Cannon, and of course there's his family to think of. He had pictures of them, too, and they look a little—well, forbidding. Yes, we must keep that secret most carefully. You like him, don't you?"

"Very much. He's a splendid chap. We had dinner together the other night. Have no fear, Mrs. Craven. The whole matter is now buried full twenty fathom deep."

Chapter VIII

I

Ludar was wakened next morning by the ringing of the telephone which had been installed in the house at the newspaper's expense. It was Allen Willing, and it was clear from his voice that something of importance had happened.

"Get on your new bike," said the editor. "There's big news this morning. A murder in town. And it's a friend of yours who's been killed."

Ludar was so surprised that he could barely achieve a whisper. "Not—not Joe Craven?"

"No. But you certainly came close. It's the other cousin, Norman Craven."

Ludar found this impossible to believe at first. Murder is a grim story which happens in the lives of other people but never comes close to oneself.

"I mean it," declared Willing. "The word came from the police station just two minutes ago. I haven't any details yet. I'm off for the office as soon as I hang up, and you had better get there, too, as fast as you can."

"Will we get out a special?"

Willing said, "I don't think so, but we'll have a busy day of it." His tone changed. "See here, why did you jump to the idea it might be Joe Craven?"

Ludar became aware suddenly that this was a serious matter and that he must step warily. "You said it was a friend of mine. Joe's my best friend, and I thought of him first."

It was difficult to make his voice sound natural when all manner of frightening possibilities filled his mind. Norman Craven murdered! Who could have done it? Joe had been angry that morning—— He shut off that train of speculation abruptly. Joe, his best friend, was not capable of violence, no matter how angry he had been.

"Yes," said Willing, "I suppose your mind might jump in that direction. It's funny, just the same, how close you got to the mark."

Ludar dressed in record time. Then he got out his bicycle, which had accumulated some mud the evening before, and rode over to the Langley Craven house. A maid came to the door and said, "I think Miss Margaret's up, but she's the only one."

Meg came down with a negligee wrapped about her. It was gray, with a long Watteau pleat, and sleeves tucked to form deep cuffs. Her hair was piled with an eye to effect on the top of her head and secured with a huge ivory pin. Even in his disturbed state of mind Ludar was conscious of the fact that Meg in a negligee was a spectacle worth seeing.

"Isn't Joe up yet?" he asked.

"That brother of mine," said Meg, "didn't bother to come home at all last night."

Ludar's face went white. "Are you sure?" he asked. "Perhaps he came in late and slipped in without anyone hearing him."

Meg shook her head. "Not a bit of it. Mother stayed up until after one o'clock waiting for him. About half an hour ago she went to his room and found the bed hadn't been slept in. She's terribly worried."

"Meg!" said Ludar. He moistened his lips before going on. "Norman was killed last night. The police think it was murder."

She let out a slight scream and then clapped a hand over her mouth. At first, it was clear, her reaction was entirely one of horrified surprise. Then, grasping the possibilities of the situation, she turned, gathered up her skirts, and ran back upstairs. He heard her rap on a door and cry, "Mother!" in urgent tones. Then she visited a second room, rapped again, and called, "Father, I'm sorry to disturb you, but . . ." Her voice sank, and Ludar did not hear in what terms she conveyed the alarming news.

In less than two minutes Joe's parents came downstairs, having hastily donned dressing gowns and slippers. Without a word they walked into the sitting room and took chairs, Mr. Craven being careful to select one at a distance from the others so they would not catch his cold. All eyes rested on Ludar, with the evidence in them of deep and tortured anxiety. For several moments nothing was said.

"See here!" exclaimed Mr. Craven finally. "I won't have it, the four of us sitting here and all thinking the same thing. I tell you, there's nothing in it. It's—it's absurd."

"Of course it's absurd," said his wife. "He couldn't possibly have had anything to do with it."

In the discussion which followed they seemed reluctant to mention names, as though doing so would turn a nebulous fear into something more tangible. Mr. Craven shook his head and stated: "That old antagonism died out long ago between them. At least it became no more than a casual dislike. As for this business of the company stock, it couldn't lead to anything—anything like this."

"But where is he?" cried Mrs. Craven suddenly. "Langley, where can our son be?"

Joe's father got up and began to pace about the room. He was obviously quite ill. His nose was red and his eyes were watery. A piece of red flannel, which undoubtedly had held a mustard plaster, showed above the top of his elegant silk pajamas. "It's one of these coincidences which can cause so much useless confusion and anxiety." Then he burst out with, "Because the two boys got to fighting years ago doesn't mean that Joe will be suspected of this."

Ludar was thinking of the bitterness Joe had displayed toward his cousin the day before. That was something he would keep buried safely away inside himself. He wondered with a sinking of the heart if his friend had expressed the same sentiment to others.

Mrs. Craven got up also and walked to one of the front windows. The gaiety of her rose-colored negligee, which had sleeves in large puffs and a butterfly bow at the waist, contrasted sharply with the pallor of her cheeks. "Joe, Joe!" she cried. "Why don't you come home and put us out of this agonizing doubt!"

After another long pause, which no one seemed able to break, Ludar got up from his chair. "I'm expected at the office. I'll let you know about anything I hear. And when Joe comes, will you have him call me up?"

Meg walked to the door with him. All the color had drained out of her face, and it was plain to see that she was badly frightened. "Please find Joe!" she whispered. "Mother will go mad if you don't."

2

Sergeant Feeny was standing in the door of the police station as Ludar rode by on his way to the *Star* office. "The chief wants to see you," he said, motioning over his shoulder with a thumb.

"I still can't believe it happened, Sarge," said Ludar, dismounting and placing his bicycle against the side of the building.

"But an easy thing it is for me to believe. I saw the body, my boy. The coroner says the neck of him was broken."

Chief Jarvis still held his post although he had not been very active the past year. A man from the west, a brisk individual named Parker, did most of the work, with the title of Chief Detective, and would undoubtedly succeed him as soon as he had proven himself. His lower jaw was so insistent that he had come to be known throughout town as Saber-tooth Parker. Both men were in the chief's office when Ludar went in.

"Hello, my boy," said Jarvis, whose hair had turned completely white. He looked across the desk at his assistant. "I remember very well the day when young Ludar was brought into court with the sign sewn on his back. He was a scared little codger and he didn't like the looks of the police court at all. Poor old Jenkinson, long since laid away in the grave of his fathers, was barking and roaring on the bench fit to frighten any small boy. That was a long time ago."

"Let's have the woman in," said Parker, who seemed in no mood for reminiscences and was scowling importantly.

Chief Jarvis raised his voice. "Mrs. Craley!"

A woman appeared in the doorway, nodded her head at Ludar, and said, "That's him."

"You identify this man as the one you saw coming out of the old lumber-yard at the north end of Diamond Street last evening?" asked Parker.

"Yes, sir. I recognized him last night. I knew he was on the *Star* because I saw him once at a lacrosse game and twice at church. He nearly ran into me on his bicycle——"

"He was in a great hurry, then?"

"I don't know about that. He didn't see me, that was all. And he apologized to me nicely. I said to him it was a nice night for a murder."

"And it turns out you were right, Grammaw. Just what time was this?"

Ludar was thinking, "Great God, do they suspect me of having something to do with the murder?" Then another and still more disturbing thought flashed through his mind. It would not be possible for him to explain what he was doing when he was seen by Mrs. Craley. To blurt out the nature of his errand would be to hurt Meg's reputation.

"What a situation this is!" he said to himself. Then he thought of Joe, and all concern for himself vanished. He must be careful not to say anything which might draw Joe into the case. He was even relieved to think that he might be suspected. Nothing would come of it, of course, and in the meantime it would perhaps keep attention away from his friend.

"I can't exactly say, sir," answered Mrs. Craley in the matter of the time.

"I think it was about nine o'clock," said Ludar. "I rode straight to the office, and it wasn't quite ten minutes after when I got there. I was in no special hurry when I rode through the lumberyard, but it's true that I nearly ran over Mrs. Craley."

"You may go, Grammaw," said Parker. "Thank you for coming to us with this information."

"Ludar," asked the old chief, "why didn't you come and tell us yourself that you were in the lumberyard at nine o'clock last evening?"

"But why should I, Chief?"

"Because," barked Parker, shoving his head belligerently forward, "Norman Craven was murdered there. And at that hour."

"This is the first time I've heard where the murder was committed." Ludar spoke in a vehement tone because it was dawning on him that he might be seriously involved. He turned to the veteran head of the force. "Chief, am I under suspicion?"

"We don't suspect anyone at this stage, my boy," said the old man. "We're trying to get all the facts from which conclusions may later be drawn. Mrs. Craley is related to one of the men on the force, and she heard what had happened early this morning. She came down at once to tell us what she knew. Now we're checking things with you. That's all." He nodded his head reassuringly. "Why do you say you know no details of the murder? Surely——"

"Allen Willing phoned me at home. He said that Norman Craven was dead and that murder was suspected. That was all he knew himself."

"I'll ask Willing about that." The brisk manner of the detective had developed a hostile edge. "I'll speak to him before he has a chance to hear from our young gentleman."

He left the office on this errand. The old man leaned forward at once and said in a low voice: "I'm sure, my boy, you had a good reason for riding through the lumberyard at that hour. It will be important, you know."

Ludar now realized that his position was one of extreme awkwardness. If he mentioned the mission which had taken him all over the North Ward he would have to give names. Meg's reputation would be badly smirched and her engagement would probably be broken off. Such an explanation, if made at all, would have to come from Mrs. Craven. In the meantime what was he to say?

"I bought a new bicycle yesterday afternoon. As soon as it stopped raining last evening I went out for a spin."

"It's most unfortunate you happened to go through the lumberyard in the course of the spin," declared the old chief. "Where else did you go?"

As Ludar told where he had been his interrogator gnawed at the ends of his white mustache and frowned. It was clear that he was not pleased.

"But didn't it occur to you," he asked at the finish, "that the path through the yard would be muddy after the rain? It seems odd that you would do that when your bicycle was new and spick and span."

Ludar answered unhappily: "It's all I can tell you, Chief."

Parker returned and nodded his head reluctantly. "He's clear on that one. Willing didn't give him any details." His sharp eyes fixed themselves on Ludar's cuff. He pointed. "What's that?"

"Blood, I expect. I fell off my bicycle and cut my hand on a piece of wood."

"Where was this?"

"In the lumberyard. It was dark in there, and my front tire slipped out from under me. I broke my fall by clutching at a block of wood. My hand was cut slightly."

"Just take the coat off, will you? We'll have the coroner make a test to be sure it *is* blood."

Ludar was not yet alarmed to the point of panic. Being innocent, it was easy to believe that the facts of the tragedy would come to light and clear him of any charge of complicity. He took off his coat with a smile and handed it to the detective.

"Did you go into the office?"

"No."

"That's where the body was found." The detective leaned forward suddenly, exposing his teeth in what was intended as a hint of friendliness. "Now, see here, we can get along without any hard feelings. But we must get at things. We want you to tell us everything. You must hold nothing back. When did you see Norman Craven last?"

"At noon yesterday."

"Outside the residence of Mrs. Gatchell?"

"Yes."

"Where you had words with him?"

"Yes."

The detective produced a notebook from an inside pocket. "Here's what his mother has told us. After dinner he put on his raincoat and cap. She protested that it would be unwise to go out in the rain. He said, 'Oh, I've got to see a man.' She suggested that he telephone the man and have him come to the house instead. Norman smiled and said, 'No, that wouldn't do.' He added when she questioned him further, *'It's someone I always quarrel with.'* He walked out then, and it was the last time she saw him alive."

This piece of evidence seemed on the surface to point to him, but the alarm that Ludar felt was not for himself. "It may have been Joe," he thought. He was remembering that Joe had told his mother he was going to speak to Norman and tell him he was beaten in his effort to get control of the stock. Had they arranged to meet somewhere?

"Are you sure you didn't see Norman Craven in the office of the lumberyard last evening?"

"I rode straight through. There was no one there at the time. Everything was dark and still."

Chief Jarvis concluded that the questioning had gone far enough. "That will be all, my boy," he said to Ludar. "For the present, that is. We may have more questions to ask you later. You mustn't think we are jumping to any conclusions. We're still in the dark and we're casting about. Everything we hear must be investigated."

"Of course, sir."

Parker produced a police sweater for Ludar to wear while the stain on the sleeve of his coat was being analyzed. It was a turtle-neck made of hard black yarn and it was large enough to swallow him up. "Our whole staff could get into this," he said, smiling. He did not feel any amusement, however, over the lack of fit. The donning of the sweater had seemed to enmesh him in the case, and his mind filled with a sudden wild alarm. The trap into which his unwary feet had stepped was not the innocent affair it had seemed at first but one with steel teeth which bit deep and from which there might be no escape.

When he had gone the old chief said to the detective, "The boy had nothing to do with it."

"Why?" There was a sharp challenge in Parker's voice.

"I watched him closely. He was genuinely surprised when he heard that the murder occurred in the lumberyard. It almost bowled him over."

"He may be a good actor."

"But not that good, Parker. It has been my experience that men are not good actors at moments like this. Art deserts them when they face stark reality. I'm positive we're going to find that this young fellow is not our man."

"Then who did do it?"

Jarvis raised his eyebrows. "That would seem to be our problem."

The note of belligerence became noticeable again in Parker's voice. "This is the only piece of evidence we've been able to stir up so far. It's not flimsy, Chief; it ties him right in with the murder. The case must be solved. It will be my first murder since I came on this job, and I can't afford a black mark on my record right off the bat."

"The truth," said the old chief, "is more important than all the police records in the Dominion of Canada."

"I know all about that. But the Police Commission will hold us responsible for clearing up *this* case. The death of Norman Craven can't go unsolved. I'm going to see that it doesn't."

Chief Jarvis was a tired old man, as the stoop of his shoulders and the deep lines about his eyes attested. It was, however, a purely physical weariness. The atmosphere of violence and shame, of cunning and cruelty, in which he had existed so long, had not robbed him of the ideals with which he had started. "Our responsibility, Parker," he said, "is to a much higher Authority than the police commissioners."

3

Willing was pacing up and down in the city room when Ludar, engulfed in the iron harshness of the police sweater, arrived. "What do those damn fools think they're up to?" he demanded.

"They think I murdered Norman Craven," said Ludar. "At least Parker does."

"Parker," said the editor, "has all the subtlety and finesse of a boar with a bullet hole in its hide. I'm sure his ideas of detection don't go much beyond the third degree. But what's all this about you being in the lumberyard at the time of the murder?"

"I happened to ride through on my bicycle." Ludar sat down abruptly in his chair. The peril in which he found himself was beginning to react on him physically. He felt weak and sick. "I'm getting worried, Allen. Parker talked to me as though I had been convicted and was on my way to the gallows."

"Aw, cheer up. What Parker thinks won't be very important in the long run. But I want to ask you an important question, straight out. Was Joe Craven with you in the lumberyard?"

Ludar shook his head quickly. "No, of course not. I haven't seen Joe since he drove me to Mrs. Gatchell's yesterday at noon."

"Where is he now?"

"I don't know."

"Well," said the editor, "that's the crucial point of the case. You realize, don't you, that everyone will think first of your friend Joe Craven when they hear of the murder? They're going to ask the same questions I did. I suppose Parker was curious on this point?"

"No. He didn't ask me about Joe."

Willing seemed startled as well as surprised. "He didn't? Well, that brings up a new train of thought. Now, why didn't he ask the most obvious of all questions? Can it mean he's already closed his mind? That he doesn't want to know anything about young Craven? I must say I don't like the looks of this."

Ludar was now so concerned over the possibility that Joe had been involved in the tragedy that he hesitated to make any suggestion for fear of consequences which could not be foreseen. Finally, however, he said: "Perhaps you would find out something if you telephoned to the Slaney house. He may have been there."

Willing shook his head. "I thought of that right away and got on the phone. No answer. Then I had inquiries made and found that Mrs. Slaney and the two girls had taken the train yesterday afternoon to some place on Lake Erie. They're spending the week end with friends. An engagement, we were told by the neighbors, of long standing. Joe drove them to the station and then went on about business of his own. No, there's no alibi for him in the Slaneys."

Ludar's fears increased. If Joe had not been with Jessie Slaney it became increasingly possible that he had seen Norman Craven at some time during the evening.

He remembered then his promise to advise the Craven family of whatever he learned. He reached for the telephone. A sharp "Don't!" from Willing arrested his hand before the receiver had been lifted from the hook.

The editor came over and drew up a chair beside him. "When a man's under suspicion," he said, "the police listen in on all his telephone conversations. I'm sure this wire is tapped already. Until this business is cleaned up they'll hear every call you make, no matter where it may be from. And of course they'll open all mail you send out or which arrives for you. For the time being you'll have to be very careful about what you say and do."

It occurred to Ludar now that to tell the Cravens of his predicament would be like a demand for succor. It would be the same as saying to Mrs. Craven, "You must get me out of this even though it means blasting the reputation of your daughter." It might come to that later on if the coils of circumstantial evidence tightened about him, but he did not want to precipitate an explanation until it became absolutely necessary.

"Besides," said Allen, "you've got to be very careful what you say to those people from now on. You're not on the same side of the fence, you know."

Ludar regarded him with a puzzled frown. "What do you mean?"

"Just this. Their interests may prove to be the exact opposite of yours. They don't want their son involved, do they?"

"Do you mean they would be willing to see me convicted if it meant shielding him?"

"I don't know." Willing was beginning to look glum over the situation. "This I do know, that strange things happen when it becomes a matter of life and death."

Ludar was silent for several moments and then he smiled. "I know them too well to think anything like that. We've been friends too long."

"Perhaps you're right. I hope so."

"Allen, I saw them after getting your telephone call this morning."

"I was sure you did. Anyone see you going into the house?"

"No. The street was deserted."

"Then keep that fact from the police. They'll suspect all sorts of things if they know you ran over there at the first hint of danger. I'm beginning to think I had better go up and have a talk with these good friends of yours."

"I wish you would," said Ludar eagerly. "I promised to tell them what's going on, and they won't know what to make of this long delay."

"The police are probably watching the house. Still, I'm at liberty to go in as a newspaperman."

"Tell me what you know about the case," said Ludar after a moment. "I'm completely in the dark about it."

"Yes, I guess you ought to know about this murder you're supposed to have committed." Willing indulged in a wry grin. "That blundering fool of a Parker! This may do one good thing; it may convince the commissioners that they've *not* found the new chief in him." He paused. "The body was discovered by old Hector Neilly. You must know him, he's a very fine gardener and he's generally called Sandy McSquash. It seems he's been in the habit of collecting what he needs in the way of wood for his kitchen stove from the abandoned supplies in the lumberyard. Being touchy and as proud as Lucifer, he always goes there early in the morning when no one is about. Today he went bright and early as usual. The rain had left everything thoroughly soaked, and so he concluded he had better pry some pieces loose in the office. As he approached the place he said to himself (as he told me in the strictest confidence) that he would be glad when it collapsed because then he would have a fine source of supply. When he came close enough he saw, through a gap which had once been a window, a body stretched out on the floor." Willing was talking now as though he were writing the story of the discovery. "This body had the white face of death! He stood stock-still for several moments, although he knew from the first moment that the man was dead.

"The old man isn't a coward. He didn't do as most men would, run out of the place and get to the nearest telephone. He walked closer and even leaned through the hole which had been a window. He tells me the body looked like 'one o' they mummies thrapped down on a board.' He knew who it was and he said to himself there would be 'gey trouble' over what had happened. Then of course he hurried to a neighbor's house and telephoned the police from there.

"The body," went on the editor, "was lying in the exact middle of the floor. On its back and with both arms stretched out. The top of the head showed evidence of a blow from a blunt instrument. There doesn't seem to have been much blood. Dislocation of the neck had caused death. At first the police inclined to the theory that the murder had been committed elsewhere and the body dumped there. Parker, the great mastermind, abandoned that theory, however, when Mrs. Craley came forward with the information that Ludar Redcraft of the *Star* had been in the yard at the time fixed by the coroner. *That* was all the proof he needed that the murder had been committed on the spot, and he's going to stand by it if the heavens open and spill evidence to the contrary like manna."

"I don't suppose I'll be allowed to do any work on the case."

"Hardly," said Willing, getting to his feet. "Jinks is up at the scene of the crime and will cover all that part of the story. Someone will have to see the mother or, at any rate, talk to the people in the house. I'll do that myself. You stay around here and read something to distract your mind. By the way, the Toronto newspapers are sending men up on the morning train. I hear Jack Beadle of the *Pilot* is coming. He's their murder expert and he always writes a sensational story. He'll tear things loose, I'm afraid. We can be dead sure he'll wade into the Craven feud and leave nothing untold. That sort of thing is his meat. Every conceivable innuendo will be woven into his story, and he'll write it in the worst possible taste. I think the fellow's a disgrace to journalism."

"Allen," said Ludar, "will you see to it that Aunt Tilly and the McGregors know what's going on? Tell them they mustn't worry. The truth will come out soon, of course."

The editor nodded and said, "Of course."

It was three o'clock in the afternoon before Ludar was summoned back by the police. In the meantime the town continued to seethe with excitement, and crowds had gathered around the station and the offices of the *Star,* where bulletins on the case had been put out. The metropolitan reporters, three in number, had arrived and were already working energetically. Beadle, a huge fellow with a red neck bulging out over a stiffly starched collar, had promptly learned of Ludar's connection with the story and had come in for a talk with him. Willing had bustled out immediately from his office and had taken part in the discussion. After the huge frame of the sensation seeker had waddled down the stairs to the street the editor had nodded approvingly. "You were cool and shrewd," he said. "You didn't let him get a hammer lock on you. I can't see that he got a lead of any kind. Ugh, he gives me the creeps!"

When Ludar reached the street he heard excited voices say, "That's him!" and there was a scramble to get a closer view of him. He had to push his way through the crowd, finally, a condition which continued until he reached the entrance of the police station.

Parker had succeeded in getting rid of Chief Jarvis. At any rate, he was alone in the office when Ludar walked in.

"Sit down," he said.

Ludar knew that he was going to be arrested for the murder of Norman Craven. He sensed it in the air, in the satisfied expression of the head detective which turned into a more than usually toothy smile as the latter began to talk.

He looked down at his wrists and wondered if they would think it necessary to put handcuffs on him.

4

"It was blood on the cuff," said Parker.

"I knew it was."

"Let me see your hand."

Ludar held it up, displaying the scraped surface of the palm and the cut made by a splinter. "Dr. Grove will have a look at that in a minute," said the detective. "In the meantime we'll keep the coat."

Ludar made an attempt to speak lightly. "As Exhibit A?"

Parker nodded his head uncompromisingly. "As Exhibit A, Redcraft," he said.

Each added hour of the day which had brought no word of Joe had confirmed the fear in Ludar's mind that his friend had been involved. That the police seemed convinced of his own guilt meant, however, that official suspicion had not yet turned in Joe's direction. A picture came into his mind: the Craven family as he had seen them in the morning, sitting in tortured silence until Mr. Craven had broken out with: "I won't have it, the four of us sitting here and all thinking the same thing."

There was some consolation in the fact that Joe had not been heard from. But it was small consolation, for the gleam in Parker's eye was that of a watching cat as the mouse emerges from its hole. The officer was ready to pounce.

"Evidence has been piling up," declared Parker. "There's that clot of blood on your cuff. We've found the marks of your bicycle tires along the path through the yard. They come to a stop in front of the office."

Ludar protested, "I told you that I fell there."

"Do you still say you didn't enter the building?"

"I did not."

"There are footprints in the mud leading to it. Impressions are being made, and I won't be surprised if we find you *did* go in. George Webster, who lives in the house where you say you stopped, is sure no one was on his veranda at any time during the evening."

"I was careful not to disturb the people in the house."

"Webster says no one could come and sit on his veranda without his knowing it." The detective shook his head with an increasing air of satisfaction. "His statement casts doubt on the whole story you told Chief Jarvis of your movements last evening."

"But I *was* on the veranda. For perhaps as long as six or seven minutes."

Parker rolled a ring with a large oval carbuncle around a fat and stubby finger. "Two men have come forward and told us all about your quarrel with the victim yesterday at noon. It seems he was very angry. He said he hated you and that the town wasn't big enough to hold you both."

"I can't help what he said. He did all the quarreling and talking. I had very little to say."

"What caused it?"

"I got ahead of him in the matter of a proxy. That was what made him so angry."

"No, it goes further back than that, young fellow. You had a fight with him once."

"Yes. He nearly ran me down in his sleigh."

"So the hatred he had for you went that far back. Isn't it true that the pair of you quarreled whenever you met?"

"No. I met him many times after the fight and nothing happened. There was no desire on my part to quarrel with him."

The questioner paused. He had been keeping his eyes on the pad in front of him which was covered with notes in a small, neat hand. He looked up now with sudden hostility. "Isn't it true that all this bad blood between you is caused by the two of you being interested in the same girl?"

Ludar answered hurriedly, "No, of course not."

"It's true. Everyone in town knows about it. Why, it's been common talk."

"I don't know, sir, what has been in Norman Craven's mind. I can speak only for myself. I haven't felt any enmity toward him."

"Actions speak louder than words," declared the detective. "Redcraft, you'll never convince me, or anyone in this town, that you and the late Norman Craven were not open and avowed enemies and that anything might happen when you got together."

There was silence for a moment, and then Parker made a switch in his tactics. "You've given no satisfactory explanation of your presence in the yard last evening. Are you prepared now to say why you were there?"

"I can only repeat what I told Chief Jarvis. I went out for a ride on my new bicycle. I not only rode through the yard, but I went up and down every street and alley in the North Ward."

"I must tell you that we consider this story highly unsatisfactory."

Ludar felt no longer the panicky urge to make a break for freedom which had swept over him at intervals since he had realized his position. He felt cool and even philosophic about it. There was no way of avoiding arrest now, and he must make the best of it, knowing that sooner or later his innocence would be established. But at this point an uneasy thought which had been at the back of his mind began to assume more alarming proportions. The Cravens were aware of what was happening to him, and so far they had not been heard from. Did this mean they had no intention of coming to his assistance? Perhaps if he told his story Mrs. Craven would deny it!

The questioning had been conducted to an accompaniment of much noise in and out of the building. The crowds standing in the streets had, quite apparently, been growing larger. There had been a continuous stamping of feet and a jumble of voices in the hall of the station. Once Ludar had heard the high, squeaky voice of Beadle say, "But aren't you going to let me talk to him?" Sergeant Feeny had answered stolidly in the negative.

Parker now got up and walked to the door. Opening it not more than an inch, he called, "Feeny!"

When the sergeant obeyed the summons by sticking his face through the door Parker said: "Clear out the place, Feeny. I want it empty and the doors closed."

"And does that mean the press too?"

"The press most of all."

"I'm telling you, they'll be mad as hops. They've all been for breaking in your door as it is."

"Sweep 'em out! If you can't I'll come and do it myself."

There was much loud expostulation, but finally the order was obeyed. Parker then escorted Ludar across the hall.

"I'm going to book you, Redcraft."

The police clerk took his place behind the high desk in the station duty room. He looked at Ludar without any hint of recognition and began to ask questions as a matter of form.

"Name?"

"Ludar Redcraft."

"Born?"

"England."

"Come," said the clerk impatiently. "Born when and where?"

"I can't tell you where. I don't know."

Parker grumbled at the clerk: "Get along with it. He's being booked for murder."

Feeny went through Ludar's pockets, placing everything he found on a large handkerchief, together with such objects as had already been taken from his coat—Uncle Billy's watch, a few keys, a handkerchief, a pad for notes, the stub of a pencil. The good-natured sergeant was talking in an undertone as he performed this duty. "It's a shame, it is. You're never the one to do a killing. I told him that, but he wouldn't listen to the likes of me. Not him. He can't wait for an inquest to get at all the facts. No, he must rush in, the great blunderhead!"

Ludar was listening to the noise from the street, which was growing louder all the time. The crowd seemed to be getting into an ugly mood. How did a crowd in an ugly mood behave? To the object of its wrath the muttering of a mob is the most frightening sound that the world can produce.

In the meantime Parker had been answering a telephone call at the desk. Ludar's interest veered hopefully in that direction when he realized that the man on the other end of the wire was Allen Willing. "No, Mr. Willing," he heard the detective say, "I can't exclude the *Star* from the general order. You can't come in, is that clear? No, Mr. Editor, I'm not trying to stand between you and the public. I'm only doing what's necessary. . . . What's that? Of course I don't want to offend the *Star* or Mr. Milner." He remained silent for several moments while Willing talked at the other end. Finally he replaced the receiver on the hook and said in an angry voice: "Willing is coming over. Let him in."

5

Willing was admitted through a narrow door off a side alley, the momentary opening of which let in a roar of sound from the street. The editor arrived in a belligerent mood. He scowled at Parker and said, "You're too new on the job to make a mistake like this."

"I know what I'm doing," declared the detective. "Keep your opinions to yourself."

"I'm going to convey them to the readers of the *Star,* who will be your employers, Mr. Detective," declared Willing. "You can't put a blot like this on a man's reputation until you're a lot surer of your ground than you are in this case. But you couldn't wait, it seems, not long enough even to let the coroner hold an inquest. You had to rush in. It won't take you much time to discover the magnitude of the mistake you've made."

"I'm doing my duty."

Willing put a hand on Ludar's shoulder and led the way down a narrow hall into the parade room. Here the members of the force had their lockers. It also contained a table at which prisoners consulted their lawyers or saw friends when the occasion rose. It was a dim room with one small grated window and a battered-looking gas jet.

"You young idiot!" Willing's anger now vented itself on the prisoner, but he was careful to speak in a low voice. "Who do you think you are, the Knight of La Mancha? Sir Philip Sidney? All the Knights of the Round Table rolled into one? I know you're keeping your mouth shut to shield your friends. It's a fool trick because it won't do any good in the end. The best thing you can do is to tell what you know and not hold back a single damned thing."

Ludar was wondering, "How far has the red car gone by this time? Is it safely over the border and lost in that great country which teems with automobiles?"

The editor let his voice sink to a whisper. "I went to the Langley Cravens' and was told Mr. Craven was too ill to see anyone. I did catch a glimpse of Mrs. Craven at the end of the hall. She was peering through a door with a frightened look on her face. I told the maid as much as I thought they should know about the situation—and came away. Ludar, those people have found out something. That, at least, is my guess. I'm sure of one thing: they're scared to death. If you think there will be any assistance offered you from that quarter, get it out of your mind." He looked cautiously over his shoulder before going on. "Do you know what those people out on the street are saying? They're saying that no one knows where Joe Craven is and that the police are doing nothing about it. Very few of them think you are guilty; at the worst they think you were in it with Joe. They're beginning to say that you're just the whipping boy. Parker, the blundering lunkhead, pays no attention. He has someone against whom a case of sorts can be made, and that satisfies him. And the worst of it is, he may succeed in pinning this on you if you don't look out." He stopped and motioned in the direction of the heavy oak door leading to the cell block. "Have you any idea what it's like back there? It isn't nice. You won't like it at all. My advice to you, young fellow, is to go right to bat."

"I can't!" said Ludar in a whisper. "I tell you, I can't!"

The argument continued, Willing acting as though he would like to shake the truth out of his reporter, Ludar saying again and again, "I can't!" Finally the editor threw up his hands.

"I've done all I can!" he said. "If you're determined to go through with it, I suppose you must." He turned to leave but came back to make one more effort. "There's hell to pay in town. The whole place is split wide open. Lots of people are saying that Langley Craven has paid Parker to put the guilt on you in order to protect his own son. They're even saying openly that he paid five thousand dollars."

"I don't believe it!" said Ludar. He spoke in a low tone, although a perfect panic of fear was taking possession of him. "It couldn't be true, Allen. I tell you, it couldn't be true!"

"I'll tell you someone who believes it. Mrs. Tanner Craven. I went up to see her. She didn't appear, but I talked with her secretary, the Bennet girl. The girl was as white as a sheet, and I expected her to break down every second. She had nothing to tell me. But this I know for a fact: Norman Craven's mother has sent to Toronto for private detectives. They'll be swarming all over the place by night. She believes one of the other family is guilty and she's determined to prove it."

Ludar asked after a painful pause, "Has there been any word of Joe?"

"No one seems to know anything about him. He was seen about town early last evening, but since then he and his red car have vanished from sight. There was a rumor on the streets a short time ago that the car had been seen near the border at Niagara."

"I hope it's true!" cried Ludar fervently.

Willing glanced at him sharply. "Does that mean you think he's guilty?"

"No. But I don't want him pulled into it."

"He had better be pulled in if you're to be pulled out." Willing's voice took on a note of angry urgency. "You like the Cravens and don't believe they would do anything to hurt you. Get this through your head, my boy: when it comes to a matter like this, friendship goes by the board. If they know Joe did it they'll move heaven and earth to protect him, even to the extent of sacrificing you. This has reached the stage where you must think of one thing only, your own safety." He made a start for the front of the station but came back for a final piece of information. "Lockie McGregor is in my office this minute. He says he's going to hire the best criminal lawyer in all Canada to defend you. When I left he was waving a check-book and talking in terms of thousands. He's sure you're being sold out. His daughter came down with him."

Ludar found this hard to believe. "She hasn't been well lately."

"Her husband wheeled her down. Your difficulties inspired her, apparently, to the effort. She talked to the mayor and the police magistrate, and I hear she didn't mince words. You have zealous advocates in the McGregors." He paused and then added: "I told Mrs. Christian. She was frightened, of course."

Ludar asked, "What does Mr. Milner think?"

"He doesn't know about it. I tried to get him on the telephone in Toronto, but he's out in his car, and no one knows where he's gone or when he'll get back. I tried to get Miss Adelicia Craven, but she's away for the week end, motoring with friends. I tried to get Pete Work."

"Yes!" said Ludar eagerly. He was thinking of the day when Pete had declared, *Anyone who picks on the Duke picks on me*. "I'd like Pete to know."

"He's out driving." Willing spoke in a vehement voice. "I want to tell you, the automobile is going to be the ruination of this country!" He looked back over his shoulder and was struck suddenly by the pallor of Ludar's face. "I guess you're frightened, kid, aren't you?"

"Yes," whispered Ludar. "I'm terribly frightened. But I've got to go on with it just the same."

6

The cell block consisted of a corridor with small windows at each end, one close up under the ceiling and so very small that it served no practical purpose whatever, and a row of five cells. The end cells contained three benches, each of them wide enough to accommodate two men for sleeping purposes. The three middle cells had two benches each. Thus, at a pinch, the block could hold twenty-four prisoners without quite justifying a comparison with the Black Hole of Calcutta. Each compartment had a well-scraped but nevertheless malodorous bucket and nothing else in the way of equipment. There was a stove in the corridor and two more benches, and such a liberal share of what is called the institutional smell that it took strength of will to enter, or a push from the determined hand of authority.

Ludar was put in the center cell and for half an hour he had the place to himself. At first he could think of one thing only, the story being told about town that Langley Craven was paying to have the crime fastened on his shoulders. His mind kept going back over the years, singling out the many and wonderful kindnesses he had experienced at the hands of the Cravens. It could not be true! Life would lose all meaning if it were, all sense of decency and loyalty. Their silence could mean one thing only: they were as much in the dark as he was. Perhaps they thought him guilty! No, Mrs. Craven, at least, had a reason to feel certain he was not responsible for the death of her nephew. She had sent him out on the search which had taken him through the dismantled lumberyard. The train of thought always ended at one point: she knew this, but she was saying nothing.

Then another and more terrifying strain of thought took possession of his mind.

The year before there had been a murder near Balfour and in course of time the day had arrived when a man was to be hanged.

The time set for the execution was six o'clock in the morning, and Ludar thought it a most inconvenient hour as he trudged the snow-packed streets to the courthouse, his heels crunching sharply in the still cold air. It seemed unlikely that people would stand in the bitter cold at such an hour because a hanging was to take place inside the walled building, but as he drew near he was surprised to find a small group on the nearest corner, their eyes fixed on the top of the jail wall. He understood when he saw that their skins were copper-colored. The man who would die that morning was an Indian, a descendant of the band which had followed Joseph Brant into exile, and these were his relatives and friends, waiting with the stoicism of their race for the drop of the flag above the wall.

Hangings had once been held in the courtyard, but this had entailed the erection of a scaffold each time, and so an economical county had decided on a permanent arrangement. A trap door had been cut in the floor of an upper hall for the purpose. Ludar heard the head jailer expatiating on the advantages of this as he joined the silent and strained group of witnesses in an anteroom.

"It's always a good, shipshape job this way," the official was saying. "They drop down into the lower hall with six feet leeway. No chance for bungling."

Someone asked, "How many have been hanged in here?"

The jailer made a mental count. "Five. We've had a dozen or more murders, but juries in this county are notoriously chickenhearted."

The witnesses stood in the upper hall, and they saw nothing more after the trap was sprung and the body fell through the swinging doors; nothing except the rope as it twisted and turned, twisted and turned. Ludar, feeling a little sick physically and much more so spiritually, encountered the hangman on his way out, an Englishman named Wrinch, who served in that capacity for all of Canada. Wrinch had seemed a little unsteady as he stepped out to officiate. Now it was only too clear that he was drunk. He clutched Ludar by the arm.

"How was it, lad?"

Ludar nodded as a sign of his belief that the execution had been satisfactorily carried out.

"I thought it was." The man staggered and then steadied himself by attaching his other hand to Ludar's sleeve. "I told you he'd have it."

"Have what?"

"The Sign. The Sign on his hand. Aren't you the reporter who bought me some drinks yesterday and got me talking? Of course you are. And you're English; you're from the land I love and will never see again." He gave an alcoholic gulp. "That's why I talked to you. That's why I told you about the Sign I've found on the hand of every man I've ever hanged. Well, this one had it. There it was, as plain as the nose on your ruddy face."

The hangman had refused the day before to say what the Sign was. Now, being mentally unstrung as well as quite drunk, he laughed and said, "Tip us your flipper, lad."

He took the palm reluctantly tendered him and said, "Now this is what you look for, a line which goes . . ." And he went on to say what the Sign of the Scaffold consisted of. "It's right here on the life line, and once you've seen it, I can tell you, lad, you never forget the look of it. Well, it isn't here. That means you'll die snugly and comfortably in bed and not at the end of a rope tied by Freddie Wrinch."

Ludar had kept this knowledge to himself, knowing the mental anxiety he would cause any who had lines on their hands resembling this particular combination. He had never really believed it himself and had since given it no more than a casual thought.

He spread out both palms and looked at them carefully, fearing that some change might have come about in them and that now he might find there the dread Sign of the Scaffold.

It was not there.

This was small relief, however, from the fears which filled him. Was it true that the neck was broken in the drop, bringing unconsciousness at once? His hand went involuntarily to his throat. No death, he was sure, could be as bad as slowly strangling, swinging back and forth, even with the boasted six feet of leeway.

Then, with a suddenness which made him gasp, a terrifying possibility entered his mind. "He may have been lying!" he thought. "Perhaps he saw the Sign of the Scaffold on my hand but didn't tell me because he knew how much it would frighten me. Perhaps—perhaps what he told me wasn't the real Sign!" If that were the case, Freddie Wrinch would know and would be waiting.

7

At the end of a half-hour a policeman led in a very fat woman in a green-and-petunia dress with a silly little saucer hat on her mop of blowzy hair. Ludar had seen her in police court several times and had written facetious paragraphs about her: Sadie Small, who kept a low-grade establishment in a lonely spot along the canal.

"There you are, Sade," said the officer, opening the door of one of the end cells and shoving her in. "The place of honor tonight. What more could a young girl ask? All the comforts of home."

"I want my lawyer. Was he telephoned to?"

"Yes, he was telephoned to. But he said he had company for dinner and he wouldn't come down if all the draggle-tail leavings of Canal Street had been hauled in."

Sadie Small tugged out a handkerchief furiously from a tiny reticule at her belt and blew her nose. "The stuck-up slob! I see where I get me a new lawyer in the morning."

"Naw, Sade," commented the officer. "Get a new profession."

When the policeman had departed, slamming the heavy door shut with a resounding thud which emphasized the thoroughness with which the cell block was closed off from the rest of the world, the woman glanced about her. Her dull gray eyes were shortsighted, and all she could make out was that she had one companion in confinement. She began to talk.

"Twict a year, reg'lar as clockwork, they haul me in," she said. "They shove me into this stinking pigpen and then they take me before the beak in the morning, and he shoots off his mouth at me and he plasters a fine on me as fat as a deacon's hind end. And do you know why? To shut the mouths of the church people, that's why. They got to make the fine Christian ladies and gentlemen think the law is on the job."

"One of these days," she went on darkly, "I'll start keeping a daybook. I'll write down names in it. And the next time they pinch me, so help me, I'll lug that book in here and I'll read out the names in court. And what a mess that will be for the fine ladies and gentlemen!"

She stopped short and peered frowningly at Ludar. "Who are you?"

"My name is Redcraft."

"Oh, *him!* Well, Mr. Reporter, you've written some snotty, smart-aleck things about me in your paper. But right now I've got to say I'm sorry for you. This is a bad fix you're in."

The door opened again with a loud clatter, and the same officer herded in a file of men of the genus hobo. There were six of them, and he put two in each of the three remaining cells.

"Neighbors of yours, Sade," said the policeman. "We rounded 'em up in that jungle down near the canal."

"It's been crawling with them for days," said Sadie.

One of the tramps was quite drunk and was singing in a most unmelodious voice "We're from the Jungles Wild, Kangaroo, Kangaroo!" Another of them, an old man whose voice had a whisky croak in it but about whom clung some indefinable suggestion of better days, peered through the bars into the end cell where Sadie sat.

"This," he said, "is what we miss in life. The refining influence of a good and beautiful woman."

"Aw, shut up!" retorted Sadie. Then she added for good measure, "You dog-faced bum!"

One of the other tramps peered into the cell also. "Hell, Medicine Hat, that's just Sadie Small. She runs that hook shop on the canal."

"Is there only one of 'em in there?" asked a third. "It seemed to me the place was filled."

Sadie got to her feet, her ugly little button nose trembling with rage.

"You scum!" she shrilled. "I'm a hooker, I know. But that doesn't mean I got to take all this guff and belly-mix from a lousy lot of horseflies like you!"

The cell block was in an uproar immediately, Sadie screeching, the tramps jeering at her and laughing at the tops of their voices. In the middle of it, however, the old tramp held up a warning finger.

"Easy," he said. "We got to behave like good, honest citizens tonight, with clean hides and money in our pockets. The young gentleman in the cell behind me might hear. What's he in for? Cribbing at examinations or slapping the wrist of a dancing teacher?"

"That," said Sadie, "is the young fellah that bashed in the head of the other one."

Ludar had been so sunk in his thoughts that the talk going on around him had been no more intelligible than the buzzing of flies. He became aware at this stage, however, of a curious feeling, a prickling along his spine. Looking up uncomfortably, he was aware at first only of faces: pale, flat faces, dirty, unshaved faces with unkempt hair, all of them pressed as close against the bars as they could get and staring at him with the consuming curiosity of the vacant mind.

"You never can tell, can ye?" said one of the tramps. "He looks like a nice young fellah."

Ludar got to his feet. Then, without any warning whatever, his knees folded up under him and he felt himself falling into darkness.

When he came to he was propped up on one of the benches and Sergeant Feeny was sitting beside him. The heavy institutional smell had been replaced by the most appetizing odor in the world. The sergeant was holding a bowl of soup, steaming hot, in one hand and a sandwich in the other.

"Get outside of this, boy," he admonished. "I got it from Farley's restaurant around the corner. Is it any food at all you've had today?"

Ludar answered in the negative. "I was just going to have breakfast when I heard. I was so excited I forgot all about eating, and I didn't think about it again."

"Well," declared Feeny as the prisoner proceeded to consume the soup, "the stomach always has to be filled. It's a surprise to me, it is indeed, that condemned men can powder into their last meals the way they do."

The allusion was not a happy one. Sergeant Feeny was a kindly man, however, and meant well.

Chapter IX

I

A morose policeman named Bailey came into the cell block at six o'clock the next morning. He carried a tray with tin cups of coffee and a pile of unbuttered toast. "Fifteen cents apiece," he said. "It'll be taken out of what they frisked off you."

Ludar waved his share aside. He did not like coffee, and his appetite of the evening before had deserted him.

"Not good enough for you?" asked Bailey with a sniff. "Better get used to it, Lord Fauntleroy. It's lots more of this kind you'll be having before you're through."

At seven o'clock Allen Willing was shown through the oak door. "Nice company," he said, looking around him. "I see we have the Grande Dame of the Canal with us. Whew! What a stench!"

Ludar was released from his cell, and the pair sat down on one of the benches in the corridor.

"Hell's busting like popcorn all around us," said the editor, keeping his voice down. This was necessary because the other occupants of the cell block were standing against the bars and staring avidly. "First, a girl was seen leaving the lumberyard. A minute or two before nine. Apparently she left about the time you arrived."

Although he had not yet connected her in his mind with the tragedy, Ludar knew that this could have been Meg. He asked anxiously, "Have they any idea who it was?"

"Not a clue yet. She was seen by Mr. and Mrs. Alfred Avery, who were returning after playing five hundred for a couple of hours with some neighbors. They were on the other side of the street, and there was an argument going on between them about certain points of the evening's play. They didn't pay much attention to the girl. This is all they've been able to tell: she came out from the north end, she turned west and walked away in a great hurry, and she impressed them as young."

"Did the police find this out?"

"The police are not finding out anything. That great genius of detection, Mr. Sherlock Parker, is sitting on his prat and doing nothing. It was one of the three detectives Mrs. Tanner Craven imported from Toronto who got onto the girl. He was in touch with the Averys last evening." Willing hesitated and seemed reluctant to go on. "Beadle heard about it and used it in his dispatch to the *Pilot.* He twisted it around so there was a direct implication that the girl was none other than Tony Milner."

"No!" cried Ludar, forgetting the listeners and using a tone of anguished protest. "She's in Toronto. With her father."

"Exactly." Willing's voice contained a note of grim satisfaction. "That was what Mr. Yellow Journalist didn't know. Apparently he didn't try to find where Tony was but just blundered in. I've talked to Mr. Milner twice. I got him on the wire last night at ten-fifteen and told him what was going on. He was stunned by the news and all he could say was that everything must be done to help and protect you."

"I'm glad to hear that."

"The boss seldom lets you know what he thinks of you. For the first year I was with him I fully expected to be fired at the end of each week. Then I discovered he had liked me all the time. I know he likes you and thinks you're turning into a crack newspaperman." There was a pause, and then Willing smiled. "When I got him this morning he had just read Beadle's stuff in the *Pilot,* and I can tell you the wires crackled and sizzled. He's going to sue the paper for one hundred thousand dollars. The *Pilot's* always on the verge of bankruptcy, and they can't afford to lose any libel suits like that. Wouldn't be surprised if they apologize and fire Beadle, the yellow hound!"

Ludar found nothing to say, his mind being full of this latest, and most unpalatable, of developments.

"I tried to see the Langley Cravens again last night," went on the editor, "and this morning before I came down to the office. All the doors are locked, and they don't answer the bell. I hear he's still sick in bed. Beadle was hanging around the grounds this morning. I'd read his filthy dispatch and I took a poke at him. He said, 'Don't hit me!' in that baby voice of his and started to run."

"Has there been any word about Joe?"

"Lots of rumors, that's all. The story was all over town last evening that you and Joe had met his cousin in the lumberyard and that he had been killed in the rumpus which followed. Everybody seems to be sure at the moment that this was what happened: that Norman was done in, that Joe left by the north entrance where he had left his car and was lucky enough not to be seen. You left by the south gate and had the bad luck to be seen by Mrs. Craley. Who would invent a story like that to begin with? And here's another that's going around. They say the Langley Cravens know where Joe is and that they've got word to him to stay under cover."

There was silence for a moment, and then Willing went on: "About lawyers. Lockie McGregor gave me the names of three and said he wanted to get one of them for you. He'd picked the three best in Ontario. Leonard Hill, of course. Even if he is a local man, I think he's the best to be found. The others were Brockett of Toronto and Phillips of Ottawa." He gave his head a shake. "We can't get any of them. It seems Mrs. Tanner Craven has already put them on retainers."

Ludar again lost all sense of caution. "But why?" he cried.

"Oh, she's looking ahead. If any member of the other family becomes involved, she doesn't want them to have the advantage of good legal help. A very shrewd and hardheaded woman, Mrs. Craven. She's fighting like a she-bear who has lost her cub."

They began then to discuss points of evidence. Ludar had not slept at all during the night, and he had turned the case over and over in his mind in the hope of finding something in his favor. He asked now if the weapon had been located.

Willing shook his head. "They've been all over the place with a fine comb. Mrs. Tanner Craven's detectives as well as the police. Not a trace of it."

"Then," said the prisoner eagerly, "how could it have been me? I didn't have any weapon with me when that woman saw me ride out of the yard. So if I did it, I must have left the thing in the yard."

Willing nodded. "It's a good point. But of course the storymakers are saying that Joe carried the weapon off with him in his car." He began to talk then about the state of public opinion. "Most people are still on your side, if that's any consolation. A mass meeting's to be held tonight. On the market square. There will be thousands out. That blatherskite of a Josh Laines—ex-Alderman Laines, I should say—is going to speak. He's been looking for an issue to get himself back into public life, and now he thinks he has it. He's going to demand, or so I hear, that Parker be fired and the whole police department be investigated, root and branch, as he'll undoubtedly put it. He's a flannel-mouthed skunk, but this time I say, 'All power to him.' He's working for you, Ludar, and I hope he succeeds in nailing down the lid on Parker's coffin."

He got to his feet. "Well, boy, I admire your courage more than your intelligence. Be of good cheer. Plenty of friends are working for you."

At ten o'clock Sadie Small was summoned to climb the steep flight of steps which led from one end of the corridor directly up into the courtroom. She had become chatty with the gentlemen of the road during breakfast and she waved a fat hand to them as she placed a foot on the first step.

"Ta-ta!" she said. "I've a notion they're going to put the boots to me proper this time."

"How much dust did they scrape off Sadie?" asked one of the hobos when the morose Bailey returned to escort them upstairs in a body.

"Two hundred and fifty simoleons."

None of his companions in crime returned, so Ludar had the cell block to himself for the next half-hour. Then the door opened and Arthur came in, looking as grave and composed as usual but secretly very uncomfortable. He walked to the door of Ludar's cell.

"My dad," he explained, "called me up last night and said I could take the morning train home. It was kind of him, wasn't it? He seemed to know how much I wanted to be here."

Ludar found himself, for the first time, on the point of tears. "I'm terribly glad to see you. It's been—pretty awful here."

"What a miserable mess the whole thing is," said Arthur. He fumbled in his pocket for his pipe and then stopped uncertainly. "I don't suppose you're allowed to smoke here."

"I suppose not. But they do. One of the bums—there were six of them and they all snored differently; it was like an orchestra—was smoking all the time he was awake. He kept reaching into a different part of his clothes and fishing out cigarettes or cigar butts. He hauled them out of the lining of his clothes and even out of his socks."

Arthur proceeded to fill his pipe. The extra care he took in the operation was an indication of the disturbed state of his mind, that he was at a loss for words.

"You're holding something back," he said finally.

Ludar nodded. He never had any doubts about telling things to Arthur or Joe. "Yes. It's a small matter, really, but it would explain how I came to be there. Small though it is, it would hurt someone a great deal if it got out."

"Has this something got to do with the killing of Norman?"

"No. And that's the tough part of it. If it had anything to do with the case it might be my duty to tell. But when it's about a person who isn't concerned at all and whose reputation would be hurt, you can't—well, it doesn't seem fair to drag them into it."

"I see," said Arthur, who now had his pipe drawing. "It does make it hard, doesn't it? But of course you can't go on keeping this secret forever. What I mean is that you can't incur too great a penalty yourself in order to protect this—this innocent bystander. Unless the truth comes out first, the time will come when you'll have to tell."

"I don't know. I've been doing a lot of thinking and I've begun to understand the—I suppose you'd call it the ethics of a case like this. The thing which got me into it, and which I haven't told, looks small now, but it would cause a lot of trouble if it got out. On that account it would be a brave thing if they came forward and told. But for me to tell, even to save my life, would be different. It might even seem cheap and mean. Arthur, I'm hemmed in. I must be brave and take the consequences or I must be a telltale and save my skin."

"You must be ready to tell just the same," declared Arthur. "At the right time."

"But when is the right time?" Ludar asked the question feverishly, passionately.

"It seems to me you should wait until you feel sure the person in question has had time to get over his or her fright and to see that a sacrifice must be made to prevent a still greater sacrifice. If they don't come forward then, you know they're too selfish to be worth protecting and that any obligation to keep them out of it has ceased. When that time comes, it's right and proper for you to tell."

Ludar looked at him thankfully. This, he knew, was good common sense. "You're right, Arthur," he said. "You're dead right. After all, this may become a matter of life and death for me." Then a new doubt took possession of him. "But how are you to know when that time has come? You sit here alone and think, and you hear all sorts of rumors, and you know that terrible things are happening which may be the end of you. And you don't know what's happening to *them*—the other people concerned. Perhaps they're in as much of a fix as you are. Perhaps they want to help you but don't know how. Perhaps—Arthur, it's sheer hell, because you sit here and think and you nearly go crazy!"

"Aren't you exaggerating in your mind the effect it would have on this other person? My opinion, Ludar, is that the time to talk is—is *right now*."

The officer put his head through the door. "You've had enough time," he said.

Arthur got to his feet. He had something more to say, however, and it was causing him trouble. He opened his mouth, stammered, and finally got out a question. "Ludar, was Meg having anything to do with Norman?"

"Not that I've ever heard." Ludar was startled. This possibility had not entered his mind, and yet it now appeared quite possible, even probable.

Meg's willingness to conquer knew no bounds at all. If she had stooped to captivate the much-hated cousin, it might have been Norman she had gone out to meet and not Rupert Corliss. She might have been the girl seen by the Averys as they walked home and argued about the fine points of five hundred.

His answer seemed to have allayed any fears which had existed in Arthur's mind. "That's all right, then," he said. "I started thinking about it on the train. I'd seen Norman looking at her at parties and I was sure he admired her." He smiled self-consciously at Ludar. "Don't worry more than you have to. You didn't do it. Joe didn't do it. The real story is certain to come out soon."

The policeman, with a wink, had crossed the corridor and had tossed a newspaper into Ludar's cell. It was lying face up on the floor, and the prisoner now saw that it was a copy of the *Pilot*. As soon as he heard the door close he picked it up, and his eyes lighted on a great black headline four columns wide:

HEIR TO FORTUNE MURDERED
LOVE RIVAL ARRESTED

He started to read with a dread which became steadily greater as he went along.

It began with a dispatch from London, telling the story of Vivien Redcraft's marriage, his misappropriation of securities, his unfortunate flier in the market, and his flight to Canada under the name of Prentice. A reference followed to his wife which caused the reader to gasp with incredulous dismay:

> Until her death the deserted wife served as a cook's assistant at Pursley Castle, the property of Cyril Anthony Tresselar. Here lived also the boy, who afterward was to be sent out to Canada with a sign sewn on his back and bearing his father's assumed name. Miss Callie Railey, the unfortunate woman's half sister, who gave this information, took the boy immediately after his mother's death and kept him with her in London until the arrangements were made . . .

The paper fell from his hands, and he stumbled blindly to his feet. His mother a cook! The dark house with the clock tower not the home of his family but the property of strangers in whose kitchen his mother had worked!

At first he could not believe it. This was no more than a particularly lurid sample of the kind of yellow journalism which many newspapers were adopting. Gradually, however, the conviction grew in his mind that it was true. He seemed to recall that the large dark rooms of the castle had been forbidden territory and that he had visited them with stealth. The tall figure of Old Mr. Cyril had stalked about in an aura which he was now able to define as mastership. On what other basis could he and his unfortunate mother have lived at this imposing seat of a family of strangers?

Many minutes passed before he mustered the courage to pick up the newspaper again. The dispatch from London was followed by a long story signed by Orley Dawes Beadle, in which that crime specialist told of Ludar's arrival in Balfour, of the money raised in his behalf (but not a hint that most of it had been most painstakingly paid back), and of the belief which had prevailed of his aristocratic origin. It made much of the fact that the Christian and Milner yards backed together and there had been a youthful romance between the English boy and the daughter of the wealthy publisher. After a lurid description of the Craven feud the story of the murder was told in sensational terms, and the evidence connecting Ludar with the tragedy was presented with a bias which verged on malice. It was full of innuendoes and sly references. Beadle had not gone to the length of stating that the girl seen leaving the scene of the crime was Tony, but he made it very easy for readers to draw that conclusion. He made it almost impossible for them not to believe that Ludar had been responsible for the killing of the Craven heir.

The paper fluttered to the floor of the cell a second time. Ludar, acutely aware of the gravity of his position, was now incapable of caring what might happen to him. His world was in ruins. He saw no chance of rebuilding the tower of self-respect without which life is unbearable.

Chapter X

I

At one minute after two Policeman Bailey came through the door. Crossing the corridor, he unlocked the center cell and motioned to Ludar. "Come on," he said.

The time had come to appear in court and answer the charge laid against him. He had been dreading this moment even more acutely since reading what the exponent of yellow journalism had written about him. What could he say when he faced his employer, whose daughter's name had been dragged into this unsavory affair? How would he take it when people laughed because they remembered he had been called Duke and now knew that his mother had been a cook?

"Hurry up, Lord Fauntleroy," grumbled the officer. "Won't do you any good to lag. Sooner up, sooner over." He looked back over his shoulder, however, to say in a better humor: "Never saw the beat of this. The court's packed, and there's a thousand people outside if there's one, and all of them yapping to get in. You ought to be the proudest prisoner that ever stood in the dock."

The policeman had not exaggerated. The courtroom was filled, and there was a loud buzz of excited comment when the prisoner's head appeared over the railing at the head of the stairs. Ludar kept his eyes down and was prepared to follow the officer to the dock when a hand took him by the sleeve and the voice of Chief Jarvis said impatiently to his guardian, "Not there, Bailey!"

He stood up then and saw that the old police head was smiling at him. "Stand here beside me," said the chief.

The tone of voice used had been so friendly that he began to feel a faint quiver of hope. He took a quick glance around the room and saw familiar faces everywhere. Mr. Milner and Allen Willing were sitting in the front row of chairs, the proprietor's face set in a scowl and one knee swinging impatiently over the other. Arthur Borden was jammed so tightly against one of the walls that if he had raised an arm in greeting he would never have been able to get it down again. Lockie McGregor was quite near him and was glowering, perhaps because of the way he had been forestalled in the matter of a lawyer. ("It's funny," thought Ludar, "that no lawyer has been provided for me.") Jinks Snider was at the press table and was scribbling rapidly on a pad of paper. On every hand he saw people he knew, and the surprising thing about it was that they did not seem openly unfriendly. Some actually smiled at him. Nowhere did he see any signs of the derisive attitude he had thought would be engendered by the article in the Toronto newspaper.

Two things leaped at him and crowded out all other impressions. The first was an unoccupied chair inside the railing, its emptiness dramatic in a room so densely packed. Had it been reserved for the mother of the victim? The second was the presence of Mr. and Mrs. Langley Craven sitting in one of the rear rows and striving to appear unaware of the interest they were causing. Even if they had not figured so prominently in all discussion of the tragedy, Mrs. Craven would have drawn every eye in the place. She was sitting proudly erect, and her own eyes, which were feverishly large and eloquent of emotional strain, were lending her a beauty she had never had in her prime. Her plain vagabond fedora was worn with an aduacity which said, "I am ready to face whatever befalls."

Ludar's heart bounded hopefully when he saw them. Their presence meant, he was sure, that they had not abandoned him.

The Crown attorney, old George Maddocks, who had withered to slippered pantaloonism in long years of perfunctory performance of his duties, rose rheumatically to his feet. He looked about the room, frowned as though in disapproval of such a display of public interest, and cleared his throat with one hand over the straggling ends of his tobacco-stained mustache.

"Your Worship," he said, crossing the enclosure on stiff pipestem legs and standing in front of the magistrate, "there has been—*ach, pugh!*—new evidence produced in this case. As a result of this additional information, it has been deemed necessary—*ach, pugh!*—to have an inquest. The coroner has selected a panel to hear the evidence."

"Quite so," said the magistrate, who had succeeded Mr. Jenkinson when a comparatively young lawyer. "What disposition do you propose of the charge laid against Ludar Redcraft?"

The Crown attorney rumbled still more explosively in his throat. "We ask, Your Honor, that the charge be dismissed."

Could it be true? Ludar looked about him in amazement. This was so different from what he had expected, so far beyond his wildest hopes, that he could neither comprehend it completely nor put full credence in what he had heard. And yet it must be true because the room had burst into such a wave of excited comment that the pounding of the magisterial gavel had no effect whatever. Ludar tugged at the black serge sleeve beside him and asked, "Have I—have I been cleared?" And the old chief nodded with no attempt to conceal his elation and formed words with his lips which Ludar construed as, "Yes, thank God, you are free!"

People came crowding into the enclosure to shake hands with him and thump him on the back. Allen Willing, grinning excitedly, said: "Well, Sir Galahad, it seems that stubbornness pays after all. I guess the saber-toothed detective will be hunting another job soon." Mr. Milner put a hand briefly on his shoulder and said, "You behaved well, my boy." Ludar was dazed at all this proof of good feeling. Did this mean that the article in the *Pilot* had not turned people against him as he expected it would, irrespective of his innocence or guilt?

The spectators were beginning to stream out, and it became easier to make oneself heard. Ludar asked the chief, "When may I leave?"

"You're as free as the air now. But I want to have a talk with you before you go. There's such a crowd outside, anyway, that you'll find it convenient, I think, to delay your leaving awhile."

Ludar realized now that he had not seen the Cravens since his eyes had lighted on them during his first quick and embarrassed survey of the courtroom. They were not among the few spectators still left at the rear. He asked the chief about them.

"Mrs. Craven had to be carried into the magistrate's office," said Jarvis. "She took a faint turn, I expect. Langley was looking pretty sick, too, and I question if they should have come. You probably don't know it, my boy, but they've been having almost as bad a time of it as you. And even now they're not free of it. Their boy hasn't been heard from yet."

"Then it wasn't true," said Ludar, smiling with relief, "that they knew where he was all the time."

The chief's face became red with anger. "That wild and infamous lie was all over town, and I'm not surprised it even reached you. Of all the things for presumably decent people to repeat and believe! Certain men who have always seemed honest citizens were saying they knew for a fact that Langley Craven had bribed Parker with five thousand dollars in cash! I tell you, your faith in human nature becomes shaken at times like this."

2

"The inquest will begin as soon as the court can be cleared," said Jarvis when they had seated themselves in his office downstairs.

Ludar asked, "What is the new evidence?" His mind was full of fresh anxieties, for his talk with the old chief in the courtroom had again raised his fear that Joe's absence was the key to the mystery.

"It was an accident." Chief Jarvis shook his head with sudden resentfulness. "If Parker hadn't been so determined to nudge me out finally by showing how smart he was you'd have been spared this ordeal and we wouldn't be in the absurd fix we find ourselves in now. The regular course is to hold an inquest and to determine the nature of the death and the direction in which the finger of guilt points. Parker was certain he had a clear case and he wanted to get a reputation for speed and efficiency. I grant you the case was strong. But it was purely circumstantial, and he shouldn't have rushed in as he did."

Ludar was winking back tears of intense relief. Joe had not been concerned in the case at all, and nothing else mattered. The damage which had been done to his own reputation seemed of small moment.

"Norman met the girl——" began the chief, and then paused. "We had two dozen phone calls giving us tips on who the girl was—and all of them wrong. This young fellow seems to have done a lot of running around. The youngest of the Tawney girls was suggested by half a dozen people. . . . Well, he met the girl in the office. It's been on the point of a final collapse for a long time. There was some scuffling, and he lost her in the pitch dark of the place. Apparently he was afraid she was going to run away, and when he heard her step he charged forward to intercept her. It was so completely black in the place that he didn't know one of the beams in the roof had become dislodged and was hanging down. He charged head-on into the end of it.

"He lurched forward and fell on the spot where the body was found later. The girl heard him fall. She called, 'Norman!' and when he didn't answer she knew he had been hurt. Her first thought was to get help, and she was on her way to the street when, according to her story, another thought occurred to her. He might have been stunned only, and she didn't want to give herself away unless it was absolutely necessary. So she went back. She confessed that she smokes a little and so she had matches in her purse. She struck one and discovered beyond any question of a doubt that Norman Craven was dead. That threw her into a panic, and she ran away. It was on her second exit from the yard that those people saw her. She went home and prayed long and earnestly, she says, that her part in the accident would never be discovered."

"You can imagine her state of mind," went on the chief, "when you were arrested on a charge of murder. There's this much to be said for Parker: his haste in acting, combined with the scurrilous nature of the story in the *Pilot,* brought the girl to a realization that she couldn't save her reputation at your expense. She came in with her story early this afternoon. A splendid girl, that! She was as white as paper and she told us what happened with an honesty and, yes, a dignity which you seldom see. She made no effort to spare herself, to hold anything back." He seemed to be considering the difficult nature of the position in which she had placed herself by her confession, for he shook his head gravely. "We've found plenty of evidence to back up her story. She was seen on her way to the yard as well as when she left. She told us she passed old Alvin Nimm as he was leaving his store, and he recalls it. The burned-out match was found on the floor of the office. The beam was completely dislodged, it seems, by the force with which he ran into it. The end, which did the damage, fell to the floor. Parker and several of the force spent hours examining every bit of wood in the yard, under water or not, looking for stains which would identify the murder weapon, but they neglected to look at the end of the beam where it touched the floor. They examined it half an hour ago. They found blood on it.

"That seems to clear up the case," concluded the chief. He nodded his head in the direction of a small office back of them which was used by the medical health officer. "She's in there, waiting for the inquest, where she'll have to tell her story. You'll want to speak to her, of course."

Ludar said, "Yes," and got to his feet. He was thinking, "My friends didn't come forward to tell what they knew, but this girl had the courage to do it."

"Before you go in," said Jarvis, "there's something else you should know. Langley and Mrs. Craven came in to see me also. It seems Mrs. Craven knew something with a bearing on the case. She had been reluctant to do anything about it but finally found herself unable to hold it back any longer. She confided in her husband, and he said they must come down and tell me at once. So they came, and I've never felt sorrier for anyone than those two good friends of mine. They were so certain that what they were about to do would bring their temple down about them. I got the impression that what Mrs. Craven knew did not concern the question of how Norman Craven came to die but rather your reason for being in the lumberyard." He nodded and smiled reassuringly. "Fortunately they arrived after the girl. As the solution was in our hands, I told them to wait, that it probably wouldn't be necessary for us to know. We were still checking on the girl's story, however, and I didn't feel I could tell them about it. I didn't see them again until the hearing opened in court. They went in, still having no inkling of what was coming."

Ludar's eyes were filled with tears of which he was in no sense ashamed. His friends had stood by him after all! The kindly family circle, of which he had counted himself almost a member, had not cast him out as soon as danger threatened. The stories told about town were nothing but malicious lies. He fumbled in his pocket for his handkerchief but did not find it because it had been taken away when he was arrested and used to wrap his possessions in. He rubbed his hand across his eyes and said, "Give me a second or two, Chief, to pull myself together."

3

Lou Bennet was sitting alone in the back office. She did not look up but said in a subdued voice, "I suppose you think badly of me."

Ludar was too astonished to speak for a moment. "I had no idea it was you," he said finally. Then he continued, "I'll be grateful to you as long as I live for this brave thing you've done."

"I'm not brave," declared the girl, still keeping her eyes fixed on the floor. "I'm such a coward that I—I had made up my mind not to tell under any circumstances. Then I found I had to. I couldn't leave you to face the punishment for something you hadn't done, when a word from me could save you. You've been holding something back, too, and I'm sure it's been a finer thing on your part not to tell than for me to tell."

Nothing more was said for several moments, and then Ludar asked, "Did Norman's mother know what you were going to do?"

Lou nodded. Her face was completely devoid of color, and an effect of drabness was created by the plain cut of her gray corduroy suit. She began speaking in a voice little above a whisper. "I told her, and for a moment she stared at me with such a strange expression on her face that I had no idea what she was going to do. I thought she hated me and was blaming me for her son's death. Then she asked me what there had been between us, and I told her the full truth. She picked up her Bible and went into her bedroom, locking the door after her. It must have been an hour before she came out. She gave me that same strange look, but I realized then that it wasn't hate. She asked, 'Did you love my son?' and I said, 'Yes.' She sat down then and after a while she said, 'He was going to make a strong man, a wonderful man. Was he going to marry you?' I answered, 'He said nothing about it.' Then she nodded her head. 'You loved him, and that makes a bond between us. You must stay here with me. We'll fight this out together.' "

"Then you're going to continue as her secretary?"

The girl nodded. "Yes. But what she meant was that she wanted me to live in the house with her. You see, I—I can't go home now. I told my mother what I was going to do, and she said it was bad enough that I had been 'carrying on' with my employer's son—I hadn't, Ludar, I swear I hadn't—but that if I publicly proclaimed my shame she would never see me again or speak to me as long as she lived. My mother is a very strong-minded woman, quite different from my father, who was gentle and a real Christian and would have stood by me if he were alive—even if they kicked him out of his church because of it. She meant what she said. So"—she looked up for the first time and smiled tremulously—"I'm to live with Norman's mother. It's lucky for me that she's such a remarkable woman."

"Did she believe I killed her son?"

The girl shook her head. "She was sure from the first that you didn't, but she was furious because she knew you were shielding the other family. Yesterday morning, after the first shock of the discovery, she was in a mood to fight the whole world. She hired the detectives and she got in touch with the best lawyers and she had telephone calls sent to all parts of Ontario to trace Joe Craven and his car. She was going to avenge Norman's death and she had no thought then of helping you. I think she was glad enough at the end that you had been cleared, but she never wants to see you again. You were on the other side." She paused. "Did you know she had five thousand copies printed of what you wrote about Mr. Tanner Craven? She was sending them out, enclosing them in letters. She thought very well of you then, but of course Norman's dislike of you wore on her. You'll find a great change in her now that both Norman and his father are gone. She's not closefisted herself. In fact, I've always found her quite generous. She's going to work harder than ever to make money in order to have more to spend on charities. She's considering a foundation to be called after Norman, but she hasn't decided what form it's to take."

Chief Jarvis came to the door. "The inquest has started, Miss Bennet. The public isn't being allowed in, and we're going to ask you only the necessary questions. We'll try not to make it too hard for you."

The girl got slowly to her feet. "Yes, Mr. Jarvis," she said.

4

Ludar's arrival at the *Star* offices was in the nature of a triumphal entry. All the printers dropped what they were doing and came to the door of the composing room, and they even raised a cheer as he hurried by with a wave of the hand. The business office staff came up in a body to tell him how happy they were. Searles, the manager, winked mysteriously and whispered in his ear: "You can get to be a rich man over this. Sue that Toronto newspaper and you'll collect one hundred thousand dollars. Yes, sir, *one hundred thousand dollars!*"

When the excitement was over, Ludar realized that he had not touched food since the night before and that he was ravenously hungry. Willing sent out for what he wanted, a fried-egg sandwich, a chicken sandwich, a double scoop of strawberry ice cream, and a piece of coconut cake. He consumed everything with an eagerness which proclaimed his complete recovery from the digestive troubles of his early youth.

He was so busy eating, in fact, that he paid no attention when Bates came into the city room and seated himself at his old desk. It was not until the last crumb had vanished that he turned in the direction of the old reporter and became aware of a remarkable change in him.

John Kinchley Bates looked alert and in very good health. His sandy hair, instead of lying in lank wisps on his freckled forehead, had been combed and brushed until not a strand was out of place. But this was only part of the transformation. He was attired in a neat suit of blue serge and was wearing a brand-new bowler. His bow tie was a natty blue with a pattern of small yellow anchors, and his shoes were just out of the store.

"Say, what's all this?" demanded Ludar.

Willing came out from his glass cage and stared. "Who have we here?" he asked. "Lord Dundreary himself? Willie off the yacht? A changeling strutting about in the shoes of John Kinchley Bates?"

The reformed man drew a pair of nose glasses out of a shiny new morocco case and fitted them on. "The *Star* has my permission to announce," he said, "that the Reverend Dr. Torrington, dominion head of the temperance movement, has appointed John Kinchley Bates of this city as a temperance speaker. The Balfourdian thus honored begins his duties at once with a speaking tour from coast to coast. You may state that the remuneration will be—well, quite handsome." He turned abruptly to Ludar. "Now, then, find the story behind the story."

"I already have it," said Ludar. "Have you forgotten that I followed you when you were running away and went with you to Dr. Torrington's and listened while you fascinated him with the story of your sufferings and struggles? But are you willing to have us use that?"

The redeemed alcoholic got slowly to his feet. "Tell the whole story, of course! Anything that will help the cause must be done, even if some people get a laugh out of the curing of that one-time drunkard, Sloppy Bates. I want you both to know this," he went on in an oratorical tone, "that I am going into the fight with my whole heart. I am redeemed. I am saved! I'm going to make the key song of all my meetings, 'Gone Are the Days of Wretchedness and Sin.' I feel so good over the miracle which has happened to me that I want all other sots and broken-down drunkards to experience the same miracle. I want to crush the demon rum with my feet!"

Willing looked in amazement at Ludar. "Do you suppose," he asked, "that sea gulls are nesting in his mains'l? Has he sprung a leak in his thatch?"

"I'll go still farther," orated Bates. "I want Bands of Hope to march along with me. I want them to raise their little voices in the Good Cause."

He paused and then declared with fierce determination, "I want you to know that I'll never touch another drop of the vile stuff as long as I live!"

While he listened Ludar had reached into his pocket, and his hand had touched a letter. He brought it out excitedly, the announcement from England of his good fortune. It had left his mind completely from the moment his troubles began.

"Here!" he exclaimed. "Here's proof. Here's the story behind the story for you. Just before I became involved in all this trouble I received a letter from England. Look at this."

Bates took the envelope and proceeded to read. "Whoosh!" he cried at the finish. "You're right, boy. This gives an edge to the story of the tragedy. It proves what I've been preaching all my life, and sometimes practicing. Mr. Editor, I'm no longer one of your wage slaves, but let me write this. It will be my last piece for the *Star*. My swan song. I'll try to make it a worthy one."

Ludar was becoming increasingly aware that he was very much in need of sleep. He watched Bates get to work with pencil and paper but had to struggle to keep his eyes open. Remembering that there was a space beneath the stairs which contained nothing but old newspapers, he went below with the intention of having a short nap until the paper started to roll off the presses. It would not make a downy couch, he decided when he saw it, but it would serve. He stretched himself out on the lumpy mass of papers and, having nothing to disturb him and no longer any need to speculate about the Sign of the Scaffold, he fell asleep immediately.

The establishment was strangely silent when he wakened. He stretched and rose cautiously in the darkness of the tiny closet, becoming painfully aware that the unnatural position in which he had been lying had taken its toll. Emerging stiffly, he realized that the whole plant was deserted. He could not tell whether the light coming through the windows was that of late evening or early morning. He had no idea how long he had slept.

He climbed the stairs slowly and found Allen Willing pacing about in an otherwise empty city room. "Where the hell have you been?" demanded the editor.

"I fell sound asleep, thinking the presses would waken me. I've just come to. What time is it?"

Willing glanced at his watch. "A quarter after eight. Where did you go?"

"That closet under the stairs."

Willing snorted with annoyance. "As close as that! And I've been sending people all over town to find one escaped jailbird. This place has been a madhouse. Mr. Milner wanted to talk with you. Miss Adelicia Craven has called three times frantically. She must see you as soon as you put in an appearance. Seventeen newspapers want photographs of you. A patent medicine company will pay you twenty-five dollars for a testimonial. The Society for the Abolition of Capital Punishment wired from Toronto to ask if you'd speak at their next meeting, all expenses paid. Lawyers have been calling up. They want to take on your case, I suppose, if you sue the *Pilot* for libel. Don't do it. But if you do, make the lawyer you employ take it on a contingency basis." He paused for breath. "The coroner's jury brought in a verdict of accidental death. That puts the official lid on the case. That prominent clubman and fancy dresser, John Kinchley Bates, wrote a masterpiece about the letter from England. He made quite a point of the way you had already paid back most of the funds. In fact, it's a beautiful answer to that malicious stuff in the *Pilot*."

He found a copy of the edition and held it out to Ludar. "Look at the heading I wrote for the Bates stuff. *Son of a Hundred Kings*. Isn't it a ripper?"

Ludar looked puzzled. "I don't think I understand."

"When you read the stuff you'll understand but there isn't time now. You see, it's from that line of Kipling's, 'Duke's son—cook's son—Son of a hundred kings . . .' I've always been sure he didn't mean there was anyone alive today who could prove descent from a hundred kings. But we're all straight from Adam and so, if we had complete records of life on this earth, we'd find that every man has a hundred kings or more on his family tree, you, me, John Kinchley Bates, everyone. I think he meant that it doesn't matter about such things because we're all just about equal anyway."

Willing tossed the paper aside. "Do you realize you have nearly the whole front page? That's something which will never happen to you again, I hope." He became more serious in his manner. "Mr. Milner isn't going on with the suit against the *Pilot*. They're publishing an apology tomorrow. In fact, they're eating great, dripping, unappetizing gobs of raw crow. Beadle is being fired." He indulged in a wink. "I expect they'll take him back after all this trouble is forgotten. He's too valuable a man for a paper of that stripe. I think the boss is sensible to let the thing drop."

"I think, Allen," said Ludar, "I had better see what Miss Craven wants."

"Yes, she seemed excited about something. I can afford to go home now, so we'll walk up together."

When they reached the street the sound of a voice raised in exhortation came to them from the direction of the market square. "The protest meeting must be getting under way," said Willing. "Ex-Alderman Laines seems to be doing the talking. The police department will get well raked over the coals, and Mr. Saber-tooth Parker will find that his goose is cooked to a cinder."

They walked up Monarch Street, which was pleasantly dark under the joined arch of its maple trees. There is nothing more peaceful in life than a shaded street of homes when dusk is falling. Lamps had finally been relegated to attics, and most of the houses they passed had gas or electricity, but this had not yet led to any general blaze of light. Few of the homes, in fact, had more than one room illuminated, and they provided small competition for the bonfires which had been started here and there; for the first nip of fall had been felt and the trees were beginning to yield their leaves. A buggy came up the street after them, passed, and went on, the *klop-klop* of the hoofs echoing on the hard surface of the road. Bicyclists flitted by as silently as ghosts.

Ludar indulged in a deep sigh of content. "It's wonderful to know I can go on enjoying this," he said.

"Boy," said Willing, "you certainly had us worried. If the girl hadn't come forward you'd be in prison this very minute under an indictment for murder. Jinks covered the inquest, and he says she behaved very well. Poor child, she has botched her life up!"

Ludar lost some of his sense of content. "I wonder if you can ever live down an experience like this."

"I don't know." Willing shook his head. "The old man was furious about all this triangle stuff, you and Norman and Tony. We discussed whether anything should be said in the paper about it, but I persuaded him that it wouldn't be wise. For the life of me, I couldn't see just what *could* be said about it. I think he wanted to discuss it with you this afternoon when you so mysteriously disappeared."

"Did Tony come back with him?"

"No. The old man thought it would be better for her to stay away until the air cleared."

All that was left of Ludar's content vanished. He said to himself: "Mr. Milner won't want me to stay on his paper. He won't make an issue of it, of course, but he'll let me see it would be better for me to leave. I'll have to go away and get a job—if I can. I don't suppose I'll ever see Tony again."

Chapter XI

I

Miss Craven whisked away a tear with one hand and patted Ludar on the back with the other. "You've been a brave boy!" she said. "I don't know yet what it was all about and I don't suppose you'll tell me." She paused hopefully, but Ludar shook his head. "Well, I'm sure of this much. You were afraid Joc had done it, seeing as he cleared out so suddenly, and you were ready to take the blame yourself. I suspect there was something else, but I haven't any notion what it could be. The point now is what's to be done for you?"

"I can answer that quickly. A hot bath. I won't feel right until I've soaked myself free of the cell block and the charming people I met there."

"I thought of that. You need something better than the old Saturday-night tub at home. I made sure there're lots of towels in my bathroom and plenty of hot water. Have your bath, Ludar, and then we'll settle down and talk. All I'll tell you now is I've got a surprise for you."

When he emerged from the tub, refreshed and comfortable, his kind mentor was sitting in a Morris chair with a plate of red snow apples on one side of her and a box of mints on the other. It was clear that she had dined well, for her face was as mottled as a piece of excellent tweed.

"It will be the best thing for you to leave Balfour, sonny," she said. "I'll miss you awful if you do. But—well, it's common sense, and you must do the best thing for yourself. People are going into loud raptures right now about how brave you were, but they'll get over that, the simple-minded idiots! They'll always remember you were once in jail and they'll get into the habit of saying, 'He never did tell how he came to be in that lumberyard. What do you suppose he was up to?' And how they'll smack their lips over your mother having worked as a cook! You could stay, of course, and live it down. I expect it would all be forgotten by the time you were sixty-five or seventy."

"I had been thinking about it," said Ludar. "And I had just about reached the same conclusion."

"There's an old Balfour boy owns a newspaper in Winnipeg." Miss Craven reached into her pocketbook and produced a yellow slip of paper. "We went to school together, and he always sees me when he comes back home for a visit. I sent him a long wire this afternoon and, as you'll see here, he knew you wrote the obituary about poor Tanner and thought it a remarkable job. He offers you a job on his paper—feature work, and twenty dollars a week to start."

"Twenty!" cried Ludar. It was colossal, an unheard-of, an unbelievable salary to a man who was getting nine dollars a week and hoped to be making fifteen by the time he was twenty-five. "Why, Miss Craven, that's—that's simply tremendous!"

"It's not as tremendous as you think. Everything's high out there. You lay down two bits when you walk up to a bar and ask for a shot of hooch. You're nicked four bits for a beefsteak. It costs a lot to live. Still, it's a mighty good offer, and you'll find Harvey Blair a decent fellow to work for and a damn good newspaperman as well."

"I accept!" said Ludar.

"I thought you would. Well, then, get on the telephone and send off a wire at once to Harvey telling him he's gone and hired himself a writing man."

"I'll have to speak to Allen Willing first, of course."

2

The elevator bumped and wheezed to a stop. Al Hanley swung the door open and Meg stepped out, followed by Arthur. She swept across the floor, crying in her high voice, "Ludar, how frightened we were." She put an arm around his neck and hugged him ecstatically. "I thought you might be here, so I telephoned, and they said you were."

Arthur shook hands, frowned severely, and said, "You've caused me to lose ten pounds."

"I saw you in court. You looked rather worried. I hoped to see you after the court was adjourned, but you had disappeared."

"I tried to stay, but Sergeant Feeny had other ideas about it. He gave me a shove between the shoulder blades and said, 'Git along, now, what d'ye mean taking up space that should be empty?' And I'm in trouble with my people because I didn't go back tonight. I had to stay and see you, Ludar, and naturally I couldn't leave without a look at Meg."

"It's a good thing you did," said Ludar. "I'm leaving for Winnipeg in a week, and it may be years before I get back."

Arthur was delighted. "Now that's what I call sensible. You can get ahead faster in the west. I'll see you out there, I expect. I'm aiming at an engineering job with one of the new towns as soon as I'm through college. We'll grow up together—the west, and you, and I."

"I will never go west," declared Meg, slipping her head out of the lacy white mantle she had worn around it. She had on a sleeveless visité of rough white cloth with a high collar embroidered in gold, and this she also discarded. *"I* am going east. Do you hear me, you two, east, east, east! As far east as London and Paris and Vienna. Anyone who goes west goes out of my life."

She turned to Ludar then, took him by the arm, and led him, matching steps, to a discreet distance. "Mother tells me you know about my engagement," she whispered. "You shall be the first, then, to hear the news. I am *not* going to marry him. I broke it gently to the poor boy this afternoon. I haven't dared tell Mother. She's so close to a breakdown now that I honestly think it would send her crazy. He's very nice and of course he'll come into the title—oh, it was tempting!—but, Ludar, I couldn't go through life with someone who thinks only of English politics and birds. Have you ever heard him on the subject of pipits? He talked about them for half an hour yesterday and he pipited himself right out of the running."

"Just about the shortest engagement in the annals of mankind," commented Ludar. "It must have been hard, Meg, to give up the chance to be the mistress of Brazen Tor."

"Not as hard as you might think. From the pictures he showed me, the only neighbors within five miles in any direction would be sea gulls. Except, of course, that the grounds would be overrun with pipits!"

"The truth about you, Meg, is that you don't want to settle down so soon. You'd like to have a thousand men fall in love with you first."

"You always have understood me better than anyone else. Yes, I suppose that *is* the reason. There's something deadly about the idea of settling down with one man for the rest of your life."

"That, of course, is what you'll never do. God help the poor devil who marries you! You'll be a flirt all your life, and he'll have to grin and bear it. I was saying that to Arthur not so long ago when he was very unhappy. I thought it would take some of the sting out of things for him."

"And did it, my sweet and helpful Ludar?"

"Not at all. He agreed I was right, but he said it wouldn't matter to the man who was lucky enough to win you, because the compensations in being married to you would be so great."

Meg looked across the room at Arthur, who was talking to Miss Craven. There was an almost pensive look in her eyes when she turned back.

"Arthur is sweet," she said in a low tone. "Of course I'll never be able to marry him. He'll never make very much, and I suspect he's the kind who would object to living on my family's money. There's one thing I'm going to do about him, just the same; I'm *not* going to let him go west."

They found, on rejoining the others, that the talk was of Joe's absence.

"Surely he's been out of reach of newspapers." Adelicia Craven's face showed that she was deeply concerned. "If he knew what has been going on he would have wired or telephoned, or done something to get in touch with us."

"The automobile's broken down," declared Arthur. "He's on a farm somewhere and out of touch with the world."

"Father and Mother are going away as soon as he turns up," said Meg. "They both need a change and a rest. Mother looks like a ghost." She turned to address Ludar. "She wants to get away without seeing anyone, particularly you, Ludar. She talked as though you had a secret between you."

"Yes," said Ludar to himself, "we have a secret between us, your mother and I." He looked at Meg, who was so lovely, so confident, so undisturbed by the crisis through which they had passed, and he thought, "It might do you good, my beauty, if you knew it was because of you that all this happened, but of course you'll never know."

He realized that the bitterness was gone. He was able to think of Mrs. Craven with all the old affection and he could understand why she had held back at first from coming to his aid. That she had reached the point of willingness to tell and had gone to the police was all that was needed to restore his sense of loyalty.

"Ludar, it's a mighty queer way Mother is behaving," declared Meg. "She acts as though she's afraid to see you."

He smiled at this. "I wish you would take a message to her for me. Tell her I want to see her very much as soon as she feels well enough. And tell her this also: I think that in a certain matter she knows about we managed between us to handle things very well. Be sure and tell her."

"I will. But I don't like mysteries. You ought to know that by this time. I'll drag it out of Mother, of course."

He shook his head. "Somehow I don't think you will."

Meg had dropped her white cloak on a couch near the door. She walked over and tossed it around her shoulders. After buttoning it at the top she called to Ludar: "These sleeveless things are a nuisance. Come and help me, please."

He proved slow at the task. Meg had been watching him with a smile he did not entirely understand, and now she said in a low tone, "That's right, stay as far away from me as you can, you big coward!"

He whispered back, "So you want to make it a thousand and one, do you? That's taking advantage of me when you look like a lovely swan, or a cygnet perhaps, in this white outfit."

"I'm more like a nice, innocent bunny trying so hard to be liked."

Ludar whispered scornfully, "Whoever heard of a rabbit with green eyes?"

She returned then to say good-by to her aunt. "I feel quite bitter about Ludar, Aunt Lish. He's my one failure. I've tried all my wiles on him. I've fairly thrown myself at his head. But it does no good. He's elusive and as cold as an icicle."

"That's because he's in love with someone else," said Miss Craven. "If he hadn't been he would never have been able to resist you, my pet."

"If I'm your one failure," said Ludar, smiling at her, "you'll never be able to forget me, Short-for-Margaret. That makes me feel very happy."

"Come on, we'll drive you home," said Meg in a tone which seemed to imply, "Well that's settled finally." She added proudly, "Father lets me have the car now. I think I can drive as well as he can; or Joe, for that matter. Are you going home?"

Ludar nodded. "I phoned Aunt Tilly this afternoon and said I'd be home for supper. She'll be getting alarmed. She may think I've gone and stumbled into more trouble."

The lift had arrived to take them back, but Miss Craven called, "Ludar, I've something to say to you."

He returned at once. She got up from her chair and put a hand on his arm. Her eyes were brimming with tears. "When you go to Winnipeg I'll have only one left of my three boys. I'm going to miss you like anything. Of course there will be plenty of things to remember, like that day when Tanner sent those two thugs to the Lester place, and you and little Joe and Arthur and that splendid boy Peter fought them off. *That's* a memory for an old woman to bring out, and dust off, and chuckle over with pride." She looked up at him. "You're different from the others. I'll watch what you do—with lots of pride and satisfaction."

"I can't thank you for all you've done for me."

She sighed deeply and shook her head. "I've done very little to make up for—for what I didn't do that day. Well, run along. That little witch is getting impatient, I can see."

3

Ludar felt a pang of guilt when he reached home and found Tilly sitting disconsolately in the kitchen beside an untouched supper. She had not thought to light a lamp.

"Where have you been?" she asked. "I was getting frightened. I began to wonder if you were ever coming back."

"I went to sleep and didn't waken until a short while ago." He turned up the wick on the kitchen lamp and struck a match. He was slow in fitting the chimney on, and it became slightly blackened in the process.

"How can you be so clumsy!" said Tilly. Then she uttered an exclamation of surprise. "Why, you don't look any different!"

Ludar laughed. "Did you expect I would? That I would have the prison pallor and a convict's haircut?"

Tilly shuddered. "I can't get over it. *You* in jail—and charged with murder! I thought I would die of fright. I said to myself a dozen times, 'Praise the Lord Billy Christian isn't here to know about *this.*' Of course it wasn't your fault. I told Uncle Alfred that." She looked at him intently. "What were you doing in that place?"

"I just happened to ride through on my bicycle. Strange as it may seem, I forgot I had one and left it at the office tonight."

The supper on the table was not appetizing. The johnnycake was cold and the tea had become lukewarm. Ludar took a slice of cold brisket of beef, however, and proceeded to eat. Tilly watched him closely, and if he had been observant he would have detected in her a distinct suggestion of uneasiness.

"I'm going back to Coldwater," she announced finally. "I didn't know how—how things were going to turn out, and so I wrote to Uncle Tower yesterday. I was pretty firm with him. I said if I went there must be two dollars a week to be added to my principal and he would have to put me in his will for at least two thousand. And I insisted on a bathroom being put in. The homestead is a grand, roomy house, and I'm proud of it. But it's very wintry at Coldwater, and I don't think I would enjoy the long trip out at nights. I guess you'll be able to find some nice place—not too expensive!—to board, Ludar."

He felt ashamed of the sense of relief which swept over him. Now, however, she was so well located—he knew she would be comfortable and happy to be living in the house of her lifelong pride—that he could cut his cables and go west with a free conscience.

She did not seem pleased when he told her his news. "Winnipeg is a long way off," she said doubtfully. "What are they going to pay you?"

"Twenty a week."

"Dollars? Twenty dollars—a week?" She stared at him with unbelieving eyes. "Ludar, you're not telling me the truth. There's no such thing as a salary of—of *twenty dollars a week!*"

"Everything's high in the west," he explained, laying down his fork and taking a last sip of the bitter tea. "The cost of living is high in proportion to what you make. Why, it costs a quarter to get a drink of whisky——"

"You'll do nothing of *that* kind, young man!" She still seemed unconvinced about the size of the salary he was to receive, her belief being that only presidents of countries and prime ministers and the managers of very large factories could receive as much as that. "What will you do with all that money? I'm afraid you'll waste it on—on women and riotous living."

"I'll add it to my principal," said Ludar.

"That's what you must do. You can save—why, you can save at least thirteen dollars a week. If you add that much to your principal every week, you'll soon be independent."

"I promise you I'll be sensible. I'll send you financial reports every now and then."

"You'd better. I was talking to Uncle Alfred last night. You don't know the bad news about him. He's going to be retired, and he's quite bitter about it. Of course he has enough laid by and he won't be in need. He was saying——"

"I'm not interested in Uncle Alfred's opinions, Aunt Tilly."

"Well, if you're going to be touchy about it! I'll have you know that some of the things he has said from time to time, some of them about *you,* have come true. It would be to your best interests to hear what a cautious man like Uncle Alfred has to say about how we should handle that money from England. When he knows you're going west he'll be more sure than ever that you ought to turn it over to me." Her voice rose to a high pitch. "I read what they wrote about it in the *Star,* and it gave me a turn. Now that everyone knows you have the money they'll come swarming around you and trying to take it away from you." When he made no response she gave her head an unreconciled shake and got to her feet. "I see being in trouble hasn't taught you anything. You're still as stubborn as a mule. I'm going to bed now. I'm all worn out after what I've been through these days."

"Yes, Aunt Tilly." Ludar was feeling contrite because he could not bring himself to say, "Come out west with me." Instead he said, "I've been responsible for a lot of worry and trouble, haven't I?"

"Yes, you have." She lighted another lamp and walked in the direction of the hall. "But I *don't* agree with what Uncle Alfred said about that." He could hear her mutter as she reached the door of her bedroom, *"Twenty—dollars—a—week!"*

4

There were definite streaks of white now in Clyde's hair, and it could not be denied that he was getting stout. He met Ludar at the foot of the dark stairs which led to the second floor of the McGregor home. "Catherine's pretty tired," he said in a low tone. "She's been so upset that she's tuckered out. Don't let on that you think she needs coddling—she doesn't like that at all."

"Perhaps I shouldn't go up."

"Oh yes. She would be terribly disappointed if she didn't see you. But we'll have to make it a short visit."

For the first time in Ludar's recollection of her Catherine looked like an invalid. Her eyes were deep-sunk and her features gave the impression of having been sharpened. There had been no change, however, in her spirit.

"Well, Stiddybuttons," she said, smiling, "you certainly had us all frightened to death. I didn't sleep a wink last night." She leaned over and pressed his hand. "Good boy!"

Clyde, hovering in the background, seemed anxious to avoid any emotional strain. He changed the subject by announcing: "We adopted a son this afternoon, Ludar. A fine solid little fellow. Father-in-law McGregor was so set up by the good news about you that he gave his consent without a moment's hesitation."

Catherine's eyes smiled. "There was another reason too. The boy's name. What do you think it was, Ludar? *James McGregor!* Father said to us: 'Now, here's a boy of the best stock in the world. Hasn't it paid us to wait?'"

"At last he'll be able to settle something about the sign," declared Clyde, who had given up hope years before of appearing on it and could be philosophic and amused by the situation. "I think he's at it already. I saw him after supper with a sheet of paper. He was writing things down and then scratching them out and growling."

Catherine broke into a laugh which seemed to have all of her old-time vivacity. "I'm going to suggest that he wait until our son has grown up. Then he can have two figures carved, one of himself and one of his grandson. They must be the same size exactly—I won't have Father slighting little Jamesy McGregor Carson! The sign will read 'The Remarkable Men.'"

"When do you get the boy?" asked Ludar.

"Tomorrow." Catherine nodded excitedly. "I'm going with Clyde to bring the little fellow back to his new home. I'm so excited!"

"I'd like to go with you."

Catherine looked up eagerly. "Would you really like to go?" A strained expression took possession of her face, and the two watchers knew that the pain in her back had started again. "I would be very proud if you did. Fancy a bent old woman like me going out with two such fine, straight, handsome men as escorts!"

Clyde winked at Ludar as a sign that he had stayed long enough. At the front door the latter informed him of his intention to go west. Clyde nodded approval but then added: "I'm glad you didn't mention it upstairs. It's going to be a blow for my poor girl. We must leave it until she's feeling better."

Chapter XII

I

The frame of the roller boat stood under the old cherry tree. There was not much of it left because Tilly had insisted on using it for kindling wood, but one wheel, minus a spoke or two, was still upright. Ludar groped his way to it in the dark. All the lift he had felt at the prospect of a new life deserted him in this graveyard of hopes and aspirations.

"Uncle Billy," he said aloud, resting one hand affectionately on the shaky wheel, "I've been in a lot of trouble. I guess you know all about it. It wouldn't have been so bad if you'd been here to help me."

After a pause he went on, still speaking as though his first benefactor stood beside the wheel with him. "I'm going away. Aunt Tilly's going away too. That means there will be other people living here at 138 Wilson Street. They'll destroy all that's left of your boat, Uncle Billy—but I don't see what I can do about it."

This prospect of a stranger coming out into the yard with an ax and demolishing the great invention which was to have made their fortunes and put the name of William Christian among the famous was the final blow. He swallowed hard and rubbed a hand across his eyes.

"Ludar!"

The voice was Tony's, from the other side of the high fence. She had come home after all! He said eagerly, "Yes!" and began to feel his way slowly in that direction.

"I can't see you," she said, "but I heard you speak to someone."

"I'm afraid I was talking to myself. You can have no idea how glad I am to hear your voice. I was told you were staying in Toronto until—until things settled down."

"Father wanted me to stay. I went to see him off and then I waited and took the next train. He was very angry when I arrived. But I couldn't stay, Ludar, not when you were in such trouble."

There was not a star in the sky, and the air was so still they could hear the wheels of a carriage turn on the "Y" where Fife Avenue swung around into Grand. Instinctively they had been speaking in low tones. He was thinking, "Does this mean she isn't angry with me and doesn't blame me because of what was printed?"

He put the question into words, and she answered with fitting scorn: "Ludar! Why should I blame you? After all, it was true, wasn't it?" There had always been a forthrightness about her, a willingness to face facts. "Norman wanted me to marry him, and it made him very angry when I kept saying no. He knew it was because of you." She went on in a tone which suggested she was smiling, "But poor Father! He's been in such a rage about it."

They were both standing close to the palings, their faces not more than a foot apart. "Do you remember, Tony, that we'd arranged I was to call on you tonight?"

"Of course I remember. There's still time, isn't there?"

He looked up at the fence. When he was a boy it had seemed as high as the walls of Jericho or the ramparts of Front de Boeuf's castle, and it had represented to him everything which divided them. It had never occurred to him that the time would come when it could be climbed. Now he realized that it was a bare foot above the level of his eyes.

She had been having the same thought about the fence. "I don't think it would be hard," she said.

He scrambled over and landed with a thud on the other side.

His eyes were becoming accustomed to the dark and he was able to discern now that she was wearing a cloak over her shoulders. She moved, and a ray of light from an upstairs window touched her face for a moment before she stepped back; but it had been long enough for him to see that she had nothing on her head and that the dress she wore was cut low enough to show the fine white line of her throat.

"Have you heard I'm going away?" he asked.

"Yes," she whispered. "Father told me a few minutes ago."

"Was he glad I was going?"

She was too honest to quibble. "I think he was relieved. I—I was shocked at first. But then I thought about it and I could see that it's the sensible thing for you to do."

"But now there won't be time for two friends who had become strangers to get acquainted all over again."

"Ludar, I didn't mean that. Yes, I did—when I said it. I really thought we had drifted so far apart that it would take a while to—— Oh, how can I put it! But I was wrong. How wrong and silly I was! I knew it as soon as I heard what had happened to you. It wasn't a stranger in trouble. It was you! It seemed as though the world was coming to an end."

They were standing close together. He reached out and drew her into his arms and kissed her. Having dared so tremendously, and realizing to his great delight that she had yielded willingly to his embrace, he continued to kiss her, at first furiously, then slowly and possessively, on her eyelids, on her ears, on the tip of the beautiful nose he had always admired so much.

"Can they see us from that window?" she asked breathlessly.

"No," he answered. "But would it matter if they could?"

There was a cement bench along one side of the white brick building which had once been a stable and was now a garage. It was used for some other purpose than as a seat, however, for it was quite high; so high that, when they ensconced themselves there side by side, Tony's feet did not touch the ground and she swung them slowly back and forth.

"I'm going in a week," said Ludar. "I know it's asking a lot, but—could you be ready to go then?" When she did not answer immediately, he was assailed by a devastating doubt. "You *are* going to marry me, aren't you? Or—or am I assuming too much?"

She placed one arm under his and rested her head against his shoulder. "I wouldn't want you to assume less, darling," she said.

After a pause, a very brief interval charged with the electricity of all the emotions which go to make up love, an ecstasic and glorious moment, he said: "When people hear the angels sing I don't suppose they understand the words. They don't try; they're content with the wonder of the music. That was the way I felt when you spoke. I don't believe I could repeat now what you said, but I think—I think—it meant you are going to marry me."

"Oh yes," she said, laughing briefly and happily. "That's what it meant. I've intended to marry you ever since the moment when I first saw you through the fence. Didn't you know?"

There was another pause, a longer one, perhaps, although neither was in a state of mind to gauge the passing of time with any accuracy. "Will you marry me before I leave?" he asked. "Or must I go out alone and exist in a state of the most terrible misery until I can come back for you?"

She sighed before answering. "It means giving up so much. All the fun of the engagement, and the showers, and the parties, and then the wedding in church. But I'm willing, Ludar. I'll go with you now. I don't want you to get so far away from me. We've been separated so much as it is."

"What is your father going to say?"

Tony gave her head a shake in the dark. "It's going to be a blow for him. He isn't going to care for it at all. Of course he likes you and he feels sure you have a real future, but he doesn't want me to marry for a long time. He'd feel exactly the same if it had been Norman. Aunt Mona May——" She laughed under her breath. "Oh, how furious Aunt Mona May is going to be!"

"Tony, you're giving up so much for me! It makes me happy, of course, but very humble. Does this mean we'll have to take things into our own hands? Even, perhaps, run away?"

Her head was so close against his shoulder that he could sense the determination of the shake she gave it. "Oh no! That won't be necessary. He can be unreasonable about things at first, but he always comes around. He'll glower and smooth his mustache and say a lot of things he doesn't mean at all—but he'll give us his blessing finally." She was so confident of the outcome that she let her mind leap far ahead. "We'll go to England as soon as we can and see that pretty little aunt of yours. I liked her so much. And now that everything has come out, you might want to visit Pursley Castle and see the clock tower and Old Mr. Cyril."

This caused him to say, "I can hardly believe yet that you're willing to marry a man whose father got into such trouble and disgrace and whose mother——"

"Now you are being silly! That's in the past, and you must believe, Ludar, that I don't care about it at all."

Before anything more could be said they heard an automobile come chugging and puffing up Wilson Street. It passed the corner and then came to a stop. The driver began to signal with his horn: "Pleep, pleep, pleep—*pleep!*"

"Four times!" cried Ludar excitedly. "That's Joe. He's come back. And he sounded the horn four times, which means he's married and has brought his wife with him. What a strange thing for him to arrive just when I've convinced the lovely lady of my dreams that we must be married at once!" He sprang to his feet and dragged her up beside him. "Come on, Tony, my darling. Let's go and see about this."

2

Standing at the curb beside the dark bulk of his car, Joe hailed the newly engaged couple as soon as they crossed the lot.

"That you, Ludar? Who's with you? Oh, hello, Tony. Well, it seems I helped stir up a hornets' nest by disappearing this way. What a time *you've* been having." He walked over and put a hand on each of Ludar's shoulders. "You thought I was the guilty party. Oh yes, you did. I know you, Master Ludar. You were trying to protect me." He paused and then said with a hint of emotion in his voice, "You old son of a gun!"

"We've other things to talk about," said Ludar. "Didn't you toot that horn four times?"

"You bet your life I did! I'm married." They had been aware of another figure in the car, and now the eyes of both newcomers turned in that direction. "We've been on our honeymoon, although you may think it a queer one when I tell you what happened." He reached into his pocket and produced a box of matches. Striking one, he held it up near his wife's face. "This is Mrs. Joseph Craven. You already know Ludar, sweetheart, but I don't think you've met Tony."

By the uncertain light of the match the new Mrs. Craven looked very small and very pretty and, if the light in her dark blue eyes was any indication, very happy. She was wearing a gray cloak with a deep cape of blue and an impudent black hat with a bunch of artificial flowers at one side.

"Isn't she lovely!" exclaimed Joe.

"How do you do, Jessie?" said Tony, leaning over the door of the car.

"How do you do—Tony?"

"We'll have to be good friends because our two men have always been that. They've been inseparable."

Joe had forgotten about the match. The flame bit his fingers and he let out a howl of pain. Everyone laughed, and this had the effect of breaking the ice. There was an excited chatter of good wishes and much thumping of backs on the part of the two men.

Joe then began to explain his absence and his failure to communicate with anyone for the two momentous days.

"I had a row with the parents and decided I was going to marry Jessie—if she'd have me—and start on my own. I found Pete, and he felt the same way about things. The family had gone down to Lake Erie for the week end, and so we drove off to find them and to talk Jessie and Mary into seeing reason. They saw reason, and we got someone out of bed to issue the licenses and then we found a minister. We started out on our honeymoon. The car ran beautifully for six or seven miles, and then in one of the loneliest spots in the whole universe it stopped dead. And there we were."

"There was a farmhouse quite close," he went on, "and they agreed to take us in until we could get the car fixed. Pete saw at once it was the carburetor which had gone on strike and he said he could fix it. Well, he did, but it was a long job with the few tools we had. We spent all Sunday and most of today on our backs under the car, with our wives sitting comfortably in chairs and laughing at us. We got as black as tar, and it took buckets of hot water to get us clean. The farmer wasn't feeling very well, so they didn't go to church, and there wasn't a telephone in the house. We were cut off from the world. We thought that was just perfect. But when we got the old bus running late this afternoon and I telephoned Father from the first place that had one, we realized that it hadn't been so perfect after all. With me missing, it had been necessary for my young friend here to play Sydney Carton."

"And now I have news for you," said Ludar. He paused and looked questioningly at Tony. "That is, if——"

"Oh yes. Tell them."

"I don't have a match." Ludar reached out a hand, and Joe located it in the dark and deposited the box on his palm. When the match was lighted Ludar held it close to Tony's face.

"We're going to be married," he announced. Then he repeated what Joe had said, and with at least equal ardor: "Isn't she lovely?"

Tony was lovely without a doubt. Her brown eyes had lighted up (no trace of gravity now, all laughter and delight and happiness), her lips were parted slightly, and a crinkle along the bridge of her nose made that excellent example of Grecian modeling seem gay and daring.

"We've only been engaged ten minutes," she said. Her voice was dancing with excitement. "But we're going to be married in a few days. Isn't it wonderful?"

"It's what I've wanted to see happen," declared Joe. "What times the four of us will have!"

This mention of the future made a second announcement necessary. "I'm going away, Joe," said Ludar. "To Winnipeg. I'm going on the staff of a newspaper there."

The match had gone out, and there was complete silence in the darkness for several moments.

"I might have known it," said Joe finally. "All the talk and all the things they printed! Father said it made his blood boil." Then he began to oppose the idea vigorously. "This is all wrong, Ludar. We've been friends so long that I can't get along without you now. And it doesn't make sense anyway. The *Star* will belong to the two of you someday. Why do you want to drag Tony off for the best years of all our lives when you'll have to come back sometime?"

"I don't think we'll ever come back, Joe."

"Now see here! When that old Balfour boy from New York, the one who's head of some big company down there——"

"Anderson Coyle?"

"Yes, that's the one. What did he say when he was visiting here a short time ago? You interviewed him, so you ought to remember."

"He said if he had it to do over again he wouldn't leave home."

"There you are. He knew what he was talking about. He wasn't young like us, imagining things about life. He'd been through the mill. Don't you remember what he said, that the friends you make in youth are the best friends? I tell you, it's true. I'll never make another friend like you, Ludar, and I don't want to lose you."

Ludar felt a lump grow in his throat. "I know that I'll never have another friend like you, Joe." He paused, wondering how he could make clear the thought in his mind. "I don't know whether I've got anything in me or not, but it will never come out if I stay here. I couldn't prevent myself from becoming so busy that I'd have no time for the things I want to try. I must see if—if I have any wings to spread."

"Do you think the west is the place to spread them?"

"That's just the first step."

"You mean you want to break into a big league somewhere."

"Yes, I want to try. I think I shall have to go to New York or London sooner or later."

Joe, still unreconciled, turned to Tony. "What do you think about all this?"

Tony showed no hesitation. "Whatever Ludar wants is what I want."

"The good old-fashioned type of wife, eh?" said Joe. "But your father isn't going to like it. He'll want you to stay right here."

"I know. Poor Father isn't going to be happy. We've become such great friends since we've traveled together so much. But it's our lives we're talking about, Ludar's and mine." She paused. "I'm excited about going. I feel just the way Ludar does. We must see—what luck we'll have with our wings."

"If you stay here," charged Joe, "you'll be the great lady of the town. You'll keep all your old friends and make plenty of new ones. You'll have a fine life."

"We'll try hard to make good," said Ludar. "I want to take Tony to"—he remembered the talk he had had with Meg—"to embassy dinners and important balls where she'll meet princes and all the great people of the world. Nothing less than that is good enough for her."

In the darkness Tony's hand found his, and they exchanged an ecstatic squeeze.

"Damnation!" exclaimed Joe. "You've cut the ground right out from under my feet. I can't say I hope you fail and have to come back home. I hope you succeed. And if you do, that means the end to our friendship. Oh, we'll see each other every now and then. But it isn't the same thing." He opened the door of the car and got in. "We must get along now. Pete and Mary are staying with Mrs. Slaney tonight, and Father insisted over the telephone that we come home. I think Jessie is a little nervous about it."

"I'm so frightened," acknowledged the bride, "that I can hardly keep my teeth from chattering."

"You'll be received with open arms." Ludar spoke with conviction, remembering when he had shared with the Craven family the fear that Joe's absence meant he had been implicated in his cousin's death. How happy they would have been had they known the reason for his absence! "I think you ought to get there as soon as possible."

So the bride and the bride-to-be exchanged kisses and the two friends clasped hands with all the reluctance of young males to show any emotion. Then Joe honked the horn four times again, and the diminutive red car chugged off into the darkness.

Tony put a hand on Ludar's arm. "And now we have a task ahead of us," she said. "We must go together and break the news to Father and Aunt Mona May."

Ludar said, "Just a moment." He walked to the front steps of the house and, striking a match, held it so high that the numbers "138" stood out in bold white against the blue glass of the fanlight. He waited until the match went out and then returned.

"What a debt I owe this house," he said. "Tony, my sweet, beloved, beautiful one, isn't it lucky that 138 Wilson Street backs on where you live? And that the man who lived here happened to be on that train from Toronto? If he hadn't, we might never have met. At any rate, I'm sure we wouldn't be standing here at this moment trying to get our courage up to tell your father we're going to be married and leave right away."

"I'm not sure," said Tony. "We'd have met sooner or later, and I think that—well, I think things would have worked out just the same."